THE ELECTROCHEMISTRY
OF LEAD

THE ELECTROCHEMISTRY
OF LEAD

Edited by

A. T. KUHN

Department of Dental Materials,
Eastman Institute of Dental Surgery, London

1979

ACADEMIC PRESS
London · New York · San Francisco
A Subsidiary of Harcourt Brace Jovanovich, Publishers

CHEMISTRY

ACADEMIC PRESS INC. (LONDON) LTD.
24–28 Oval Road,
London NW1

United States Edition published by
ACADEMIC PRESS INC.
111 Fifth Avenue
New York, New York 10003

British Library Cataloguing in Publication Data

The electrochemistry of lead.
  1. Electrochemistry
  2. Lead
  I. Kuhn, Anselm Thomas
  546′.688′57   QD553   78-52097

ISBN 0-12-428350-0

PRINTED IN GREAT BRITAIN BY
J. W. ARROWSMITH LTD, BRISTOL BS3 2NT

Contributors

F. Beck, Elektrochemie, Fachbereich 6, Gesamthochschule Duisburg, 4100 Duisburg 1, Lotharstrasse 65, Federal Republic of Germany.

S. M. Caulder, International Lead Zinc Research Organization, Naval Research Laboratory, Washington DC 20390, U.S.A.

W. D. Davis, Research Laboratories, Associated Lead Manufacturers Ltd., Wadsworth Road, Perivale, Greenford, Middlesex, England.

J. Dawson, Corrosion Centre, UMIST, Sackville Street, Manchester, England.

A. N. Fleming, Electrochemistry Research Laboratories, School of Chemistry, The University, Newcastle on Tyne NE1 7RU, England.

J. A. von Fraunhofer, University of Louisville, School of Dentistry, Health Sciences Centre, P.O. Box 35260, Louisville, Kentucky 40232, U.S.A.

N. A. Hampson, Department of Chemistry, Loughborough University of Technology, Loughborough LE11 3TU, England.

H. Hamsah, Department of Chemistry and Applied Chemistry, Salford University, Salford M5 4WT, England.

J. A. Harrison, Electrochemistry Research Laboratories, School of Chemistry, The University, Newcastle-upon-Tyne NE1 7RU, England.

M. Hayes, Ecological Engineering Ltd., Hurdsfield Estate, Macclesfield, Cheshire, England.

A. T. Kuhn, Department of Dental Materials, Eastman Institute of Dental Surgery, University of London, 256 Gray's Inn Road, London WC1X 8LD, England.

A. J. Owen, Research Laboratories, Associated Octel Ltd., P.O. Box 17, Oil Sites Road, Ellesmere Port, Wirral, Merseyside L65 4HF, England.

T. H. Randle, Department of Chemistry, Swinburne College of Technology, P.O. Box 218, Hawthorn 3122, Australia.

L. L. Shreir, Department of Metallurgy and Materials, Sir John Cass School of Science and Technology, City of London Polytechnic, Central House Annexe, Whitechapel High St., London E1 7PF, England.

A. C. Simon, International Lead Zinc Research Organization, Naval Research Laboratory, Washington DC 20390, U.S.A.

Preface

This book arises from a Symposium of the same title, held at the University of Salford in 1974. The contents are, however, much expanded both in their depth and scope. One need not look far for the various fields which make the subject one of real importance. Of all the different types of battery made, the lead–acid is the most important. What is more, that situation will continue. As long as internal-combustion engined vehicles are made—and that is certainly to the end of the century and possibly much longer, whether they run on fossil-formed petroleum or on man-made fuels—it seems probable that a lead–acid battery will be used to start such engines. The layman knows that all manner of sophisticated batteries are now being developed with electric vehicles in mind. Those in the field have never seriously suggested that such batteries would replace the lead–acid in its present role, where its low cost and great simplicity, its tolerance of abuse, all make it an ideal candidate. It should not be thought, however, that this battery is now a fully developed device, and as this is written, there is a flurry of research activity in battery laboratories around the world, examining new alloys and their behaviour. Indeed, those who are developing new types of battery would be making a great mistake if they assumed the lead–acid battery (in its role as a traction battery for electric vehicles) is a "stationary target". A lighter battery is, by and large, a cheaper battery, and for this reason alone, efforts will continue to be made aimed at producing more kWh kg^{-1}.

Batteries apart, we presently use lead as an anode material in the electrowinning of many different metals. In this role, some forecasters might predict a less secure future. The massive lead anode is being challenged on the one hand by titanium-based lead dioxide, and on the other by non-plumbous materials. The machinery set up to develop new anodes for the chlor-alkali industry, to replace the previously used graphite, has been extremely successful. First platinum, then platinum-iridium alloys then ruthenium oxide, finally mixtures of oxides in which precious metal forms less than 20% of the total—these catalysts deposited on titanium substrates have totally eclipsed the graphite universally used only ten years ago. The developers of such anodes, with the chlor-alkali market now assured, have understandably turned their sights to similar goals and markets. Lead anodes used in corrosion protection have been largely replaced by bielectrodes, and these in turn are being severely challenged by precious-metal containing anodes either on titanium or on niobium. So far the high acidity of metal-winning solutions has excluded the same type of technology from this, the largest of all anode markets. But it is no secret that hundreds of scientists are at work on this problem. Whether they will succeed remains to be seen. Even if they do, it seems quite likely that their recommended coatings will contain lead dioxide, which is one of a mere handful of satisfactory acid-resistant conductive oxides.

Though these are the best known areas of activity in the electrochemistry of lead, we have been fortunate in securing the collaboration of experts in other fields. Electrochemistry of lead in molten salts; the question of electrodeposition of lead which, in the editor's opinion, is likely to be one of growing importance; the use of lead-dissolving electrolytes whether in refining of the metal, in reclamation of scrap lead or as the basis of a novel battery—we are fortunate in having been able to call on authorities in these fields to produce definitive statements as to present knowledge in the fields concerned.

At this point, it is appropriate that the scope of this book be defined. This work was not intended to be a book about the lead–acid battery. In this field we have long had Vinal's excellent book* on which to lean, and the recently published work by Bode‡ updates the subject. We have therefore excluded the technological side of batteries fairly rigorously. Included, however, is the electrochemistry of the lead–water–sulphuric acid system and inevitably, much of this is "battery-orientated". Then too, there are ideas from the battery industry which may well have "spin-off" to other fields. We have thus tried to strike a balance between exclusion of battery electrochemistry on the one hand, and inclusion of the fundamental electrochemistry of that system on the other. A second field which has been deliberately excluded, is free corrosion of lead. Though this process is electrochemical in nature, the literature describing it is so peculiar, by and large, to the individual circumstances—maybe Welsh peat bogs or Transylvanian clays—that we felt

* G. W. Vinal, "Storage Batteries". Wiley (1955).
‡ H. Bode, "Lead–Acid Battery". Wiley–Interscience (1977).

nothing was to be gained in the cataloguing of the literature of the subject. Driven anodic corrosion, however, we have treated at a very considerable length, and this will be elaborated below. A final field which has been deliberately excluded is the electroanalysis of lead, and here again, one has a specialized area based on scientific principles which are expounded in this volume, but branching out into a multiplicity of specialized conditions and details of interest only to the analyst. In the field of electrosynthesis, both organic and inorganic, we have focused mainly on the anodic syntheses, since, as the reader of these chapters will discover, we feel there is at least a probability that electrochemistry of lead is involved. If the emphasis is on inorganic rather than organic reactions, it is because too little is known mechanistically of anodic organic oxidations. On the cathodic side, hydrogen evolution has been dealt with in some detail. Cathodic reductions, whether organic or inorganic, have not been treated again for two reasons, namely that it is doubtful whether the lead is actually involved on a reaction level, and also because so little is known of such reactions other than the products and the conditions required to promote them.

One of the aims of Ken Peters and myself in organizing the Salford Symposium was to try to bring together the knowledge which, it seemed to us, lay in very separate pools, with minimal interaction between the battery scientists on the one hand and the electrowinning anode technologists on the other, not to mention the other fields which are described in the book. At that time we were happy to have seen with our own eyes evidence that some cross-fertilization had taken place, and one hopes very much that this book will continue that process. All the contributors were aware of the foundations on which we were building. Apart from the books mentioned previously, one should refer to Allen Bard's excellent series "Electrochemistry of the Elements" with its chapter on lead, Noel Hampson's superb review of the "Lead Dioxide Electrode" and Jean Burbank's review of the lead–acid battery in "Advances in Electrochemistry and Electrochemical Engineering" (Vol. 8). It says much for the importance of the subject that a book of this size has been written with scarcely any overlap or repetition of material contained in these earlier works. In conclusion, I would like to use the "editorial privilege" to comment on two areas of the book in which I have been intimately involved. Speaking of the "Corrosion" chapters (Chapters 12–15) in which John Dawson and I have been involved, I think both of us would acknowledge that the subject is a very confused one. With little difficulty, one can find categoric assertions that alloy A is superior to B, with equally emphatic contradictions of the same. Why this is so, is considered in detail. Speaking with some experience, I hope, I suggest that this area is one of the most confused in the whole field of electrochemistry. After reading the chapters concerned, it will be apparent why this is so. Nevertheless, John Dawson and myself were faced with the invidious choice of trying to present the subject in a falsely simplified form or revealing it as it actually is—more than a little confused and apparently self-contradictory. We chose the latter, and the reader will see the subject in this form, and will have to be content with using published knowledge as a guide to the solution of his/her own problems, rather than as a firm prescription. This, together with intuition and analogy, aided perhaps by a limited number of well-planned experiments, will enable one to deduce as efficiently as possible the optimal answer. We believe that sometime a monograph may be written devoted entirely to this topic. However, we rest content in having written the first review chapters on the subject. The other area on which I wish to comment is the Bibliography section. Some readers may deride this as a "junk-room", though others may—and I hope they predominate—welcome this as a treasure-house of odd ideas. These references are only classified in the most rudimentary way, and even then, it could be argued as to how they are best listed. They are waiting to be "mined" and their inclusion constitutes no guarantee on our part as to their usefulness, indeed the majority of the references have not been read in their original form. Thus I, as editor, remain ignorant as to the manner in which solar activity affects battery activity, or how double-layer measurements have been conducted in methanol. However, it is our hope that by scanning through these, even though the reader may draw some blanks in the process, he or she will find a quick and efficient method of knowing what scientific precedents exist for whatever problem is of interest to him/her.

Finally, I would like to record my gratitude to the many friends and colleagues both in academic and industrial laboratories, who have been so helpful in various ways. The sheer arduousness of writing a work such as this is relieved by the pleasure of working with such a diversity of authorities in their respective fields and one of the rewards of the exercise lies in the sharing of the effort with them, and the prospect of looking back on the achievement with them. I would also like to extend my thanks to the staff of Academic Press for their patience and efficiency during the production of this book.

October, 1978   A. T. KUHN

Contents

6. Electrodeposition of Lead
J. A. VON FRAUNHOFER

7. Electrochemistry of Organolead Compounds
A. J. OWEN

To Sylvia, Martin and Andrea

1   The Aqueous System Pb^{2+}/Pb

A. N. FLEMING and J. A. HARRISON

Electrochemical Research Laboratories, The University, Newcastle-upon-Tyne

The Electrochemistry of $Pb/PbSO_4/H_2SO_4$

The measurement[1,2] of cells indicates that E^0 for the reaction

$$PbSO_4 + 2e \ \rightleftharpoons \ Pb + SO_4^{2-} \tag{1}$$

is $0 \cdot 357$ V. Thermodynamic data[3,4] indicate a value $0 \cdot 356$ to $0 \cdot 359$ V. E^0 for the reaction

$$Pb^{2+} + 2e \ \rightleftharpoons \ Pb \tag{2}$$

is $0 \cdot 126$ V from cell measurement.[5] The value agrees with thermal data.[6]

The state of complexation of Pb^{2+} in various electrolytes has been investigated by polarography. Simple electrolytes ClO_4^- and NO_3^- show no shift of $E_{1/2}$ with concentration. On the other hand, Pb^{2+} in SO_4^{2-} (with excess ClO_4^-) shows a shift[7] from which β_1 and β_2 for the complexing reactions

$$Pb^{2+} + SO_4^{2-} \ \rightleftharpoons \ PbSO_4 \tag{3}$$

$$Pb^{2+} + 2SO_4^{2-} \ \rightleftharpoons \ Pb^{2+}(SO_4^{2-})_2 \tag{4}$$

have been deduced. However, solid $PbSO_4$ which is also present in these particular experiments may be reduced and other equilibria besides (3), (4) are possible.

The solubility of $PbSO_4$ in a range of concentrations of H_2SO_4 has been measured by a dithizone method.[8] The solubility *vs* concentration curve is similar to that of Hg_2SO_4. It has a fall at concentrations up to 5×10^{-2} M, a rise to 1 M and then a further fall; $CaSO_4$

and Ag_2SO_4 have similar features. The authors suggest the fall in solubility is caused by the common ion effect and depression of the Pb^{2+} ion concentration according to the solubility relation

$$K_s = a_{Pb^{2+}} a_{SO_4^{2-}} \qquad (5)$$

They attribute the rise in solubility to possible complexes $PbSO_4$, $Pb(SO_4)_2^{2-}$, $Pb(HSO_4)^+$ and $Pb(HSO_4)_2$. A decision was difficult as the ionic composition of H_2SO_4 was not known; this is now known from Raman spectroscopy.[9] Solubility measurements have also been made with radio tracers.[10] Values of β_1 given in the literature are 420,[11] 8500[12] and 500[13] 1 mol^{-1}.

Stationary i–E curve

When current flows through a Pb electrode in 1 M H_2SO_4, the $i-E$ relation is shown in Fig. 1. The curve was measured[14] by standing at

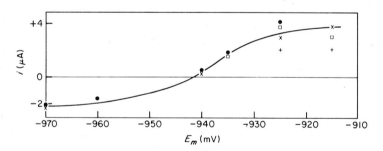

Fig. 1. The approximate steady-state current-potential curve measured (●) 5 min, (×) 30 min, (□) 1 h, (+) 5 h from the start of the experiment for a mechanically polished Pb rotating disc electrode (area = 0·2 cm^2) in 1·0 M deoxygenated H_2SO_4. Rotation speed = 17·25 rev s^{-1}.

each potential for periods up to 5 h. Current flows at potentials positive to the $Pb/PbSO_4/H_2SO_4$ potential and then reaches a steady value. The steady value does not depend on rotation of the electrode. Figure 1 shows clearly that low currents are involved. This is a difficulty in investigating the Pb/H_2SO_4 system and is a unique situation among battery reactions.

Structural and thermodynamic considerations (see Pourbaix diagram in ref. 20) suggest that under the condition of Fig. 1, the only possible solid product is $PbSO_4$.

The anodic dissolution of Pb to form soluble species

The times to reach the steady state in Fig. 1 show that the formation of $PbSO_4$ to form a thick layer is slow. This fact has been utilized to observe Pb dissolution into solution and to investigate the kinetics of this reaction.

(i) *Linear potential sweep on solid Pb.* Figure 2(a) shows a reduction peak, following an anodic current, which disappears on rotating the

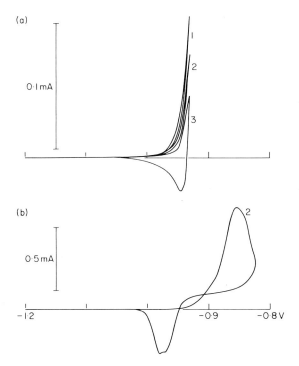

Fig. 2. Current voltage curves showing: (a) Rotation speed dependence of (1) 40, (2) 16, (3) 0 rev s^{-1}. Sweep rate = 30 mV s^{-1}. (b) Potential sweep at 30 mV s^{-1}. Rotation speed = 24 rev s^{-1}. The measurements were made on a mechanically polished Pb rotating disc electrode (surface area = 0·2 cm^2) in 1·0 M deoxygenated H_2SO_4. The potential scales on the abscissa are against E_m.

electrode. This demonstrates conclusively that during the anodic process a solution species can be formed.[15] The shape of the sweep curve is also similar to that calculated by computer.

Figure 2(b) shows that if the anodic current is higher another product can be reduced. This has the characteristics of solid $PbSO_4$ reduction on the surface of the electrode.

The results depend on the treatment of the electrode. Mechanical polishing with β-alumina enhances $PbSO_4$ solid formation; electropolishing delays it and the solution species can be observed more readily.

(ii) *Linear potential sweep on Pb(Hg)*. Similar measurements[16] on concentrated Pb(Hg), which is, in comparison with solid Pb, free of mechanical surface defects, showed the relation between $PbSO_4$ formation and soluble species. In Fig. 3, as $PbSO_4$ increases soluble Pb species decreases. This shows that Pb dissolves into solution between the $PbSO_4$ crystals and not through them. At high sweep rates when $PbSO_4$ does not interfere the dissolution has the characteristics of a reversible $2e$ reaction.[17]

(iii) *Potentiostatic potential step on solid Pb*. The measurements confirm the linear potential sweep results.

(iv) *Rotating ring-disc*. Figure 4 gives the disc current as a function of time and the answering ring current. When the disc current rises due to solid $PbSO_4$ formation, the ring current falls as less soluble species reach it.

The collection efficiency shown in Fig. 5 depends on the history of the electrode as the hydrodynamics depend on the electrode remaining planar. However, newly polished electrodes which are anodically polarized for short times give[18] an observed collection efficiency which approaches the calculated value.

(v) *Rotating disc*. It has been demonstrated in (i) to (iv) that experimentally, dissolution and film formation occur together in the same potential range. However, they can be separately investigated. This allows a more detailed investigation of the kinetics of dissolution. Figure 6 shows stationary $i–E$ curves derived from sweep measurements under conditions where no $PbSO_4$ is formed. The $i–E$ relation

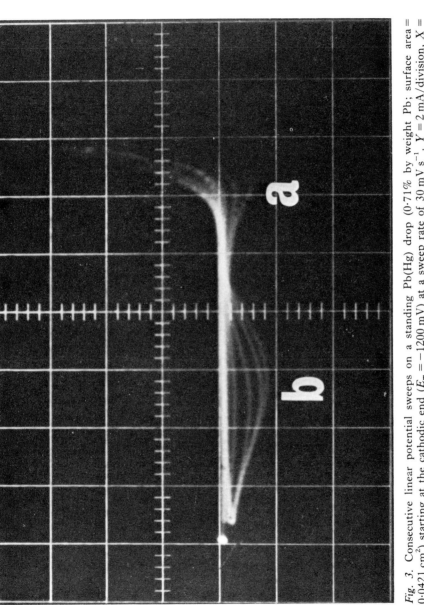

Fig. 3. Consecutive linear potential sweeps on a standing Pb(Hg) drop (0·71% by weight Pb; surface area = 0·0421 cm²) starting at the cathodic end ($E_m = -1200$ mV) at a sweep rate of 30 mV s⁻¹. $Y = 2$ mA/division, $X = 50$ mV/division. As reduction peak (a) decreases, the corresponding peak (b) increases. Potential goes anodic left to right. Solution is 1 M deoxygenated H_2SO_4.

corresponds to a reversible $2e$ electrode reaction (30 mV Tafel slope). The curve depends on H_2SO_4 concentration, suggesting that a complex species is formed. This could either be $Pb^{2+}(SO_4^{2-})_x$,

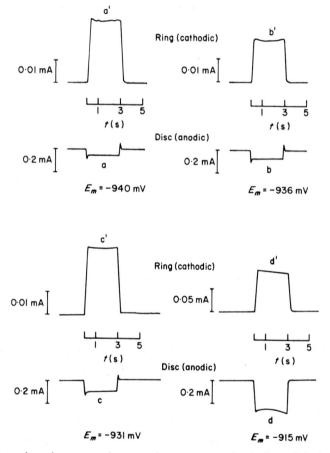

Fig. 4. Potentiostatic current-time transients on a mechanically polished ring-disc electrode, $N_{calc} = 0.57$. Rotation speed = 24 rev s^{-1}. The solution used is 1 M deoxygenated H_2SO_4. The species reaching the ring were under diffusion control.

$Pb^{2+}(HSO_4^-)_x$, where x is 1 or 2, or even small particles of $PbSO_4$ diffusing from the electrode. A definite identification is impossible as the ionic activities of SO_4^{2-} and HSO_4^- are not known. The reaction cannot be investigated[19] in the presence of an inert electrolyte as most common ions, possibly even ClO_4^-, complex more strongly with Pb^{2+} than SO_4^{2-}.

(vi) *a.c. impedance*.[14,62] The electrode impedance, measured under conditions where a film of solid PbSO₄ is absent and where it is present, shows only diffusion characteristics. An example is shown in

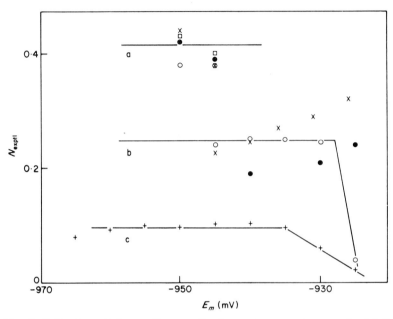

Fig. 5. Collection efficiency, N_{exptl} vs E_m, for various rotation speeds. (a) Freshly mechanically polished electrode, $N_{calc} = 0.57$, (×) 12, (●) 36, (□) 18, (○) 48 rev s⁻¹. (b) Mechanically polished electrode but having been exposed to potentials anodic to $E_m = -935$ mV, (×) $N_{calc} = 0.57$, 24 rev s⁻¹; (●) $N_{calc} = 0.49$, 7.5 rev s⁻¹; (○) $N_{calc} = 0.45$, 15 rev s⁻¹. (c) Electropolished $N_{calc} = 0.48$, 15 rev s⁻¹. The solution is 1.0 M deoxygenated H₂SO₄.

Fig. 7. Before film formation a solution species diffuses from the Pb surface and after film formation, Pb²⁺ or SO₄²⁻ migration and diffusion occurs (PbSO₄ has low electronic conductivity).

The anodic formation and reduction of PbSO₄

Potentiostatic Potential Step

Figure 8 shows an oxidation *i–t* curve to form a thick PbSO₄ layer (of the order of mC cm⁻²) and the corresponding reduction.[14] Only a portion (900 μC cm⁻²) of the anodic charge is recovered. The form of the reduction transient and its potential dependence suggests that a thin layer of PbSO₄ is being reduced at nucleation sites which are determined by defects in the substrate. The layer may be next to the Pb surface. What consequences this has for subsequent formation and reduction of PbSO₄ are not known.

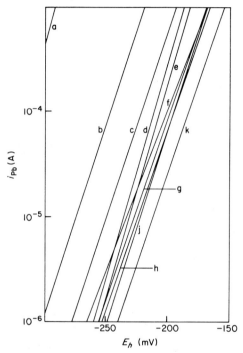

Fig. 6. Tafel plot for deoxygenated (a) 5 M H₂SO₄, (b) 1·0 M H₂SO₄, (c) 0·5 M H₂SO₄, (d) 1·5 M HClO₄/0·5 M NaHSO₄, (e) 1·0 M HClO₄/1·0 M NaHSO₄, (f) 1·0 M HClO₄, (g) 1·9 M HClO₄/0·1 M NaHSO₄, (h) 0·1 M HClO₄, (j) 0·1 M HClO₄, (k) 1·99 M HClO₄/0·01 M NaHSO₄, plotted on a standard hydrogen scale taken from sweep measurements at a constant sweep rate of 10 mV s⁻¹. The Pb rotating disc electrode is 0·21 cm² in area, mechanically polished and rotating at 17·25 rev s⁻¹. The scatter in plots (d) to (k) represents the experimental accuracy.

Shorter times of anodic polarization show that a greater proportion of the charge can be recovered, although the cathodic amount remains fixed.

A significant difference in response is observed if the potential is advanced manually[14] instead of electronically. During the double layer charging time, solid PbSO₄ is obviously nucleated when electronic

$$1. \text{ THE AQUEOUS SYSTEM } Pb^{2+}/Pb$$

potential switching is used. The "manual" i–t curve can be interpreted as evidence for precipitation of $PbSO_4$ from the dissolving Pb complex.

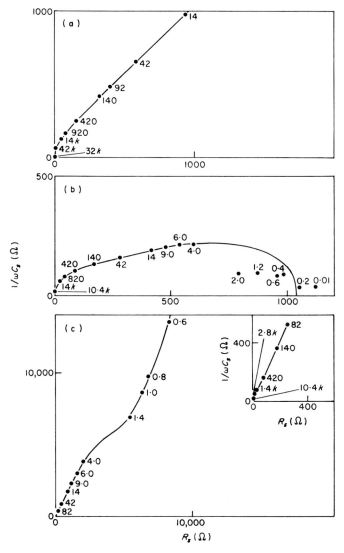

Fig. 7. Complex impedance plane plots at potentials (a) $E_m = -940$ mV, (b) $E_m = -920$ mV, (c) $E_m = -850$ mV, for a mechanically polished Pb rotating disc electrode (surface area $= 0.196$ cm²) in 1.0 M deoxygenated H_2SO_4. Rotation speed $= 17.25$ rev s⁻¹. The frequencies on the complex impedance plots are quoted in Hz.

The effect of dissolved oxygen

The previous sections have dealt with the kinetics of oxygen-free solution. The presence of oxygen has a marked effect[19] on the kinetics of dissolution of the Pb in H_2SO_4. A complete understanding of this mechanism is not yet available.

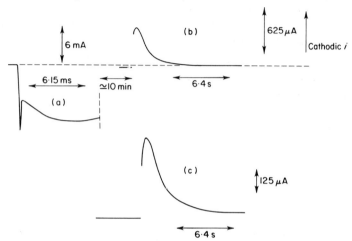

Fig. 8. Curve (a) shows part of a typical growth transient for $PbSO_4$ caused by anodically polarizing a mechanically polished Pb rotating disc electrode (surface area = 0.196 cm^2) in 1 M deoxygenated H_2SO_4 from $E_m = -1200$ mV to $E_m = -850$ mV for 10 min. Curve (b) shows the corresponding reduction transient of this $PbSO_4$ layer by cathodically pulsing from $E_m = -850$ mV to $E_m = -980$ mV. The current zero for curves (a) and (b) is defined at $E_m = -1200$ mV. Curve (c) shows the reduction transient of a similar $PbSO_4$ layer formed at $E = -850$ mV for 20 min and then cathodically pulsing from $E_m = -850$ mV to $E_m = -980$ mV. Rotation speed = 17.25 rev s^{-1}.

Conclusion

The electrochemical measurements demonstrate that formation of solid $PbSO_4$ formed by a solid state nucleation mechanism is a slow process. Until the surface is covered with a complete layer of this film the Pb dissolves as a soluble Pb complex. The formation of the solid $PbSO_4$ probably follows without any appreciable overpotential, although the formation is so slow at low potentials that this is difficult to decide with any accuracy.

A solution precipitation of the soluble Pb complex can be observed under conditions where solid state $PbSO_4$ formation is minimized.

Implication for the mechanism of the negative plate of a battery

The formation of solid state $PbSO_4$ seems to be undesirable for the operation of a battery. However, the foregoing measurements demonstrate that for short times, oxidation of Pb in H_2SO_4 can occur by dissolution of the true surface (estimated by the B.E.T. method). The current per unit area in a battery plate is low (10^{-5} to 10^{-6} A cm^{-2}) so that the Pb plate is actually oxidized at the foot of the curve in Fig. 1. This means that for times of minutes or perhaps hours virtually no solid state $PbSO_4$ would be formed. However, the $PbSO_4$ would be stored as a precipitate from the dissolving species until exhaustion of SO_4^{2-} occurs. This situation is made more favourable when inhibitors (expanders) are added and solid state nucleation of $PbSO_4$ is suppressed further.[17] It is also probable that dissolved O_2 has some effect on battery operation; however, this has yet to be confirmed.

The Kinetics of Metal Dissolution

A number of simple models have been calculated.[22–24] These will be briefly presented in this section. Only the situations which most commonly occur during dissolution of metals will be discussed. The most useful electrochemical method for following the kinetics is the potential step method as the reactions are non-linear in time and potential.

Another method is the small signal a.c. impedance method. This is a useful method as the steady state of the system can be investigated. However, it can be difficult to measure and interpret if the processes at the electrode are complex and the system reaches a steady state slowly.

Constant potential measurements

Dissolution from a Plane ($v_3 > v_2$ in Figure 10)

If the metal dissolves smoothly and rapidly (reversibly) the situation of Fig. 9 can be envisaged. The metal dissolves smoothly, the resulting ions leave the surface by diffusion. The problem is similar to the corresponding problem in heat, i.e. the melting of a solid.[21] It can be shown (see Appendix 1) that under most conditions the current time curve for an anodic dissolution will be

$$i = nFC_1^s \left(\frac{D_1}{\pi t}\right)^{1/2} \tag{6}$$

Fig. 9. Axis system for Appendix 1. The dashed line is the original position of the surface and full line is the position of the surface at time t.

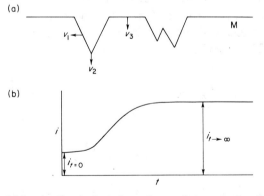

Fig. 10. (a) Model for the dissolution of a surface to form etch pits. (b) i–t transient to a potential pulse corresponding to the model shown in (a). M is the metal.

where C_1^s is the Nernst concentration of the dissolving species at the moving interface. This is the well-known equation for the current determined by the diffusion of one species from the electrode. The potentiostatic control of potential means that the double-layer charging current can be neglected and the ohmic resistance R_Ω reduced to a minimum or compensated by positive feedback.

If a particular metal dissolution follows equation (1), then C_1^s can be calculated and its potential dependence investigated.

A reaction controlled by the rate of the interfacial reaction and diffusion will give the well-known response to a potential step, height η,

$$i = i_0 \left[\exp \left(\frac{\alpha n F \eta}{RT} \right) - \exp \left(\frac{-(1-\alpha)nF\eta}{RT} \right) \right] \exp \lambda^2 t \,.\, \mathrm{erfc} \lambda t^{1/2} \quad (7)$$

where

$$\lambda = \frac{\dfrac{i_0}{nF} \exp \dfrac{\alpha n F \eta}{RT}}{C_1^b D_1^{1/2}} \quad (8)$$

and the species 1 diffuses away from the metal, which is assumed to have a stationary surface. This probably introduces little error.

Dissolution with the Formation of Etch Pits ($v_3 < v_2$ in Figure 10)

Figure 10(a) shows diagrammatically the formation of the etch pits. The model can be described mathematically[25] by assuming that the etch pits are circular pyramids formed at random. It will be assumed that the electrochemical reaction is slow with the various rates given by v_1, v_2, v_3. The calculation is reproduced in Appendix 2. The result is

$$i = i_{t \to \infty} \left[1 - \exp \left(\frac{-t^2}{\tau_0^2} \right) \right] + i_{t=0} \left[\exp \left(\frac{-t^2}{\tau_0^2} \right) \right] \quad (9)$$

where

$$\frac{1}{\tau_0^2} = + \pi N_0 v_1^2 \left(1 - \frac{v_3}{v_2} \right)^2 \quad (10)$$

Equation (9) is illustrated in Fig. 10(b). By fitting, $1/\tau_0^2$ can be found and hence one of N_0, v_1, v_2, v_3 if the others are known. A corresponding equation has also been calculated[25] for the case where the growing centres are located in rows with the average distance between rows different from the average distance within a row. However, this introduces a further parameter into the equation. Further calculations can be made, assuming the surface grows via a two-dimensional growth mechanism.

Dissolution to Form a Surface Film

If the film-forming reaction can be considered to form circular pyramids in the early stages of Fig. 11(a) then the i–t curve will be formally described by equation (9) and again by an i–t curve as in Fig. 11(b), which is of the same form as Fig. 10(b). It is assumed that the slow

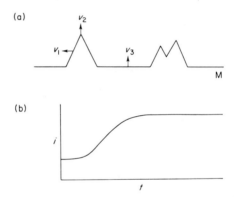

Fig. 11. (a) Model for the formation of a deposit on an electrode surface. (b) Corresponding i–t transient to a potential pulse. M is the metal.

step is at the growing edge and there is no hindrance from supply of material through the growing nuclei to the edge.

Perhaps a more realistic model situation is to consider[26] that the edge grows by diffusion of material through the nuclei. A model of this type can only be calculated approximately, assuming that the radius [Fig. 12(a)] has a particular time dependence $r \propto t^{1/2}$. The i–t transient can be approximately described by

$$i = ct^{-1/2}(1 - \exp(-kt)) \tag{11}$$

where k is a constant which characterizes the growth of separate nuclei and c characterizes the transport of the growing edge. The form of equation (11) is shown in Fig. 12(b).

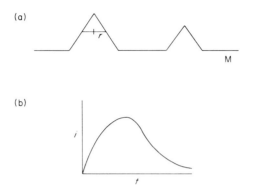

Fig. 12. (a) Model for the formation of a deposit growing under transport control, $r \propto t^{1/2}$. (b) Corresponding i–t transient. M is the metal.

Dissolution and Transport in a Film

If the electronic conductivity of the film is high, then the diffusion of a single ionic species can be the rate-determining mode of transport and

$$i = \frac{nFD_1 C_1^s}{x} = \frac{nF\rho}{M}\left(\frac{dx}{dt}\right) \tag{12}$$

hence

$$i = nF\left(\frac{\rho D_1 C_1^s}{2M}\right)t^{-1/2} \tag{13}$$

and C_1^s is given by the Nernst equation.

If a potential gradient can exist and the rate-determining reaction is at the metal film boundary, then for an i–E reaction

$$i = i_0\left[\exp\left(\frac{\alpha nF\eta}{RT}\cdot\frac{a}{x}\right) - \exp\left(\frac{-(1-\alpha)nF\eta}{RT}\cdot\frac{a}{x}\right)\right] \tag{14}$$

When the exponentials can be expanded, i.e. $(\alpha nF/RT)\cdot a/x < 1$,

$$i = i_0\frac{nF\eta}{RT}\cdot\frac{a}{x} = \frac{nF\rho}{M}\left(\frac{dx}{dt}\right) \tag{15}$$

and

$$i = \left(\frac{i_0 nF\eta}{2RT}\right)^{1/2} \cdot \left(\frac{nF\rho}{M}\right)^{1/2} \cdot t^{-1/2} \tag{16}$$

Other more complex situations have been discussed by a number of authors, for example ref. 27 for Fe dissolution.

a.c. impedance measurements

Dissolution from a Plane ($v_3 < v_2$ in Figure 10)

Similar equations to (6) and (7) may be derived for the a.c. impedance measured by imposing a small signal on the steady state. The most useful situation is to use the method pioneered by Epelboin and co-workers in which measurements are made at a rotating disc so that diffusion processes in the solution are in the steady state. When $(D/\omega)^{1/2} < \delta$, the a.c. measurement is independent of δ. If the d.c. process is reversible, then for an a.c. reversible process corresponding to equation (6)

$$Z_{j\omega} = R_\Omega + (\sigma\omega^{-1/2}) - j(2\sigma^2 C_d + \sigma\omega^{-1/2}) \tag{17}$$

The ohmic resistance R_Ω and the double layer capacity C_d enter directly into the measurement.

A reaction controlled by the rate of the interfacial reaction and diffusion gives the response

$$Z_{j\omega} = R_\Omega + (\theta + \sigma\omega^{-1/2}) - j(2\sigma^2 C_d + \sigma\omega^{-1/2}) \tag{18}$$

for small θ and

$$Z_{j\omega} = R_\Omega + \frac{\theta}{1 + \omega^2 C_d^2 \theta^2} - j\frac{\omega C_d \theta^2}{1 + \omega^2 C_d^2 \theta^2} \tag{19}$$

for large θ. Equation (19) is the well-known Cole–Cole semicircle.

The charge transfer resistance, θ, is the slope of the i–E curve at the potential of measurement and $\sigma = RT/n^2 F^2 \sqrt{2} \cdot (1/C_i^s D_i^{1/2})$ in this case, for the diffusion of a single species.

Equations (17), (18) and (19) are limiting cases of the equation

$$Z_{j\omega} = R_\Omega + \cfrac{1}{j\omega C_d + (\theta + \sigma\omega^{-1/2} - j\sigma\omega^{-1/2})^{-1}} \qquad (20)$$

which can be used if on-line computing facilities are available.

The case when the d.c. process is not reversible has also been treated in some detail.[30] See Appendix 3 for the basis of equations (17) and (18) and the general definition of θ and σ.

Dissolution with the Formation of Etch Pits ($v_3 < v_2$ in Figure 10)

The small signal a.c. impedance corresponding to the situation of Fig. 10(a) can easily be calculated as in Appendix 4. The result plotted in the complex impedance plane is shown in Fig. 13. Similar diagrams have been observed in a number of metal depositing systems.[32]

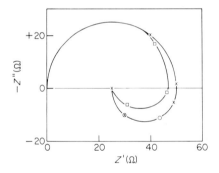

Fig. 13. Calculated impedance diagram $-Z''$ against Z' according to equation (61) with a value of double layer capacity 10^{-5} F added in parallel. The values

$$\left(\frac{\partial i}{\partial E}\right)_f = \frac{1}{50}\,\Omega^{-1}; \qquad \left(\frac{\partial i}{\partial f}\right)_E \cdot \left(\frac{\partial f}{\partial E}\right)_{\omega \to 0} = \frac{1}{50}\,\Omega^{-1}$$

were used to demonstrate the behaviour. □, $k = 100\ \mathrm{s}^{-1}$, frequency = 10, 400, 1000 rad s⁻¹. ○, $k = 10\ \mathrm{s}^{-1}$, frequency = 10, 30 rad s⁻¹. ×, $k = 1\ \mathrm{s}^{-1}$, frequency = 10^{-2}, 1, 10, 100, 1000 rad s⁻¹.

Conclusions

The equations for etch pit formation and film formation are identical, which means that evidence for the actual process must be obtained from electronic diffraction or S.E.M. measurements. Similarly, equations (6), (13) and (16) have the same form. Supplementary evidence can also be obtained from electrochemical reduction measurements, ring disc measurements and (for large changes) weight loss. It is possible in suitable cases to obtain *in situ* identification of a film by ellipsometry or light reflectance techniques.

The Effect of Expanders

It is well known that the capacity of a negative battery plate falls off at low temperatures and the addition of expanders both raises the capacity of the plate and also its recycling life. Some loss of capacity in the plate is due to formation of thick $PbSO_4$ layers which are apparently resistant to recharge ("hard sulphate"). Most commercial batteries use a mixture of carbon black, $BaSO_4$, lignin derivatives and trace amounts of various metals, the composition of which may be varied to give the plates differing properties. An attempt[34,35] was made to group expanders by the effect they are expected to produce in the battery.

It seems probable[36-38] that the lignin is adsorbed on to the surface of lead; Pierson *et al.*[39] have shown that it is capable of altering the geometry of lead crystals formed both on charge and discharge. Russian workers[40,41] have shown that the organic expanders can modify $PbSO_4$ crystals in the absence of $BaSO_4$ and they postulate that the expander is adsorbed on to the growing faces of the $PbSO_4$ nuclei and thus yields a porous layer.

Zachlin[42] studied the effect of adding an expander and $BaSO_4$ to a battery plate and concluded that $BaSO_4$ was only necessary for maintaining capacity during prolonged cycle life. It may also reduce passivation during discharge by providing alternative nucleation sites for $PbSO_4$ (as $BaSO_4$ is isomorphous to $PbSO_4$).

Anodic Corrosion of the $PbSO_4$ Layer

At low dissolution rates, Pb passivates as $PbSO_4$ only (see section on anodic dissolution of lead). However, on further anodic oxidation and subsequent thickening of the corrosion layer, basic lead sulphates and

PbO have been detected. Burbank[43] has studied the composition of such films as a function of potential in great detail. The first passivating species formed on a clean electrode is $PbSO_4$;[44] this appears to be permeable to H^+ and H_2O but not SO_4^{2-}. Lander,[45] using potentiostatic control and X-ray analysis, established that corrosion to PbO occurred below this film of $PbSO_4$.

To explain the corrosion of Pb and subsequent formation of basic lead sulphates, Pavlov[46] introduced the idea of differing fluxes of Pb^{2+} and SO_4^{2-} which resulted in the alkalinization of the metal surface. In a later paper, Pavlov and Popova[47] interpreted the alkalinization process as the anodic corrosion acting as a perm-selective membrane.

Ruetschi et al.[48] studied the thin film formation of $PbSO_4$ at low overpotentials on freshly prepared Pb–Ca alloys under potentiostatic control. They discussed the formation of basic lead sulphates in terms of high local pH values in the film interior and critical thicknesses of the $PbSO_4$ layer. Stabilization of these high local pH values is assumed to occur by decomposition of H_2O molecules which can penetrate the corrosion layer.

In a further study, Ruetschi[49] likened $PbSO_4$ corrosion layers on Pb in H_2SO_4 to those of Pb^{2+} and SO_4^{2-} interdiffusing into a cellophane foil. The perm-selective $PbSO_4$ (i.e. permeable to H_2O and OH^-) precipitation membranes in cellophane thus produced were compared with the situation occurring at the Pb/H_2SO_4 interface and, allowing for the diffusion potential in the film, gave good agreement.

Current Distribution in Flooded Porous Electrodes

Flooded porous electrodes by virtue of their practical importance in batteries, fuel cells and electrodes in organic synthesis have been studied in detail since 1948.[50] The obvious advantages of increased surface area for such an electrode is offset by limited transport of reacting species in the pores which may be reduced by pumping electrolyte/reactants through the electrode. However, this situation will not be considered.

De Levie[51] has considered the problem of current distribution for uniform interconnecting pores. An alternative approach,[52] which is probably more valid for battery plate application in that it can allow for a distribution of pore sizes in the electrode, is to describe the plates as being made up of two continua; one the electrode matrix and the other, the solution, filling all the unoccupied space within the matrix. A review[53] has appeared which uses this method to highlight the problems associated with porous electrodes used in batteries.

Describing porous electrodes in terms of a mathematical model which relates measurable parameters such as porosity, surface area, current distribution etc. is an active field of research as it lends itself to optimizing electrode design. In this respect,[54–58] many authors, of which only a few have been cited, have shown how this may be achieved in practical battery applications.

List of Symbols

a	parameter determining the amount of potential drop at the interface
a_1	activity of species 1
c	a constant in equation (11)
C_d	double layer capacity
C_1^s	surface concentration of species 1
C_1^b	bulk concentration of species 1
C_s	series capacitance
D_1	diffusion coefficient of species 1
E	potential of working electrode
E_h	potential against the hydrogen scale
E_m	potential against the $Hg/Hg_2SO_4/1\cdot0$ M H_2SO_4 reference scale
$E_{1/2}$	half wave potential
f_e	equilibrium surface area about f_0
f_0	a fixed value of surface area
F	Faraday's constant
i	current
i_0	exchange current
j	$\sqrt{-1}$
k	a constant in equation (11)
k_1	a rate constant in equation (59)
K_s	thermodynamic solubility constant
M	molecular weight
n	no. of electrons transferred
N_0	no. of growth centres
p	Laplace variable
r	radius of nuclei formed
R	gas constant
R_s	series resistance

R_Ω	solution resistance
S	area of surface covered
t	time
T	absolute temperature
v_1, v_2, v_3	rates of growth defined in Fig. 10
x	distance
Z	electrode impedance
Z'	real impedance component
Z''	imaginary impedance component
α	transfer coefficient
β_1	conventional stability constants for equilibria
δ	diffusion layer thickness
ΔE	a potential change
Δf	a surface area change
Δf_e	an equilibrium surface area change
η	overpotential
θ	charge transfer resistance
λ	a constant defined in equation (8)
λ_1	a constant in equation (27)
ρ	density
σ_1	Warburg coefficient of species 1
τ_0	a constant defined by equation (10)
ω	frequency of applied a.c. signal
$\overline{dC_1^s}, \overline{dC_2^s}$	Laplace transforms of complete differential of concentration of species 1 or 2 at the interface
$\overline{dE}$	Laplace transform of complete differential of potential
$\overline{di}$	Laplace transform of complete differential of current.

References

1. H. S. Harned and W. J. Hamer, *J. Amer. Chem. Soc.* **59**, 33 (1935).
2. J. Shrawder and I. Cowperthwaite, *J. Amer. Chem. Soc.* **56**, 2340 (1934).
3. W. A. Latimer, "Oxidation Potentials". Prentice-Hall, Englewood Cliffs, New Jersey (1952).
4. T. F. Sharpe *in* "Encyclopedia of Electrochemistry of the Elements" (A. J. Bard, ed.), Vol. 1, p. 235. Dekker, New York (1973).
5. W. R. Carmody, *J. Amer. Chem. Soc.* **51**, 2905 (1929).
6. A. J. de Bethune and N. S. Loud *in* "The Encyclopedia of Electrochemistry" (C. A. Hampel, ed.), p. 414. Reinhold, New York and London (1964).
7. A. M. Bond and G. Hefter, *J. Electroanal. Chem.* **34**, 227 (1972).
8. D. N. Craig and G. W. Vinal, *J. Res. Nat. Bur. Stand* **22**, 55 (1939).
9. T. F. Young, L. F. Maranville and H. M. Smith *in* "The Structure of Electrolyte Solutions" (W. J. Hamer, ed.). Wiley, New York (1959).

10. K. H. Lieser, G. Beyer and E. Lakatos, *Z. Anorg. Chem.* **339**, 208 (1965).
11. B. Van't Riet and I. M. Kolthoff, *J. Phys. Chem.* **64**, 1045 (1960).
12. I. M. Korenman, *Izv. Vyssh. Uchebn. Zaved. Khim.* **4**, 554 (1961).
13. G. Gardner and G. H. Nancollas, *Anal. Chem.* **42**, 794 (1970).
14. A. N. Fleming and J. A. Harrison, *Electrochim. Acta* **21**, 905 (1976).
15. G. Archdale and J. A. Harrison, *J. Electroanal. Chem.* **34**, 21 (1972).
16. G. Archdale and J. A. Harrison, *J. Electroanal. Chem.* **43**, 321 (1973).
17. G. Archdale and J. A. Harrison, *J. Electroanal. Chem.* **47**, 93 (1973).
18. G. Archdale and J. A. Harrison, *J. Electroanal. Chem.* **39**, 357 (1972).
19. A. N. Fleming, J. A. Harrison and J. Thompson *in* "Power Sources 5" (D. H. Collins, ed.), pp. 1–14. Academic Press, London and New York (1975).
20. "Atlas d'Equilibres Electrochemiques" (M. Pourbaix, ed.), pp. 286–293. Gauthier–Villars, Paris (1963).
21. H. S. Carslaw and J. C. Jaeger, "Conduction of Heat in Solids", 2nd edn. Oxford University Press, Oxford (1959).
22. M. Fleischmann and H. R. Thirsk *in* "Advances in Electrochemistry and Electrochemical Engineering" (P. Delahay, ed.), Vol. 3, p. 123. Interscience, New York (1963).
23. J. A. Harrison and H. R. Thirsk *in* "Electroanalytical Chemistry" (A. J. Bard, ed.), Vol. 5, p. 67. Dekker, New York (1971).
24. R. D. Armstrong, J. A. Harrison and H. R. Thirsk, *Corrosion Sci.* **10**, 679 (1970).
25. W. Schaarwächter and K. Lücke, *Z. Phys. Chem.* **53**, 367 (1967).
26. W. Davison and J. A. Harrison, *J. Electroanal. Chem.* **44**, 213 (1973).
27. C. Lukac, J. B. Lumsden, S. Smialowska and R. W. Staehle, *J. Electrochem. Soc.* **122**, 1571 (1975).
28. M. M. Clarke, J. A. Harrison and H. R. Thirsk, *Z. Phys. Chem.* **98**, 153 (1975).
29. J. A. Harrison and S. K. Rangarajan, *Far. Soc. Symp.* No. 12. Chemical Society, London (1977).
30. M. Sluyters-Rehbach and J. H. Sluyters *in* "Electroanalytical Chemistry" (A. J. Bard, ed.), Vol. 4, p. 32. Dekker, New York (1970).
31. J. H. Sluyters, *Rec. Trav. Chim.* **79**, 1092 (1960).
32. W. Davison, J. A. Harrison and J. Thompson, *Far. Soc. Disc.* **171**, No. 56. Chemical Society, London (1973).
33. P. Delahay, "New Instrumental Methods in Electrochemistry". Interscience, New York (1954).
34. E. J. Ritchie, *Trans. Electrochem. Soc.* **92**, 227 (1947).
35. E. J. Ritchie, *Trans. Electrochem. Soc.* **100**, 53 (1953).
36. G. Archdale and J. A. Harrison, *J. Electroanal. Chem.* **47**, 93 (1973).
37. T. F. Sharpe, *Electrochim. Acta* **14**, 635 (1969).
38. T. F. Sharpe, *J. Electrochem. Soc.* **116**, 1639 (1969).
39. J. R. Pierson, P. Gurbisky, A. C. Simon and S. M. Caulder, *J. Electrochem. Soc.* **117**, 1463 (1970).
40. Y. B. Kasparov, E. G. Yampol'skaya and B. N. Kabanov, *Zh. Prkl. Khim.* **37** (9), 1936 (1964).

41. E. G. Yampol'skaya, M. I. Ershova, I. I. Astakhov and B. N. Kabanov, *Elektrokhimiya* 2, 1327 (1966).
42. A. C. Zachlin, *J. Electrochem. Soc.* 98, 325 (1951).
43. J. Burbank, *J. Electrochem. Soc.* 106, 369 (1959).
44. D. Pavlov and N. Iordanov, *J. Electrochem. Soc.* 117, 1103 (1970).
45. J. J. Lander, *J. Electrochem. Soc.* 98, 213 (1951).
46. D. Pavlov, *Electrochim. Acta* 13, 2051 (1968).
47. D. Pavlov and R. Popova, *Electrochim. Acta* 15, 1483 (1970).
48. P. Ruetschi and R. T. Angstadt, *J. Electrochem. Soc.* 111, 1323 (1964).
49. P. Ruetschi, *J. Electrochem. Soc.* 120, 331 (1973).
50. V. S. Daniel'-Bek, *Zh. Fiz. Khim.* 22, 697 (1948).
51. R. De Levie, "Advances in Electrochemistry and Electrochemical Engineering" (P. Delahay, ed.), Vol. 6, p. 329. Interscience, New York (1967).
52. J. Newman and C. W. Tobias, *J. Electrochem. Soc.* 109, 1183 (1962).
53. J. Newman and W. Tiedemann, *Amer. Inst. Chem. Eng. J.* 21(1), 25 (1975).
54. D. Simonsson, *J. Electrochem. Soc.* 120(2), 151 (1973).
55. J. S. Dunning, D. N. Bennion and J. Newman, *J. Electrochem. Soc.* 120(7), 906 (1973).
56. T. Katan and H. F. Bauman, *J. Electrochem. Soc.* 122(1), 77 (1975).
57. W. Tiedemann and J. Newman, *J. Electrochem. Soc.* 122(1), 70 (1975).
58. K. Micka and I. Rousar, *Coll. Czech. Chem. Commun.* 40(4), 921 (1975).
59. D. E. Smith *in* "Electroanalytical Chemistry" (A. J. Bard, ed.), Vol. 1, p. 1. Dekker, New York (1966).
60. S. K. Rangarajan, *J. Electroanal. Chem.* 55, 297, 329, 337, 363 (1974).
61. J. A. Harrison, D. R. Sandbach and P. J. Stronach, *Electrochim. Acta* (in press).
62. A. N. Fleming, J. A. Harrison and J. M. Ponsford, *Electrochim. Acta* 22, 1371 (1977).

Appendix 1

The effect of a moving boundary as in a metal dissolution, on the rate of a reversible electrochemical reaction can be easily calculated from the similar problem in heat transfer (Neumann's solution).[21] If a species with concentration C_1 diffuses then

$$\frac{\partial C_1}{\partial t} = D_1 \frac{\partial^2 C_1}{\partial x^2} \tag{21}$$

If the boundary is at position X then

$$t > 0 \qquad x = X \qquad C_1 = C_1^s \text{ (given by the Nernst equation)} \tag{22}$$

$$x \to \infty \qquad C_1 \to 0 \tag{23}$$

and

$$t = 0 \qquad X = 0 \qquad C_1 = 0 \tag{24}$$

If the current is given by

$$i = -nFD_1 \left(\frac{\partial C_1}{\partial x} \right)_{x=X} \tag{25}$$

The solution of equation (21) is

$$C_1 = A \operatorname{erfc} \left(\frac{x}{2(D_1 t)^{1/2}} \right) \tag{26}$$

where A is a constant and erfc is the complementary error function. Assuming that

$$X = 2\lambda_1 (D_1 t)^{1/2} \tag{27}$$

then λ_1 is a constant, to be determined. In addition to conditions (22)–(24)

$$-D_1 \left(\frac{\partial C_1}{\partial x} \right)_{x=X} = \frac{\rho}{M} \left(\frac{dX}{dt} \right) \tag{28}$$

then from (26), (27) and (28)

$$D_1 A \exp(-\lambda_1^2) = \frac{\lambda_1 \rho \pi^{1/2} D_1}{M} \tag{29}$$

condition (22) gives A and hence

$$D_1 \frac{C_1^s}{\operatorname{erfc}(\lambda_1)} \exp(-\lambda_1^2) = \frac{\lambda_1 \rho \pi^{1/2} D_1}{M} \tag{30}$$

λ_1 can be determined from the constants ρ, M, C_1^s and the current determined by equation (25)

$$i = nF \frac{\rho \lambda_1}{M} \sqrt{\frac{D_1}{t}} \tag{31}$$

This is to be compared with equation (6)

$$i = nFC_1^s \sqrt{\frac{D_1}{\pi t}} \tag{32}$$

for dissolution from a stationary plane. It is clear that i is always proportional to $t^{-1/2}$. When C_1^s is small, equation (31) becomes (32). However, when C_1^s is large the current from (31) is larger or smaller than that predicted from (32), depending on the direction of movement of x.

Appendix 2

The effective radius of a slice of the growing circular pyramid under the conditions of Figs 10(a) or 11(a) is

$$r = v_1\left(1 - \frac{v_3}{v_2}\right)t \tag{33}$$

The cross-sectional area fraction of the surface covered, S, can be corrected for overlap

$$dS_t = (1 - S_t)N_0 2\pi r \, dr \tag{34}$$

which on substituting for r in terms of t and integrating from 0 to t becomes

$$S_t = 1 - \exp\left(-\frac{t^2}{\tau_0^2}\right) \tag{35}$$

where τ_0 is defined previously for N_0 centres. The current i is then

$$i = \frac{nF\rho}{M}(v_2 S + v_3(1 - S)) \tag{36}$$

$$i = i_{t\to\infty}\left(1 - \exp\left(-\frac{t^2}{\tau_0^2}\right)\right) + i_{t=0}\left(\exp\left(-\frac{t^2}{\tau_0^2}\right)\right) \tag{37}$$

Other models can be considered based on a two-dimensional nucleation and growth mechanism (*see* refs 28 and 29 for recent calculations). The reader is advised to consult the literature for details.

Appendix 3

The general Faradaic response for a small amplitude a.c. potential signal superimposed on d.c. can be derived as follows. Given a redox reaction where 1 and 2 are electrochemically active species, i.e.

$$1 \underset{i}{\overset{ne}{\rightleftharpoons}} 2 + e, \tag{38}$$

then in general

$$\overline{di} = \overline{dE}\left(\frac{\partial i}{\partial E}\right)_{C_1^s, C_2^s} + \overline{dC_1^s}\left(\frac{\partial i}{\partial C_1^s}\right)_{E, C_2^s} + \overline{dC_2^s}\left(\frac{\partial i}{\partial C_2^s}\right)_{E, C_1^s} \tag{39}$$

for a perturbation $\overline{dE}$ and its response $\overline{di}$. The quantities $\overline{di}$, $\overline{dE}$, $\overline{dC_1^s}$ and $\overline{dC_2^s}$ are Laplace transforms.

Solution of the Laplace transforms of Fick's second law

$$\frac{\partial C}{\partial t} = D\left(\frac{\partial^2 C}{\partial x^2}\right) \tag{40}$$

gives

$$\overline{dC_2^s} = -\frac{\overline{di}}{nF\sqrt{pD_2}} \quad \text{and} \quad \overline{dC_1^s} = +\frac{\overline{di}}{nF\sqrt{pD_1}} \tag{41}$$

Hence

$$\left(\frac{\overline{di}}{\overline{dE}}\right) = \frac{1}{Z_{(p)}} = \frac{\left(\frac{\partial i}{\partial E}\right)_{C_1^s, C_2^s}}{1 + \frac{1}{nF\sqrt{pD_2}}\left(\frac{\partial i}{\partial C_2^s}\right)_{E, C_1^s} - \frac{1}{nF\sqrt{pD_1}}\left(\frac{\partial i}{\partial C_1^s}\right)_{E, C_2^s}} \tag{42}$$

which on substituting $p = j\omega$ gives the familiar Faradaic a.c. impedance[30,31,33]

$$Z_p = \frac{1}{\left(\frac{\partial i}{\partial E}\right)_{C_1^s, C_2^s}} + \frac{1}{nF\sqrt{pD_2}} \cdot \frac{\left(\frac{\partial i}{\partial C_2^s}\right)_{E, C_1^s}}{\left(\frac{\partial i}{\partial E}\right)_{C_1^s, C_2^s}} - \frac{1}{nF\sqrt{pD_1}} \cdot \frac{\left(\frac{\partial i}{\partial C_1^s}\right)_{E, C_2^s}}{\left(\frac{\partial i}{\partial E}\right)_{C_1^s, C_2^s}} \tag{43}$$

and for the diffusion of a single species

$$Z_{j\omega} = \frac{1}{\left(\frac{\partial i}{\partial E}\right)_{C_1^s, C_2^s}} + \frac{1}{nF\sqrt{j\omega D_2}} \cdot \frac{\left(\frac{\partial i}{\partial C_2^s}\right)_{E, C_2^s}}{\left(\frac{\partial i}{\partial E}\right)_{C_1^s, C_2^s}} \tag{44}$$

$$= \theta + (\sigma_2\omega^{-1/2} - j\sigma_2\omega^{-1/2}) \tag{45}$$

This equation defines the Warburg coefficient σ and charge transfer resistance θ which can then be inserted into equation (17) or (18) or, in the general case, equation (20).

Similar considerations can be applied to a 2e reaction, which is more appropriate for the case of Pb oxidation. If[59,60,61]

$$1 \underset{i_1}{\overset{n_1 e}{\rightleftharpoons}} 2 \underset{i_2}{\overset{n_2 e}{\rightleftharpoons}} 3 \tag{46}$$

then

$$\overline{di_1} = \left(\frac{\partial i_1}{\partial E}\right)_{C_1^s, C_2^s} \overline{dE} + \left(\frac{\partial i_1}{\partial C_1^s}\right)_{E, C_2^s} \overline{dC_1^s} + \left(\frac{\partial i_1}{\partial C_2^s}\right)_{E, C_1^s} \overline{dC_2^s} \qquad (47)$$

$$\overline{di_2} = \left(\frac{\partial i_2}{\partial E}\right)_{C_2^s, C_3^s} \overline{dE} + \left(\frac{\partial i_2}{\partial C_2^s}\right)_{E, C_3^s} \overline{dC_2^s} + \left(\frac{\partial i_2}{\partial C_3^s}\right)_{E, C_2^s} \overline{dC_3^s} \qquad (48)$$

and the solution of Fick's second law in terms of the Laplace transforms and the Laplace variable gives:

$$\overline{dC_1^s} = \frac{\overline{di_1}}{n_1 F \sqrt{pD_1}} \qquad (49)$$

$$\overline{dC_2^s} = \left(\frac{\overline{di_2}}{n_2 F} - \frac{\overline{di_1}}{n_1 F}\right) \frac{1}{\sqrt{pD_2}} \qquad (50)$$

$$\overline{dC_3^s} = -\frac{\overline{di_2}}{n_2 F \sqrt{pD_3}} \qquad (51)$$

Hence

$$y_1 = \frac{\overline{di_1}}{\overline{dE}} = \frac{1}{\alpha_1}\left(\frac{\partial i_1}{\partial E}\right)_{C_1^s \cdot C_2^s} + \frac{1}{\alpha_1}\left(\frac{\partial i_1}{\partial C_2^s}\right)_{E, C_1^s} v \qquad (52)$$

$$y_2 = \frac{\overline{di_2}}{\overline{dE}} = \frac{1}{\alpha_2}\left(\frac{\partial i_2}{\partial E}\right)_{C_2^s \cdot C_3^s} + \frac{1}{\alpha_2}\left(\frac{\partial i_2}{\partial C_2^s}\right)_{E, C_3^s} v \qquad (53)$$

where

$$\alpha_1 = \left[1 - \frac{1}{n_1 F \sqrt{D_1 p}}\left(\frac{\partial i_1}{\partial C_1^s}\right)_{E, C_2^s}\right] \qquad (54)$$

$$\alpha_2 = \left[1 + \frac{1}{n_2 F \sqrt{D_3 p}}\left(\frac{\partial i_2}{\partial C_3^s}\right)_{E, C_2^s}\right] \qquad (55)$$

$$v = \frac{\dfrac{1}{n_2 F \alpha_2}\left(\dfrac{\partial i_2}{\partial E}\right)_{C_2^s \cdot C_3^s} - \dfrac{1}{n_1 F \alpha_1}\left(\dfrac{\partial i_1}{\partial E}\right)_{C_1^s \cdot C_2^s}}{\left[\sqrt{pD_2} + \dfrac{1}{n_1 F \alpha_1}\left(\dfrac{\partial i_1}{\partial C_2^s}\right)_{E, C_1^s} - \dfrac{1}{n_2 F \alpha_2}\left(\dfrac{\partial i_2}{\partial C_2^s}\right)_{E, C_3^s}\right]} \qquad (56)$$

The impedance is then given in general by

$$Z_p = \frac{1}{y_p} = \frac{1}{(y_1 + y_2)} \qquad (57)$$

Equation (57), with (52), (53), (54), (55) and (56), has been cal-culated,[61] by computer, for the case of a small amplitude signal on an imposed d.c. condition by substituting $p \to j\omega$ and the actual values of the partial derivative coefficients. The theory and experiments[61] have been compared for the case of the deposition of Cu, Zn and Cd.

Appendix 4

Suppose a surface state, characterized by the surface area f_0, exists. Perturbation by a potential change ΔE gives

$$\Delta f_e = f_e - f_0 = \left(\frac{\partial f}{\partial E}\right)_{\omega \to 0} . \Delta E \tag{58}$$

If $\Delta f = f - f_0$ then

$$\frac{d}{dt} \Delta f = -k_1(\Delta f - \Delta f_e) \tag{59}$$

Assuming that the current is a function of area and potential

$$i = i_{(f,E)} \tag{60}$$

then expanding equation (60), taking the Laplace and substituting $p = j\omega$ gives the small signal impedance $Z_{j\omega}$

$$\frac{1}{Z_{j\omega}} = \left(\frac{\partial i}{\partial E}\right)_f + \left(\frac{\partial i}{\partial f}\right)_E \left(\frac{\partial f}{\partial E}\right)_{\omega \to 0} \left[\frac{k(k - j\omega)}{k^2 + \omega^2}\right] \tag{61}$$

This equation is plotted in Fig. 13 for some trial values of the parameters. It shows how the surface relaxation due to electrocrystal-lization could appear in an impedance measurement as an inductance.

2   The Aqueous System Pb^{4+}/Pb^{2+}: Electrochemical Aspects

N. A. HAMPSON

Department of Chemistry, Loughborough University, Leicestershire

Introduction

The whole of the electrochemistry of this system in aqueous solution involves the compound lead dioxide PbO_2, because only this Pb(IV) compound is stable. It is the purpose of this section to review the important features of the PbO_2/Pb(II) systems insofar as they have been investigated. A number of previous reviews have dealt with this area in some depth and the reader is referred to these.[1-7]

The Preparation and Properties of Lead Dioxide

Lead dioxide is readily produced as an insoluble phase when lower valency states are subjected to strong oxidizing conditions.[8,9] Chemically, this is effected by the oxidation of Pb(II) salt solutions using, for example, chlorine, bromine or hydrogen peroxide. The most convenient method of production is electrochemically by the anodic oxidation of Pb(II) in the form of alkaline solutions of plumbite or acidic solutions of perchlorates, nitrates, fluoborates or fluosilicates. Lead sulphate is of course readily oxidizable to lead dioxide (lead acid cell).

The substance appears to have no solubility in aqueous solution although it has been reported that it dissolves in very strong alkali to form plumbates. The Pourbaix diagram[10] for Pb in sulphate environments shows the regions of thermodynamic stability of the various possible lead species and has a relatively high electrode potential for the reaction

$$PbO_2 + SO_4^{2-} + 4H^+ + 2e = PbSO_4 + 2 H_2O \qquad (1)$$

It is also evident from Pourbaix diagrams that PbO_2 and Pb are not thermodynamically possible in contact with each other in acid media, so that if PbO_2 is to be stable on a Pb base in acid solutions it must be in contact with the Pb intimately enough to exclude the electrolyte solution. In the case of lead-dissolving acids, e.g. HNO_3, $HClO_4$, such duplex structures are unstable and it is impossible to anodically deposit PbO_2 from highly acid Pb^{2+} solutions on to Pb, for the metal dissolves at all practical potentials. The presence of SO_4^{2-} ions enables PbO_2 to form a stable phase on Pb either from an overlaying sulphate deposit:

$$PbSO_4 + 2 H_2O = PbO_2 + 4 H^+ + SO_4^{2-} + 2e \qquad (2)$$

or from a lead basis:

$$Pb + 2 H_2O = PbO_2 + 4 H^+ + 4e \qquad (3)$$

The oxidation of Pb to PbO_2 is also achieved in alkali at moderate oxidizing potentials. The reaction goes through a Pb(II) stage by both solid state and solution reactions. The form of the lead dioxide produced by the oxidation of the base metal is not a very desirable surface on which to study electrochemical reactions; however, such surfaces provided Planté with enough energy storage capacity to make the first viable lead–acid cell. For the most scientifically useful electrochemical measurements it is necessary to form PbO_2 as a dense deposit on a non-reacting basis. Conventionally, Pt, Au or Ta are chosen as bases and PbO_2 is electrodeposited from a Pb(II) solution, usually the perchlorate, acetate or alkaline plumbite.

Lead dioxide does not exactly conform to the stoichiometry of PbO_2; there is always an oxygen deficiency[9] and compositions of $\sim PbO_{1.98}$ are typical.* The analysis of such compounds is difficult and the presence of very small quantities of hydrogen as water in such compounds cannot be entirely ruled out.

There are two forms of lead dioxide,[11] α (orthorhombic) and β (tetragonal). Various chemical methods of oxidation have been used to form exclusively α-PbO_2, thus, for example, yellow lead monoxide is oxidized by a fused sodium chlorate–sodium nitrate mixture, and sodium plumbite may be oxidized by chlorine dioxide. Electro-

---

* In this chapter, PbO_2 is taken to represent the compound called lead dioxide, although its formula does not exactly conform to this stoichiometry.

chemically α-PbO_2 is readily prepared by the electro-oxidation of lead acetate in alkaline solution. β-PbO_2 is prepared by the electro-oxidation of lead (II) from acid solutions, usually of the perchlorate or nitrate. To form exclusively β-PbO_2, careful control of the current density is necessary.

A close relationship exists between the lattices α- and β-PbO_2.[12] In both cases each metal ion is at the centre of a distorted octahedron; however, the octahedra are differently packed in each polymorph. The Pb–O distances (2·15 Å–2·16 Å) are almost certainly the same in both polymorphs. Qualitatively, X-ray analysis readily distinguishes between the two forms. There are, however, some problems with the X-ray quantitative analysis, the intensity of the diffraction pattern of α-PbO_2 being weaker than would be expected relative to the known amount of this phase present. Reasons for this effect have been discussed by several authors. A γ-form of PbO_2 pseudotetragonal has been suggested; however, the unequivocal identification of this structured PbO_2 has not yet been made.

Lead dioxide has a high conductivity[13] (equivalent to that of the metal bismuth) due to electron mobility. Carrier concentrations of 10^{20}–10^{21} electrons cm^{-3} are typical. It has been suggested that the conductivity is associated with the excess lead present in the non-stoichiometric compound; other explanations have been advanced but it seems that the balance of all the evidence is that PbO_2 is a highly doped semiconductor with excess Pb and a band gap of about 1·5 eV.[14] α-PbO_2 is a somewhat better conductor than β-PbO_2.

The mechanical properties of PbO_2 have been studied in connection with lead–acid battery investigations (*see* p. 291 *et seq.*). The mechanical properties of electrodeposited PbO_2 (so-called smooth PbO_2) have been less well investigated. The stress in the PbO_2 electrodeposit has received the most attention and it is clear that the stress (sign and magnitude) depends strongly on the deposition conditions and electrolyte composition.[15]

The Standard Electrode Potential

The most reliable determination of E^{θ} for the most important system $PbO_2/PbSO_4,H_2SO_4/H_2(Pt)$, corresponding to the reaction:

$$PbO_2 + SO_4^{2-} + 4\,H^+ + 2e = PbSO_4 + 2\,H_2O \tag{4}$$

appears to be that of Beck *et al.*[16] These authors found a value of
1·687 V (at 25°C). The electrode potential accurately conformed to
the Nernst equation when used with the activity data of Stokes. The
temperature dependence of e.m.f. from Beck *et al.* provided enthalpies
which were in agreement with calorimetric data. There is a small
difference between E^θ values corresponding to α- and β-PbO$_2$; this
difference has been variously reported but is probably $\sim$10 mV.[17]
Other thermodynamic data of PbO$_2$ have been given in an important
paper by Duisman and Giaque.[9] The E^θ value corresponding to the
reaction

$$PbO_2 + 4\,H^+ + 2e = Pb^{2+}_{aq} + 2\,H_2O \qquad (5)$$

can readily be found to be 1·456 V.

The Structure of the PbO$_2$–Aqueous Solution Interphase

Following the original conclusion of Kabanov[18] that diffuse double
layer theory can be applied to PbO$_2$ electrodes, there have been a
number of examinations of the structure of the PbO$_2$–aqueous solution interphase. The results have been summarized. There are some
difficulties with the identification of the point of zero charge (p.z.c.) in
sulphuric acid solutions because of the adsorption of H$_2$SO$_4$ on to the
eléctrode surface[19] (this difficulty might well also extend to perchloric
acid; early work with this electrolyte indicated a p.z.c. which seems
too positive). The electrolyte most suitable for the study of the interphase structure is potassium nitrate. Differential capacitance data for
α- and β-PbO$_2$ have been obtained over a range of KNO$_3$ concentrations and calculations of the surface charge density made assuming
that the p.z.c. coincides with the capacitance minimum in dilute
solution (diffuse layer minimum). The data conform to theoretical
expectations around the point of zero charge when likely roughness
factors are taken into account. At highly charged (+ or −) electrodes,
deviations occur which are to be entirely expected for surfaces having
large roughness factors. The p.z.c. of PbO$_2$ can be fixed with some
confidence at 1·06 ± 0·01 V for α-PbO$_2$ and 1·15 ± 0·01 V for β-PbO$_2$.
    In sulphate electrolytes adsorption of ions is a complication to
which has been ascribed time-dependent capacitances, and certain
oxygen evolution effects. The magnitude of the electrode capacitance

is much greater than observed with the non-interacting nitrate electrolyte and is strongly pH dependent. It has been suggested that surface reactions of the type given below occur

$$H^+ + SO_4^{2-} = HSO_{4,ads}^-$$ (6)

$$4\,HSO_{4,ads}^- + PbO_2 + 2e = PbSO_{4,ads} + 3\,SO_{4,ads}^{2-} + H_2O$$ (7)

which give rise to a reaction pseudocapacitance. Here, there was no difference reported between α- and β-PbO$_2$.

The behaviour of the electrode capacitance in phosphate electrolyte was similar to that in sulphate. There was a pronounced pH dependence of the magnitude of the capacitance: an explanation of these effects has been given in terms of a redox reaction involving Pb(II) phosphate adsorbed at the PbO$_2$ along similar lines to that suggested for the sulphate.

There was, however, a significant difference between the behaviour in sulphate and phosphate electrolyte. Forced beyond the negative limit of the experimentally polarizable region in sulphate electrolyte, the electrode capacitance decreased due to the presence of a thickening film. In phosphate electrolyte the capacitance reduction was nothing like so marked. This behavioural difference suggested that either the products of the faradaic reaction were able to leave the electrode or formed only a poorly adherent layer.

The only other electrolyte solution for which capacitance data has been reported is sodium hydroxide. The results are rather difficult to interpret, for whereas β-PbO$_2$ apparently has a minimum in dilute solution, the α-polymorph does not. More work in this area would be of great interest.

Other determinations of the p.z.c. on PbO$_2$ have been made[19] using the hardness technique, the potential of maximum hardness being found by microtechniques. These data showed that the maximum occurred at different potentials in sulphuric acid solutions of different concentrations. The results are generally considered to confirm that the electrolyte is specifically adsorbed.

The electrode capacitance of PbO$_2$ in propionate electrolytes has recently been studied by Sekine;[19a] the results are of interest and represent an interesting extension of the range of electrolytes in which interphase data for PbO$_2$ are available. Further extensions are awaited.

Exchange Reactions, $PbO_2/Pb(II)$

These may be conveniently divided into two types with the electrode behaving as one of the first (PbO_2/Pb_{aq}^{2+}) or of the second kind (e.g. $PbO_2/PbSO_{4,s}, H_2SO_4$). For those of the first kind the systems are simpler than those of the second kind because difficulties due to partial coverage of the PbO_2 with solid films are avoided.

Exchange reactions in alkaline solution

It was found that electrodeposited β-PbO_2 was not stable enough in alkaline solution to allow reliable kinetic measurements to be made; mechanical strength and adhesion deteriorated to such an extent that the results of experiments became erratic. For α-PbO_2 the concentration dependence of both the faradaic current and the exchange current density suggested the mechanism of exchange was:[21]

$$PbO_2 + H_2O + 2e = PbO_{ads} + 2OH^- \qquad (8)$$

$$PbO_{ads} = PbO_{aq} \qquad (9)$$

$$PbO_{aq} + OH^- + H_2O = Pb(OH)_3^- \qquad (10)$$

where PbO_{ads} and PbO_{aq} represent respectively PbO adsorbed at the electrode and in solution.

The variation of the faradaic current density with potential could be interpreted in terms of a one-step two-electron charge transfer at low overpotential and two consecutive one-electron transfers at high overpotential.

Exchange reactions in perchlorate solutions

The discharge of both α- and β-PbO_2 in acid perchlorate solutions is relatively simple since the reduced species is an uncomplexed solution species.[22] The concentration dependencies of the rates of the reaction and the variation of the exchange current densities with concentrations have been investigated using conventional electrochemical methods. The data are consistent with the following mechanisms:[23]

$$PbO_2 + 2 H^+ = PbO_2(H^+)_{2,ads} \qquad (11)$$

$$PbO_2(H^+)_{2,ads} + e = HO.Pb.OH^+ \qquad (12)$$

$$HO.Pb.OH^+ + e = Pb(OH)_2 \qquad (13)$$

$$Pb(OH)_2 + 2 H^+ = Pb^{2+} + 2 H_2O \qquad (14)$$

An analysis of the current density-potential data suggested that the slow step in the reaction sequence was likely to be that leading to the intermediate reaction (11). At potentials near the equilibrium (low overpotential) it was not possible to detect the discrete steps (11) and (12), rather a simultaneous two-electron transfer appeared to occur. Some justification for this behaviour has been given in terms of the regions of stability of the intermediate species.

Results for α- and β-PbO$_2$ were generally similar but some differences were observed which could be explained in terms of α/β equilibration processes where the stable form of PbO$_2$ in the acidic media is the β form, the α material undergoing a changeover to β because of the exchange reaction occurring at equilibrium.

An alternative treatment of the PbO$_2$ electrode has been recently given by Pohl and Rickert[24] who considered it as a non-stoichiometric oxide electrode. Basically the PbO$_2$ electrode is considered as both an oxygen and a lead electrode in which the oxygen and lead contents can change independently of each other. The main evidence for this conception rests on the fact that during the anodic polarization of PbO$_2$, lead dissolves initially; during the cathodic polarization initially oxygen only dissolves. It will be interesting to see how far this theory can be taken, for if it is correct, a reassessment of all the experimental data in this area reported so far may be needed; developments are awaited.

Other anionic solutions

Some measurements have been reported using Pb(II) salts of anions other than those already noted;[6,7,25] these include nitrate, fluoborate, fluosilicate and acetate. Results have been broadly similar to analogous cases.

Enthalpies of activation of exchange

The apparent enthalpy of activation of the exchange reaction for both perchlorate[23] and hydroxide[21] reactions calculated from the temperature dependence of the exchange current is about $30\,\mathrm{kJ\,mol^{-1}}$ for β-PbO_2. For α-PbO_2, $\Delta H^{\ddagger}$ is $\sim 40\,\mathrm{kJ\,mol^{-1}}$; however, at temperatures below $\sim 40°C$, the slope of the Arrhenius plot changes to $\sim 10\,\mathrm{kJ\,mol^{-1}}$. There is no satisfactory explanation for this change.

The $PbO_2/PbSO_{4,s}$,H_2SO_4 Electrode

The PbO_2 electrode in sulphuric acid is a typical electrode of the second kind conforming to

$$PbO_{2,s} + 4\,H^+ + SO_4^{2-} + 2e = PbSO_4 + 2\,H_2O \tag{15}$$

and for which[16]

$$E_\beta = 1 \cdot 6871 - 0 \cdot 1182\,\mathrm{pH} + 0 \cdot 0295\,\log a_{SO_4^{2-}} \tag{16}$$

and

$$E_\alpha = 1 \cdot 6971 - 0 \cdot 1182\,\mathrm{pH} + 0 \cdot 0295\,\log a_{SO_4^{2-}} \tag{17}$$

In $4 \cdot 62\,\mathrm{M}$ H_2SO_4 $(\mathrm{d}E_\beta/\mathrm{d}T) = -0 \cdot 20\,\mathrm{mV/°C}$, $(\mathrm{d}E_\alpha/\mathrm{d}T) = -0 \cdot 36$ mV/°C. A kinetic study of reaction (15) is complicated because unless electrodes are very carefully controlled, varying porosity and composition impose a considerable degree of uncertainty on the experimental results. Fleischmann and Thirsk[26] studied the anodic oxidation of $PbSO_4$ to PbO_2. Nucleation centres for PbO_2 were identified at local sites of crystal imperfections at which electrodes could be ejected. Although a certain amount of kinetic data concerning nucleation and growth processes (see later) were obtained, the system was too complicated for a reaction mechanism to be abstracted from the electrometric data. The exchange between lead nuclei in α- and β-PbO_2 electrodes and Pb(II) in solutions of $PbSO_4$ has been examined by Wynne-Jones et al.[27,28] using radiochemical and electrical methods. It was found that although the electrode reaction (15) was undoubtedly the solution/solid, Pb(II)/Pb(IV) exchange was concerned in charging only the double layer.

The Nucleation of PbO_2 from a Pb(II) Solution on to an Inert Substrate

Fleischmann and Liler[29] have investigated the electrodeposition of α-PbO_2 on to a platinum basis from a lead acetate solution. The concentration dependence of the rate constants controlling the reaction were studied. It was suggested that the formation of α-PbO_2 in the non-steady state was determined by the rate of formation of nuclei, an induction period before nucleus formation and the growth of formed centres. The complexity of these component processes made the extraction of a rate equation difficult. By preforming nuclei using an initially high potential pulse of current, the growth process may be sufficiently separated in time from the nucleation process to obtain a limiting value of current above which the current at the lower overpotential becomes independent of the time of the preformation step, i.e. the "growth of deposit current" was obtained. At low overpotential it was shown that the current varies as the square of the time

$$i = N_0 B_3 t^2 \qquad (18)$$

where N_0 is the original number of nuclei on unit surface and B_3 a growth constant. This relationship indicated three-dimensional growth of the electrodeposit. It was suggested that an induction period t_0 is necessary before nucleation occurred and that the nucleation current–time relationship (18) represents the behaviour.

$$i = \frac{AB_3 N}{3}(t - t_0)^3 \qquad (19)$$

A is the nucleation constant and N is the number of nuclei on unit surface. From rather complex orders of reaction, Fleischmann and Liler suggested that the slow stage in the reaction

$$Pb^{2+} + OH_{ads} + OH^- - e = Pb(OH)_2^{2+} \qquad (20)$$

$$Pb(OH)_2^{2+} = PbO_2 + 2\,H^+ \qquad (21)$$

was (20). Fleischmann and Liler were able to argue that the surface free energy of lead dioxide increased with increasing potential.

Later work by Fleischmann and co-workers[30] with β-PbO_2 when electrocrystallization was on a more secure base showed that for a potentiostatic step into the deposition region

$$i = \frac{2F\pi MlAN_0k^2t^2}{\rho} \exp -\frac{\pi M^2N_0Ak^2t^3}{3\rho^2} \qquad (22)$$

where M = molecular weight, l = height of nucleus, A = area, N_0 = number of nucleation centres, k = constant and ρ = density of nucleus.

Equation (21) indicates progressive two-dimensional nucleation growth of three-dimensional cylindrical centres. It was found possible to change the nucleation law to conform to two-dimensional growth with an instantaneous nucleation

$$i = \frac{2F\pi MlN_0k^2t}{\rho} \exp -\frac{\pi M^2N_0k^2t^2}{\rho^2} \qquad (23)$$

by the use of a current pre-pulse.

The Nucleation of PbO₂ on to PbSO₄

It is necessary, as with the deposition of PbO_2 from a solution phase, to form lead dioxide nuclei which act as centres for the spreading of PbO_2 through the mass of the $PbSO_4$ phase. Fleischmann and co-workers[25,30] have considered this reaction in some detail. The transient current–time data under potentiostatic conditions were found to obey a cubic law of the form

$$i = ABt^3/3 \qquad (24)$$

where B is a potential-dependent constant and A is the nucleation constant. The rate of the oxidation was found to reach a maximum and then decay exponentially. It was assumed that the growth of a PbO_2 centre was confined to discrete single lead sulphate crystals and that the decay was due to the completion of the reaction. The relationship between the number of nuclei and time was shown to be

$$N = N_0(1-\exp(-At)) \qquad (25)$$

where N is the number of nuclei on unit surface and N_0 the maximum number of nuclei on unit surface. It was found possible to derive two rate equations:

when $t < t_{max}$

$$i = B_3 N_0 \left(t - \frac{2t}{A^2} + \frac{2}{A^2} - \frac{2}{A^2} \exp(-At) \right) \tag{26}$$

when $t > t_{max}$

$$i = B_3 N_0 \exp\left[-A(t - t_{max})\right] \times \left(t_{max}^2 - \frac{2t_{max}}{A} + \frac{2}{A^2} - \frac{2}{A^2} \exp(-At_{max}) \right) \tag{27}$$

more or less in agreement with the observed facts.

Using similar reasoning to that used for the deposition from homogeneous solutions, it was concluded that the surface free energy of β-PbO$_2$ increased with potential. It is interesting that for both the deposition of PbO$_2$ from homogeneous solution and the oxidation of PbSO$_4$ to PbO$_2$, the variation of the nucleation rate with overpotential η was found to obey the relationship

$$A = k i_0 \exp\left(-K \sigma^3 / \eta^2\right) \tag{28}$$

where k is a frequency factor, i_0 the exchange current, K a constant determined by the shape of the nucleus and σ the surface free energy.

A recent interesting extension of the de Levie[32] treatment of porous electrodes has been made by Simonsson[33] who proposes a theoretical model for spongy PbO$_2$ electrodes on the basis of the macro-homogeneous model of porous electrodes. The structural effects of pore plugging and gradual increase in the (insulating) PbSO$_4$ content of the electrode are considered. A numerical simulation has been made which shows that discharge capacity is linked with these structural factors in addition to the transport restrictions. The final passivating process is confirmed as the expected physical blocking of the electrode with PbSO$_4$. At low rates of discharge the simulation procedure has been used to show that provided that the initial porosity (v/v) exceeds 50%, the electrode is uniformly discharged, otherwise structural factors cause the inner parts of the electrodes not to be fully utilized.

References

1. G. W. Vinal, "Storage Batteries". Wiley, New York (1965).
2. P. Ness, *Electrochim. Acta* **12**, 161 (1967).
3. C. K. Morehouse, R. Glicksman and G. S. Lozier, *Proc. IRE* **46**, 1462 (1958).

4. J. P. Hoare, "The Electrochemistry of Oxygen". Interscience, New York (1969).
5. J. Burbank, A. C. Simon and E. Willihnganz *in* "Advances in Electrochemistry and Electrochemical Engineering" (P. Delahay, ed.), Vol. 8, p. 157. Interscience, New York (1971).
6. J. P. Carr and N. A. Hampson, *Chem. Rev.* **72**, 687 (1972).
7. T. F. Sharpe *in* "Encyclopedia of Electrochemistry of the Elements" (A. J. Bard, ed.), Vol. 1, p. 235. Dekker, New York (1973).
8. N. V. Sidgwick, "The Chemical Elements and their Compounds". Oxford University Press, London (1950).
9. J. A. Duisman and W. F. Giauque, *J. Phys. Chem.* **72**, 562 (1968).
10. S. B. Barnes and R. T. Mathieson *in* "Batteries" (D. H. Collins, ed.), p. 43. Pergamon Press, New York (1963).
11. N. Kameyama and T. Fukumoto, *J. Chem. Soc. Ind. Jap.* **46**, 1022 (1943); **49**, 155 (1946).
12. W. Mindt, *J. Electrochem. Soc.* **117**, 615 (1970).
13. U. B. Thomas, *Trans. Electrochem. Soc.* **94**, 42 (1948).
14. F. Lappe, *J. Phys. Chem. Solids* **23**, 1563 (1962).
15. C. J. Bushrod and N. A. Hampson, *Brit. Corrosion J.* **6**, 129 (1971).
16. W. H. Beck, R. Lind and W. F. K. Wynne-Jones, *Trans. Far. Soc.* **50**, 136 (1954).
17. P. Ruetschi, R. T. Angstadt and B. D. Cahan, *J. Electrochem. Soc.* **106**, 547 (1959).
18. B. N. Kabanov, I. G. Kiseleva and D. I. Leikis, *Dokl. Akad. Nauk SSSR* **99**, 805 (1954).
19. D. I. Leikis and E. K. Venstrem, *Dokl. Akad. Nauk SSSR* **112**, 17 (1957); **112**, 97 (1957).
19a. J. Sekine, *Denki Kagaku* **41**, 339 (1973).
20. J. P. Carr, N. A. Hampson and R. Taylor, *J. Electroanal. Chem.* **27**, 109, 201, 466 (1970).
21. J. P. Carr, N. A. Hampson and R. Taylor, *Ber. Bunsenges Phys. Chem.* **74**, 557 (1970).
22. H. B. Mark, *J. Electrochem. Soc.* **109**, 634 (1962); **110**, 945 (1963). H. B. Mark and W. C. Vosburgh, *J. Electrochem. Soc.* **108**, 615 (1961).
23. N. A. Hampson, P. C. Jones and R. F. Phillips, *Can. J. Chem.* **45**, 2045 (1967); **46**, 1325 (1968).
24. J. P. Pohl and H. Rickert, *in* "Power Sources 5" (D. H. Collins, ed.), p. 15. Academic Press, London and New York (1975).
25. L. M. Bonzunova, V. F. Lazarev and A. I. Lenin, *Zh. Prikl. Khim.* **8**, 864 (1972). D. D. McDonald, E. Y. Weissman and T. S. Roemer, *J. Electrochem. Soc.* **119**, 660 (1972). V. A. Volgins, E. A. Nechaev and N. G. Bakhchisaraits'yan, *Elektrokhimiya* **9**, 984 (1973). L. I. Lyamina, N. I. Koral'kova, E. K. Oshe and K. M. Gorbunova, *Elektrokhimiya* **10**, 841 (1974). N. A. Hampson and C. J. Bushrod, *J. Appl. Electrochem.* **4**, 1 (1974).
26. M. Fleischmann and H. R. Thirsk, *Trans. Far. Soc.* **51**, 71 (1955).

27. S. J. Bone, K. P. Singh and W. F. K. Wynne-Jones, *Electrochim. Acta* **4**, 288 (1961).
28. S. J. Bone, M. Fleischmann and W. F. K. Wynne-Jones, *Trans. Far. Soc.* **55**, 1783 (1959).
29. M. Fleischmann and M. Liler, *Trans. Far. Soc.* **54**, 1370 (1958).
30. As reported in "Advanced Instrumental Methods in Electrode Kinetics", p. 92. Dept. of Chemistry, University of Southampton (1975).
31. M. Fleischmann and H. R. Thirsk, *Electrochim. Acta* **1**, 146 (1959).
32. R. de Levie *in* "Advances in Electrochemistry and Electrochemical Engineering" (P. Delahay, ed.), Vol. 6, p. 329. Interscience, New York (1967).
33. D. Simonsson, *J. Appl. Electrochem.* **3**, 261 (1973); **4**, 109 (1974).

3   The Aqueous System Pb^{4+}/Pb^{2+}: Morphological Aspects

S. M. CAULDER and A. C. SIMON

*International Lead Zinc Research Organization,
Naval Research Laboratory, Washington D.C., U.S.A.*

The Oxidation of Lead Sulphate

On a lead surface

When lead is placed in sulphuric acid, Pb^{2+} ions tend to go into solution. These react with SO_4^{2-} ions to form $PbSO_4$, and because of the extreme insolubility of this product, it is formed directly at the lead surface. At the same time, the conversion from lead to lead sulphate is accompanied by an increase of 168% in volume, so that the result is a very densely packed, but thin, film of lead sulphate that completely passivates the lead from further reaction. An equilibrium condition is thus set up that can persist indefinitely. Several investigators have studied this $Pb/PbSO_4$ electrode equilibrium.[16,22,27,30,39,45,59,76,83]

When an anodic current is applied to this electrode, however, this equilibrium is destroyed. At increasing potential, lead dioxide is nucleated on the lead sulphate surface and the lead sulphate film is converted to lead dioxide. Conversion of the thin, passivating layer of $PbSO_4$ to PbO_2 does not result in an equally passivating layer of the latter. In the first place, the conversion to PbO_2 results in a 48% decrease in volume. This undoubtedly results in some porosity, so that sulphate ions can reach the metal surface and cause further formation of $PbSO_4$ which, in turn, will be converted to PbO_2. There is also the possibility that oxygen ions diffuse through the PbO_2 layer[46] to cause reaction at the interface with the metal, producing more PbO_2 or PbO.

As a consequence of these various factors, the layer thickness increases, and as it increases, it will become increasingly difficult for the acid to diffuse into the mass. Thus the pH at the interior can be expected to increase with the possible formation of basic products. These must then be oxidized to PbO_2, but by a somewhat different mechanism.

The actual conditions existing within the PbO_2 and at its surface therefore become very complex, and have been studied by a large number of investigators, whose findings form the basis for our present concept of the oxidation process at a lead electrode.

In the early years, devices for automatic control of potential, independent of current, were not available, so that most of this early work[7,32,39,49,53,58–60,73] and even some of the more recent work[4,30,31,37,61,72] was performed with galvanostatic techniques, and the potential change that occurred with the completion of each reaction at the electrode surface, together with the extent of the resulting potential plateau, gave useful information about the electrode reactions. On the other hand, when the potential is held constant, the current and consequent rate of reaction will vary as films are built up in thickness or dissolved. From a study of these current excursions and their duration, information can be obtained as to the rate-controlling mechanisms.

The very early investigations[39,59] were unable to detect any intermediate arrests between the equilibrium potential of the $Pb/PbSO_4$ electrode and that of $PbSO_4/PbO_2$. Indeed, a number of the more recent investigators have also been unable to detect arrests for some of the intermediates that will be discussed here. However, some of the recent investigators have used a combination of experimental techniques such as electron diffraction, X-ray, and electron and optical microscopy, along with strictly electrochemical analysis, so that considerable information about these intermediates has been collected.

Most real progress in characterizing the composition of the multiphase oxidation films on lead has been made during the last decade,[21,64–67,69,75,78,79,86] although a considerable portion of this work also confirms the findings of the earlier investigators. The multiphased structure of the film formed by the continued oxidation of the $PbSO_4$ film first formed on lead is now generally recognized. Tetragonal PbO has been found[4,9,30,45,54,55,63] as well as $PbO \cdot PbSO_4$,[30,61,66,75] α-

PbO_2[4,10,44,57,66,73,74,78,86] and $3\,PbO.PbSO_4.\,H_2O$,[68,69] and it is very possible that $5\,PbO.2\,H_2O$ and $2\,PbO.PbSO_4$ also may appear at some stage of the process.

From the foregoing investigations, it is possible to put together a credible description of the chemical and physical changes that occur during the oxidation of a lead electrode in sulphuric acid.

Upon simple immersion of lead in sulphuric acid, lead sulphate crystals are nucleated. These do not homogeneously cover the whole surface but first appear as exceedingly small dendrites[27] epitaxially growing upon the lead, the density of growth depending upon the orientation of the lead sulphate, as well as the time. In this initial stage, the surface is not covered uniformly and certain favoured crystals can grow at the expense of the uncovered portions of the lead. It has been observed that the crystal size increases with time,[41] and, in general, a crystal grows more rapidly in directions perpendicular to the lattice planes having the highest reticular density. Because of this varied rate of growth, the fastest growing planes increase in area and, given sufficient time, will disappear from the crystal surface. The remaining faces have a slower growth rate, so that the growth rate varies with time. As the crystals grow in size, the area of uncovered lead decreases. However, available photomicrographs[16,65,83] seem to indicate that relatively few of the larger crystals appear in a background matrix consisting of many smaller ones. Thus it would appear that while crystals with advantaged positions continue their growth away from the lead surface, less favoured dendritic crystals must continue their practically two-dimensional growth into the available uncovered lead areas, leading finally to a non-porous dielectric layer about 1 μm thick[31] of $PbSO_4$, which has been described[83] as amorphous to X-rays, but which has been found to give well-defined electron diffraction patterns.[21] This layer separates the Pb^{2+} and SO_4^{2-} ions, presumably becoming impermeable to SO_4^{2-}, HSO_4^- and Pb^{2+}, but remaining permeable to H^+, OH^- and probably H_2O.[75]

If this process continues long enough, there will result a higher Pb^{2+} ion concentration within the dielectric layer which will be compensated by the dissociation of water. Since the mobility of H^+ is considerably higher than that of Pb^{2+} and not hindered by the dielectric layer, the H^+ ions will migrate through the layer, while the pH within the layer will increase, and reactions between Pb^{2+} and OH^- ions are favoured.[65] At this point, the film will consist of a layer of PbO against the metal surface, an intermediate layer of $PbO.PbSO_4$, and a layer of $PbSO_4$ next to the electrolyte.

Upon application of increasing potential steps, these products will appear with increasing rapidity up to about $+300 \, mV$ vs the Hg/Hg_2SO_4 electrode. Above this potential step $5 \, PbO.2 \, H_2O$ also begins to appear, and between $+900$ and $+950 \, mV$, $\alpha\text{-}PbO_2$. Above $1000 \, mV$, $\beta\text{-}PbO_2$ also appears.[65] Between $+900$ and $+950 \, mV$, PbO becomes hydrolysed, or the $5 \, PbO.2 \, H_2O$ dissociates, to form $PbOH^+$ and OH^- ions. The $PbOH^+$ ions diffuse through the PbO layer to the metal surface and become oxidized to $\alpha\text{-}PbO_2$. At this point, H^+ or OH^- ions must be carrying the current since the overlying $PbSO_4$ is impermeable to Pb^{2+} or SO_4^{2-} ions.[64] At about $+900$ to $+950 \, mV$, $\alpha\text{-}PbO_2$ begins to nucleate.[66]

Tetragonal PbO, Pb and $\alpha\text{-}PbO_2$ have close similarities in structure[8] so that $\alpha\text{-}PbO_2$ could form either at the lead surface or on the PbO. The initiation of PbO_2 growth has been described[31,41,86] as the process of nucleation at the interfaces of $PbSO_4$ crystals, and a growth of very small filaments of PbO_2 across the lead sulphate crystal surfaces. In the present description, however, it appears that $\alpha\text{-}PbO_2$ forms in the PbO or at the Pb surface. Filaments must then grow outward through the basic lead sulphate zone, towards the lead sulphate crystals. On the other hand, $\beta\text{-}PbO_2$, which begins to appear at $+1000 \, mV$, will form on the $PbSO_4$ crystal layer, so that it is possible to have a layer of $\alpha\text{-}PbO_2$ next to the Pb and separated from the $\beta\text{-}PbO_2$ by a layer of $PbSO_4$.[62] Between $+1200$ and $+1400 \, mV$, the sulphate formed at the beginning is oxidized to $\beta\text{-}PbO_2$ very rapidly, although $\alpha\text{-}PbO_2$ and tetragonal PbO continue to form. However, above $+1300 \, mV$, tetragonal PbO does not form and the coating then consists of only α- and $\beta\text{-}PbO_2$.[66]

While the theory of ion-selective permeable films has been very plausibly developed to explain this process,[66,75,86] physical defects in the film will also occur and it would seem that this might change the situation from that described above.

With the formation of each phase in the film, a considerable change in the specific volume occurs, which can cause changes in crystal size and perhaps introduce stress in the layers. The growth of tetragonal PbO beneath the $PbSO_4$ crystal layer will create stresses that may displace some of the $PbSO_4$ crystals, opening cracks in the formerly impervious layer through which sulphuric acid can penetrate, thus lowering the pH at the interior so that monobasic lead sulphate and PbO are dissolved. However, further growth of $PbSO_4$ will then occur to seal such cracks, so that the process remains much as originally stated. As long as $PbSO_4$ crystals are able to continue their growth, the impermeable film will be self-healing.

Oxidation of PbSO₄ in a porous electrode

In the lead–acid battery, the oxidation of $PbSO_4$ assumes very great importance, as the rechargeable nature of the cell depends upon it. Two essentially different oxidation processes are involved, (i) the manufacturing process known as electrolytic formation, and (ii) the oxidation of the $PbSO_4$ formed during discharge of the cell to return the active material to the PbO_2 form or to "recharge" or "store" energy in the system.

Electrolytic Formation

The original electrolytic formation is required to place the cell in an operable position. At present, no satisfactory and economical process has been developed to place PbO_2 of an active form directly into the plates. The method used at present is to paste a plastic mixture of PbO, sulphuric acid, water, and perhaps other compounds of lead, into a grid cast from lead–antimony or other suitable lead alloy. This material must then be allowed to dry out or "cure" at a definite rate, presumably to ensure the proper crystallization and equilibration of a mixture of 3 $PbO.PbSO_4.H_2O$, $PbO.PbSO_4$, PbO and $PbSO_4$. The dried plate is then placed in an electrolyte in the range 1·050–1·150 sp.gr. and, after a prescribed soaking period, is electrolytically oxidized to PbO_2. A number of studies have been made of this oxidation or formation process[3,6,19,40,43,49,67,71,76,82,83] in porous electrodes.

The most comprehensive of these studies[3,43,67] assist materially in the determination of the most important phase changes that occur in the electrode. The original mixing of the paste used for filling the grid uses insufficient sulphuric acid to react completely with the original PbO. As a consequence of this, as well as of the tendency for lead sulphate to form acid impervious films around pellets of the original oxide, a non-equilibrium condition exists even after extensive mechanical mixing, the pasting and the rather slow drying (curing) processes. As a result, the overall condition of the porous electrode is one of high pH. The practice of soaking the plate in the electrolyte before beginning the electrolytic process does not greatly change this condition for the following reasons:

(i) The soaking process does not continue for a prolonged period and is accompanied in its initial period by a release of CO_2 that has been absorbed in the alkaline paste and converted into carbonates. The continuous release of gas prevents penetration of the acid electrolyte very deeply into the pores and the process responsible for the gas formation partially neutralizes the acid.

(ii) The initial condition of the plate following curing is such that pores are numerous but very small, and these become rapidly blocked by lead sulphate crystals.

(iii) The well known tendency of $PbSO_4$ to form films impervious to sulphuric acid isolates large areas of the active material from further penetration.

As a result of these factors, only a thin surface layer is converted to $PbSO_4$, with the interior remaining essentially unchanged. It has been shown[67] that 10 h of chemical reaction with sulphuric acid of 1·15 sp.gr. were required to completely remove basic sulphates, and 12% of PbO remained even after 18 h treatment. With normal soaking times of less than an hour, the interior of the plate will contain large quantities of basic sulphates and PbO. In contrast, when electrolytic formation was used, basic sulphates disappeared within the first two hours of formation, and PbO had disappeared by the sixth hour. It has been shown microscopically[6,67,82] that oxidation begins at the grid wires and usually from the surface of the plate. The growth of the PbO_2 is inward, away from the surface, and is preceded by a zone of lead sulphate. The reason for this is that the transfer from lead sulphate to lead dioxide at the surface is accompanied by a considerable reduction in volume, about 50% in fact. This considerably increases the porosity in the reaction zone, allowing increased penetration of electrolyte. At the same time, the conversion of $PbSO_4$ to PbO_2 releases additional SO_4^{2-} ions that combine with the PbO or the basic sulphates to produce more $PbSO_4$. It would be expected that, under these conditions, the oxidation of the $PbSO_4$ would produce only β-PbO_2, and in some plates this appears to be the case. In other cases, however, in the first stages there is an initial formation of α-PbO_2 as well as β-PbO_2, indicating a condition of high pH in the reaction zone. This formation of α-PbO_2 appears to be mainly around the grids and in the internal portions of the plates between the grid wires.[3,67] During later stages of the formation process, basic lead sulphates are not observed and the process apparently is by the direct oxidation of $PbSO_4$ to β-PbO_2. Under these conditions, only β-PbO_2 is formed, although any α-PbO_2 formed at lower potential will remain.

In some respects this oxidation during plate formation is very similar to that occurring at a lead surface. The principal difference is the porous nature of the medium to be oxidized and that sulphuric acid can enter to change the pH from high to low. In general, the entrance of this acid results in the dissolution of tetragonal PbO and of the basic sulphates, and generates $PbSO_4$ and water. The rate of this reaction will depend on the acid concentration of the forming electrolyte, the current density, and the size of the pores in the plate. It has been pointed out[67] that the geometry of the pores is important. Small pores are filled with a volume of electrolyte that is small compared with the surface area of the pore walls. As a consequence, hydration and neutralization occur rapidly in such small pores. While small pores may hinder mobility, as has been pointed out, they possess a much larger surface area in their entirety than do the large pores, and therefore probably control the electrochemical processes that occur within the plate. Pores also act as current conductors between the metal grids and the bulk electrolyte, and the shortest pores possess the least resistance so that it is natural for reaction to begin above the grid bars next to the bulk solution. It has already been mentioned that the pores in a cured plate are very small.

On the other hand, the formation of PbO_2 increases the porosity, and the PbO_2 now forming the walls of the pores no longer reacts with the H_2SO_4, so that the original high pH of the interior is gradually destroyed as the zone of reaction advances into the interior. The basic sulphates become converted to $PbSO_4$, and at this point, only β-PbO_2 is formed. While α-PbO_2 may remain in the plate from its earlier formation in high pH areas, its location and extent is not so clearly defined as was the case with the $PbSO_4$ oxidation on a lead sheet.

Oxidation in the Charge Process Following Discharge

Depending upon the current density and depth of the previous discharge, there will be either scattered lead sulphate crystals of fairly large size upon a matrix of uncovered PbO_2, or a densely packed layer of $PbSO_4$ crystals entirely covering the underlying PbO_2. In cases of extremely slow discharge, it is even possible that practically no PbO_2 will be present and the entire mass will consist of $PbSO_4$. In each of these cases, finite crystals of lead sulphate of more or less uniform habit have been formed. In two of the cases, the crystals rest upon already present PbO_2, and there is at least a small quantity of PbO_2 in the third case. Although the discharge process has depleted the acid, the pH is still low (battery operation normally occurs in acid that ranges from $1\cdot250$–$1\cdot280$ sp.gr. when the battery is charged) and the three-dimensional crystals are surrounded by it.

In the first case, PbO_2 is exposed so that upon application of a sufficiently high potential, $PbSO_4$ will go into solution and the exposed PbO_2 will have additional PbO_2 deposited on it, fed by the dissolving $PbSO_4$ crystals in the vicinity. In the second case, if sufficient $PbSO_4$ crystals are present to completely cover the remaining PbO_2, nucleation of additional PbO_2 will have to take place on $PbSO_4$ crystals in the manner that has been described.[31,41,86] In the third case, it is often found that the lead sulphate crystals contain small amounts of PbO_2 which they have either surrounded during their growth or which have fallen upon their surfaces during growth. It has been found that these small amounts of PbO_2 can act as nuclei and that growth of additional PbO_2 during oxidation of the $PbSO_4$ may begin at these points and spread through the space that the crystal occupied before its solution.[81]

Nothing is known about the electrochemical mechanism of the charge reactions taking place on $PbSO_4$ in the absence of a lead substrate. Attempts to find a suitable substitute substrate have led to difficulties. Efforts to analyse the electrode kinetically have been somewhat more successful and considerably more is known about the morphology of the nucleation and growth of PbO_2 on $PbSO_4$ than about the charge exchange reactions. It is generally recognized that the morphology of the porous electrode is extremely important for its proper functioning. Important properties such as surface area, porosity and passivation of active surfaces will depend upon crystal shape and size. Dissolution processes will be incomplete for crystals exceeding a certain size. Better cohesion within the crystal mass will be achieved by certain crystal growth habits. The above are but a few of the reasons for studying morphology. It has been shown that certain configurations of relatively large euhedral crystals, as seen with the optical microscope, have imparted benefits to batteries with grids of lead calcium,[80,82] when held at float potential. Other reports have stressed the importance of the morphology seen with the electron microscope.[11–13,15] More recently, the morphology has been studied with the scanning electron microscope.[42,81,90]

The positive active material, as well as preparations of α- and β-PbO_2, have been studied by X-ray and neutron diffraction[51] and the conclusion reached that the interior of the active material was well crystallized, whereas the outer layers were poorly crystallized. A shape factor of 1·2–1·3 (whereas 0·5 is the shape factor for spherical particles) was determined for this Pb, but it was impossible to determine whether the particles were present as platelets or rods, although the average particle size was given as between 0·38 and 0·56 μm. Other investigators, using electron microscopy, concluded that compound spikes of 0·5 μm were intermixed with sessile crystallites of about 0·1 μm or less diameter, along with rod-like whiskers. During the course of charge and discharge cycling, the PbO_2 particles were observed to increase in size to about 0·55–0·6 μm,[17] while the shape factor was reduced to 0·9, indicating an anhedralization of the PbO_2 crystals.

The ability of such small crystals to form a porous yet rigid structure that still remains serviceable after repeated cycles between $PbSO_4$ and PbO_2 is truly amazing, and one wonders how the crystals remain attached to one another. It has been found that antimony, introduced from corrosion of the lead–antimony alloy grid, has a stabilizing effect on the PbO_2 structure. Electron microscope studies of the morphology of PbO_2[11–13] have shown that interlocked prismatic crystals produce a firmer and better retained PbO_2 structure than do nodular crystals. The presence of antimony caused the formation of the prismatic forms, or at least increased the percentage of them in the active material.

Apparently it is also important that the lead sulphate crystals, from which the PbO_2 is formed, fall within a definite size range.[52] An investigation was made of the effect of $BaSO_4$ in the positive plate and it was shown that the smaller the particles of $BaSO_4$, the greater was the loss of service life. In other words, increasing the number of the $BaSO_4$ crystals, which form the nuclei for the lead sulphate crystals, increased the number but decreased the size of the latter, with adverse effect on service life.

There is no question that repeated cycling between PbO_2 and $PbSO_4$ produces changes in the active material structure that reduce its reversibility and lead to eventual failure. It is apparent that the structure must consist of a porous mass of PbO_2, the crystals of which are firmly welded together to provide rigidity to the structure and the necessary electrical conductivity. The above is equally true when the active material is in the form of $PbSO_4$. Enclosed porosity is of no value and a high surface area of the crystals is a necessity. As an electrode continues to cycle, some of these conditions are no longer met. As mentioned previously, PbO_2 particles were observed to increase in size during continued cycling.[17] It has also been found that the structure produced by the original electrolytic formation has very small and numerous pores of uniform shape. After a period of cycling, this structure is opened out into a matrix containing much larger, more irregular and less numerous voids, so that the overall appearance is that of brain coral.[81] This coralloid structure remains rigid, interconnected, and has pores that are open and interconnected, but it is less satisfactory than the active material present at the beginning. The reason for this appears to lie in the relatively large diameter and dense interconnecting PbO_2 branches of the coralloid structure. A large proportion of the total PbO_2 remains in these arms, the arm diameter preventing access of the electrolyte to the interior portions of the contained PbO_2 and also allowing passivation to occur before a major portion of the interior PbO_2 can be reached by the reaction. In addition, it is found that a gradual lack of cohesion occurs, and more and more PbO_2 particles are disengaged from the mass as cycling is prolonged.

Techniques for Structural Characterization of the PbO_2 System

Lead dioxide is a non-stoichiometric compound that contains sufficient adsorbed water[5,36] as well as lattice water or hydroxyl ions to provide for a $Pb:O$ ratio approaching $1:2$. Since the energy content of the two polymorphs of PbO_2 $(\alpha + \beta)$ are nearly the same,[2] it appears that deviations from the ideal crystal lattice could substantially influence the electrochemical properties of the PbO_2 electrode.

Within the past decade, a number of new instrumental techniques have become commercially available. Recently, these techniques have found application in the structural characterization of the Pb^{2+}/Pb^{4+} electrochemical system.

Structural characterization of PbO_2 using X-ray and neutron diffraction

Information obtained from X-ray and neutron diffraction experiments can be used in conjunction to highlight the scattering of the lighter oxygen and hydrogen atoms. X-ray diffraction is primarily sensitive to the heavy Pb atom scattering, whereas neutron diffraction is sensitive to lead, oxygen and hydrogen. The total neutron scattering for hydrogen is practically all incoherent. If hydrogen, in the form of hydroxyl ions or water, plays an important role in electrode activity, as has been proposed by various authors,[2,20,24,25,32,34,35,51,87] then differential X-ray and neutron analysis may verify this.

The basic question of electrochemical activity as related to atomic arrangement of the active constituents in the PbO_2 electrode can be investigated using X-ray and neutron diffraction. Direct structural information can be derived from a radial distribution function (RDF), which is the Fourier sine transform of the experimental total intensity after the atomic background scattering has been determined and removed. The RDF, which is well approximated by a series of over-lapping Gaussian functions, is a distribution of the interatomic distances in the sample, and also yields information about disorder parameter and coordination numbers. One can systematically analyse the interatomic distance distribution and then attempt to correlate structural changes with electrochemical activity changes as the PbO_2 electrode is oxidized and reduced.

The theoretical RDF for a substance whose crystal structure is known can be calculated from the relation:[29]

$$4\pi\rho_c(r) = \sum_A \sum_B \frac{N_{AB}C_{AB}}{(2\pi)^{1/2}l_{AB}r_{AB}r} \exp\left[\frac{(r - r_{AB})^2}{2l_{AB}^2}\right] \qquad (1)$$

where $4\pi\rho_c(r)$ is the probability, weighted by the average scattering power of atom pair AB, C_{AB}, of finding atoms B from a bulk sample in a shell of thickness dr and radius r_{AB} from origin atoms A; N_{AB} is the coordination number: l_{AB} is the disorder parameter. If there is only thermal disorder, l_{AB} is the root mean square amplitude of vibration.

The RDF reflects the distribution of distances arising from the bonding topology in a spherical sample of a three-dimensionally periodic atomic arrangement. In calculating the RDF from a finite spherical region, the double sum in equation (1) need only be taken over the asymmetric unit, which consists of one Pb and one O atom for β-PbO$_2$. The bulk density, ρ_0, can be obtained from the limiting value of $4\pi\rho_c(r)$ at large r. The resultant distribution function, $G(r)$, can be determined by the expressions:

$$G(r) = 4\pi r(\rho_c(r) - \rho_0) * \varepsilon(r) \tag{2}$$

$$\varepsilon(r) = 1 - \frac{3}{2} \cdot \frac{r}{t} + \frac{1}{2}\left(\frac{r}{t}\right)^3, \qquad r \leqslant t \tag{3}$$

where $\varepsilon(r)$ is a spherical particle function and corrects the relative frequency of occurrence of an interatomic vector in a structure of unbound dimensions to that which pertains within a sphere of diameter t.

At the top of Fig. 1, the atomic arrangement of the Pb and O atoms in β-PbO$_2$ is shown. These atomic positions were obtained by X-ray diffraction analysis[85] and the oxygen parameters were verified by neutron diffraction.[56] The Pb atoms are the smaller circles (i.e. atoms A and D) and the O atoms are the larger circles (i.e. B, C, E, F and G). The unit cell is outlined at the top of Fig. 1.

In the first column of the table in Fig. 1, the first resolved peaks observed in $G(r)$ are listed. Columns 2 and 3 describe the types of atom pairs which give rise to these distances. N is the coordination number and corresponds to the number of neighbouring atoms about each origin atom in the asymmetric unit.

For example, the distance at $2 \cdot 157$ Å is composed of four Pb–O distances between origin atom Pb(A) and target atoms of type O(B), in addition to two O–Pb distances between origin atom O(C) and target atoms of type Pb(A). Since the unit of composition is PbO$_2$, the distances arising from origin atom O(B) need to be doubled and we then have eight contributing distances to the peak at $2 \cdot 157$ Å. The

2·157 and 2·165 Å Pb–O distances are too close to be resolved and consequently the first peak in the $G(r)$ curve is clustered with a total coordination of 12. The second resolved peak is due to an O–O distance between origin atom O(B) and target atom O(G).

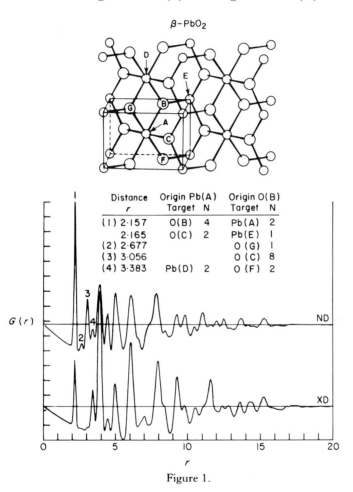

β-PbO$_2$

Distance r	Origin Pb(A) Target	N	Origin O(B) Target	N
(1) 2·157	O(B)	4	Pb(A)	2
2·165	O(C)	2	Pb(E)	1
(2) 2·677			O (G)	1
(3) 3·056			O (C)	8
(4) 3·383	Pb(D)	2	O (F)	2

$G(r)$

ND

XD

Figure 1.

The disorder parameters l_{AB}, used in Figs 1 and 2, were estimated from experimental results[28] on chemically prepared β-PbO$_2$, since thermal parameters are not reported in the literature.

In the lower part of Fig. 1, two $G(r)$ curves are shown. In the neutron diffraction (ND) theoretical $G(r)$ curve, the first four resolved distances are indicated. These numbers correspond with the numbers in the table above the curve and allow one to identify the contributing atom pairs.

As can be seen by comparing the two curves, the X-ray diffraction theoretical $G(r)$ (XD) is missing the second and third peaks of the ND curve. These peaks are due to O–O distances, as seen in the table. The reason these two curves differ, and a discussion of how we can use this to our advantage, is given below. The area A_{AB} of a peak in $G(r)$ is a function of the average scattering power of atom pair AB, C_{AB}, the coordination number, N_{AB}, and the internuclear distance r_{AB}:

$$A_{AB} = \frac{C_{AB}N_{AB}}{r_{AB}} \qquad (4)$$

$C_{AB}(s)$, however, is a function of the type of radiation used,

$$C_{AB}(s) = \frac{f_A(s) \cdot f_B(s)}{\sum_k^{UC} f_k^2(s)} \qquad (5)$$

where $f_A(s)$ is the coherent atomic scattering factor for atom A as a function of the scattering variable s, and the sum in the denominator is over the unit of composition (UC), i.e. PbO_2. C_{AB} is the mean value of $C_{AB}(s)$ over the experimental data range. For X-rays $f_{Pb} \cong 10f_O$, and the lighter atom information is lost because the area is relatively very small. For neutrons, the scattering factor is referred to as b and $b_{Pb} \cong 2b_O$. Hence, the area ratios are more nearly equal. For this reason, one can determine information concerning the positions of lighter atoms (O and H) in the presence of the much heavier Pb atoms. The neutron scattering factors do not exhibit the dependence on s that is observed for X-rays. These constant scattering factors improve the accuracy of the $G(r)$ calculation.

Another polymorph of lead dioxide, α-PbO_2, has been observed in porous PbO_2 electrodes. The $G(r)$ for α-PbO_2, along with an illustration of the spatial atomic arrangement, with unit cell outlined, is shown in Fig. 2. Also given is a table listing the atoms involved in the first three distances in the α-PbO_2 sample. The atomic coordinates obtained from Wyckoff[91] yield 12 Pb–O distances at 2·17 Å. Zaslavsky[93] has published slightly different atomic parameters for α-PbO_2. Calculations based on these parameters yield a range in the Pb–O distances from 2·16–2·24 Å. As in the β-PbO_2, the second peak, which is due to the O–O distances, is unobservable in the XD $G(r)$ curve.

The atomic spatial geometry and theoretical $G(r)$ curves in Figs 1 and 2 illustrate how the structure of α- and β-PbO$_2$ differ. Although the bonded Pb–O distances agree to ± 0.01 Å in the α- and β-forms, the $G(r)$ distance distributions are appreciably different.

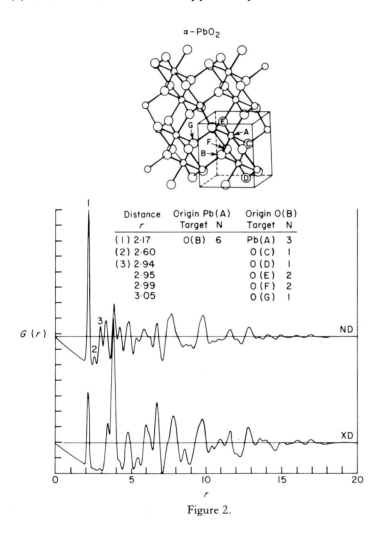

Figure 2.

Work done to date has shown that the structure of PbO$_2$ plays an important role in the kinetics of the PbO$_2$ electrode. Chemically prepared β-PbO$_2$ has been characterized by X-ray and neutron-diffraction to be a highly crystalline well-ordered structure,[28,51,56] while material obtained from the porous PbO$_2$ electrode has been found to be a disordered structure and with poor crystallinity. Recent neutron diffraction results[28] have shown that chemically prepared β-PbO$_2$ gives a minimal amount of incoherent scattering, while material obtained from a porous PbO$_2$ electrode gives a much greater percentage of incoherent scattering. This greater incoherent scattering may be due to a more disordered structure containing a greater amount of hydrogen and/or water.

Thermoanalytical techniques

Thermoanalytical techniques such as differential analysis (DTA) and thermogravimetric analysis (TGA) have been used extensively to study the lead–oxygen system. Thermal analysis, when used in conjunction with other techniques such as X-ray diffraction and nuclear magnetic resonance, can provide valuable information about nucleation on lead dioxide. Hopefully, information gained from thermal decomposition studies can be correlated with electrochemical decomposition studies, since the PbO$_2$ nucleation sites are probably the same.[36,44]

It has been found that the thermal decomposition mechanism depends upon the chemical or electrochemical method of preparation of PbO$_2$.[1,2,18,25,26,33,36,47,62,70,88,89] Several authors have observed[25,51] that as one heats lead dioxide, an exothermic enthalpy change occurs at approximately 185°C. This peak is associated with a reordering of the disordered PbO$_2$ lattice with the evolution of adsorbed and lattice water. Continued heating to higher temperatures produces two distinct oxide phases before obtaining Pb$_3$O$_4$. These intermediate oxides, designated α-PbO$_x$ and β-PbO$_x$ by Bystrom,[20] have similar crystal structures. The crystallographic system to which these oxides belong is still subject to controversy. The Pb:O ratio over which α-PbO$_x$ exists is generally accepted as 1·60–1·51, and that for β-PbO$_x$ as 1·50–1·44. These compounds, α-PbO$_x$ and β-PbO$_x$, have been postulated[44] to exist as intermediate materials during the discharge of the PbO$_2$ electrode.

Nuclear magnetic resonance (NMR)

Pulsed nuclear magnetic resonance has recently been used to study the hydrogen bonding in lead dioxide.[24] The results of this study show that two different types of hydrogen bonding exist in PbO_2. One of these is attributed to water while the other is attributed to some other hydrogen species, possibly hydroxyl ion. These results are in agreement with the ESCA results[48] reported below.

ESCA (electron spectroscopy for chemical analysis)

Electron spectroscopy produces a high-resolution energy spectrum of electrons ejected from a molecule after interaction with a monoenergetic beam of X-rays, photons or electrons. Each ionizing source provides a different type of spectrum, all of which can be used for chemical analysis. ESCA is unique as a surface technique (depths of 10–100 Å) because it gives information about elemental composition, oxidation state and molecular structure. Kim *et al.*[48] used ESCA to study the lead–oxygen system. They obtained ESCA spectra of seven different lead–oxygen species and found that the binding energies of the Pb $4f$ electrons in PbO_2 were lower than those of the Pb $4f$ electrons in PbO. The apparent reversal is rationalized in terms of a relaxation effect. Two kinds of water were found on PbO_2. These were explained in terms of a surface adsorbed water species and a water or hydroxyl species which occupies anion lattice vacancies near the surface. These spectra were used to characterize electrochemically produced PbO_2. ESCA has proven to be a valuable tool for identifying "active" surface species, and as such provides needed structural information to be used in conjunction with kinetic studies.

Auger spectroscopy

Auger spectroscopy is another technique which probes surface composition by means of secondary electron emission. The sample is bombarded with electrons having a few keV energy, and the energy distribution of secondary electrons is measured. The location of the Auger peaks in this energy distribution identifies the nature of the surface species while the intensity of these peaks is a measure of the abundance of the species on or near the surface. The technique is capable of uniquely identifying each element and is essentially a surface probe as the data come from about the top five atomic layers. It is quite sensitive to trace contaminants. Concentrations as low as

10^{12} atoms cm^{-2} have been detected under ideal conditions; this corresponds to 10^{18} atoms cm^{-3} bulk composition. The Auger spectra for each element is only slightly dependent on the chemical environment of the atom. This makes for relatively easy identification of the elements but has the disadvantage of saying nothing about the chemical bonding.

Caulder et al.[23] used Auger spectroscopy to investigate the elemental composition of the PbO_2 active material surface as well as the composition of the corrosion product of a Pb/Sb/Sn grid alloy in a lead–acid battery as a function of cycle life. They were able to correlate the appearance and amount of Sb and Sn in the active material surface with morphology changes and capacity loss.

References

1. R. T. Anderson and M. Sterns, *J. Inorg. Nucl. Chem.* **11**, 272 (1959).
2. R. T. Angstadt, C. J. Venuto and P. Ruetschi, *J. Electrochem. Soc.* **109**, 177 (1962).
3. J. Armstrong, I. Dugdale and W. J. McCusker *in* "Power Sources, 1966" (D. H. Collins, ed.), pp. 163–176. Pergamon Press, Oxford (1967).
4. I. I. Astakhov, E. S. Vaisberg and B. N. Kabanov, *Dokl. Akad. Nauk SSSR* **154**(6), 1414 (1964).
5. N. E. Bagshaw, R. L. Clark and B. Halliwell, *J. Appl. Chem.* **16**, 180 (1966).
6. N. E. Bagshaw and K. P. Wilson, *Electrochim. Acta* **10**, 867 (1965).
7. W. H. Beck, R. Lind and W. F. K. Wynne-Jones, *Trans. Far. Soc.* **50**, 147 (1954).
8. H. Bode and E. Voss, *Z. Elektrochem.* **60**, 1053 (1956).
9. J. Burbank, *J. Electrochem. Soc.* **103**, 87 (1956).
10. J. Burbank, *J. Electrochem. Soc.* **104**, 693 (1957).
11. J. Burbank *in* "Batteries" (D. H. Collins, ed.), p. 43. Pergamon, New York (1963).
12. J. Burbank, *J. Electrochem. Soc.* **111**, 765 (1964).
13. J. Burbank, *J. Electrochem. Soc.* **111**, 1112 (1964).
14. J. Burbank, *J. Electrochem. Soc.* **113**, 10 (1966).
15. J. Burbank *in* "Power Sources, 1966" (D. H. Collins, ed.), p. 147. Pergamon, New York (1967).
16. J. Burbank, *J. Electrochem. Soc.* **118**, 525 (1971).
17. J. Burbank and E. J. Ritchie, *J. Electrochem. Soc.* **116**, 125 (1969).
18. G. Butler and J. L. Copp, *J. Chem. Soc.* **1956** (3), 725.
19. W. O. Butler, C. J. Venuto and D. V. Wisler, *J. Electrochem. Soc.* **117**, 1339 (1970).
20. A. Bystrom, *Arkiv. Kemi, Min. Geol.* **20A**, No. 11 (1945).

21. J. P. Carr, N. A. Hampson and R. Taylor, *J. Electroanal. Chem.* **33**, 109 (1971).
22. E. J. Casey and K. N. Campney, *J. Electrochem. Soc.* **102**, 219 (1955).
23. S. M. Caulder, J. S. Murday and A. C. Simon, *Extended Abs., Battery Div., Electrochem. Soc., Abs. 40*, p. 87, Oct. 13–17 (1974).
24. S. M. Caulder, J. S. Murday and A. C. Simon, *J. Electrochem. Soc.* **120**, 1515 (1973).
25. S. M. Caulder and A. C. Simon, *J. Electrochem. Soc.* **121**, 1546 (1974).
26. P. Chartier and J. Brenet, *C. R. Acad. Sci.* **255**, 100 (1962).
27. T. Chiku and K. Nakajima, *J. Electrochem. Soc.* **118**, 1395 (1971).
28. P. A. D'Antonio, S. M. Caulder and A. C. Simon, Annual Report, Dec. 1977. International Lead Zinc Research Organization Report LE-255.
29. P. A. D'Antonio, P. Moore, J. Konnert and J. Karle, *Trans. Amer. Crystall. Ass.* **13**, 43 (1977).
30. K. Ekler, *Can. J. Chem.* **42**, 1355 (1964).
31. W. Feitknecht, *Z. Elektrochem.* **62**, 795 (1958).
32. W. Feitknecht and A. Gaumann, *J. Chim. Phys.* **49**, 136 (1952).
33. D. Fogue, P. Fouilloux, P. Bussiere, D. Weigel and M. Prettre, *J. Chim. Phys.* **62**, 1088 (1965).
34. A. B. Gancy, *J. Electrochem. Soc.* **116**, 1496 (1969).
35. S. Ghosh, *J. Electrochem. Soc., Japan, Overseas Edition* **34**, 1 (1966).
36. M. I. Gillibrand and B. Halliwell *in* "Power Sources" (D. H. Collins, ed.), p. 179. Pergamon Press, New York (1966).
37. V. I. Goncharov, F. I. Kukoz and M. F. Skalozubov, *Issled. Obl. Khim. Istchnikov Toka* **1966**, 193.
38. R. H. Greenberg and B. P. Caldwell, *Trans. Electrochem. Soc.* **80**, 71 (1941).
39. H. F. Haring and U. B. Thomas, *Trans. Electrochem. Soc.* **68**, 293 (1935).
40. J. E. Hatfield and O. W. Brown, *J. Electrochem. Soc.* **72**, 361 (1937).
41. S. Hisano, *Kogyokogaku Zasshi* **62**, 376 (1959).
42. T. J. Hughel and R. N. Hammar *in* "Power Sources 3" (D. H. Collins, ed.), p. 35. Oriel Press, Newcastle-upon-Tyne (1971).
43. S. Ikari, S. Yoshizawa and S. Okada, *Denki Kagaku* **27**, 487 (1959); *J. Electrochem. Soc., Japan, Overseas Edition* **27**, E186 (1959).
44. S. Ikari and S. Yoshizawa, *J. Electrochem. Soc., Japan, Overseas Edition* **28**, E138, E192 (1960).
45. B. N. Kabanov, *Dokl. Akad. Nauk SSSR* **31**, 581 (1941).
46. B. N. Kabanov, E. S. Weisberg, I. L. Romanova and E. V. Krivolapova, *Electrochim. Acta* **9**, 1197 (1964).
47. T. Katz and R. Faivre, *Ann. Chim.* **5**, 5 (1950).
48. K. S. Kim, T. J. O'Leary and N. Winograd, *Anal. Chem.* **45**, 2214 (1973).
49. D. F. A. Koch, *Electrochim. Acta* **1**, 32 (1959).
50. J. Konnert and J. Karle, *Acta Cryst.* **A29**, 702 (1973).
51. D. Kordes, *Chem. Ing. Tech.* **38**, 638 (1966).
52. I. I. Koval and V. I. Barilenko, *Trudy Chetvertogy Soveshchaniya Elektrokhim., Moscow* **1956**, 748 (published in 1959).

53. E. V. Krivolapova and B. N. Kabanov, *Trudy Soveshchaniya Elektrok-him.*, *Akad. Nauk* **1950**, 539 (1953); *Chem. Abstr.* **49**, 12161 (1955).
54. J. J. Lander, *J. Electrochem. Soc.* **98**, 213 (1951).
55. J. J. Lander, *J. Electrochem. Soc.* **98**, 220 (1951).
56. J. Leciejewicz and I. Padlo, *Naturwissenschaften* **49**, 373 (1962).
57. L. M. Levinzon, I. A. Aguf and M. A. Dasoyan, *Zh. Prikl. Khim.* **39**, 525 (1966).
58. M. Maeda, *J. Electrochem. Soc. Japan*, 197 (1957); *Overseas Edition* **26**, E21, E183 (1958).
59. W. J. Müller, *Kolloid-Z. Z. Polym.* **86**, 150 (1939).
60. K. Nagel, R. Ohse and E. Lange, *Z. Elektrochem.* **61**, 795 (1957).
61. R. W. Ohse, *Werkst. Korros.* **11**, 220 (1960).
62. E. M. Otto, *J. Electrochem. Soc.* **113**, 525 (1966).
63. H. S. Panesar *in* "Power Sources 3" (D. H. Collins, ed.), p. 79. Oriel Press, Newcastle-upon-Tyne (1971).
64. D. Pavlov, *Berichte der Bunsengesellschaft* **71**, 398 (1967).
65. D. Pavlov, *Electrochim. Acta* **13**, 2051 (1968).
66. D. Pavlov and N. Iordanov, *J. Electrochem. Soc.* **117**, 1103 (1970).
67. D. Pavlov, G. Papazov and V. Iliev, *J. Electrochem. Soc.* **119**, 8 (1972).
68. D. Pavlov and R. Popova, *Electrochim. Acta* **15**, 1483 (1970).
69. D. Pavlov, C. Poulieff, E. Klaja and N. Iordanov, *J. Electrochem. Soc.* **116**, 316 (1969).
70. G. Perrault and J. Brenet, *C. R. Acad. Sci.* **250**, 2921 (1960).
71. J. R. Pierson, *Electrochem. Technol.* **5**, 323 (1967).
72. A. Ragheb, W. Machu and W. H. Boctor, *Werkst. Korros.* **16**, 676 (1965).
73. P. Ruetschi and B. D. Cahan, *J. Electrochem. Soc.* **104**, 406 (1957).
74. P. Ruetschi and B. D. Cahan, *J. Electrochem. Soc.* **105**, 369 (1958).
75. P. Ruetschi, *J. Electrochem. Soc.* **120**, 331 (1973).
76. A. I. Rusin, M. A. Dasoyan and N. N. Federova, *Elektroteknika* **38**, 4 (1967).
77. T. F. Sharpe, *J. Electrochem. Soc.* **116**, 1639 (1969).
78. T. F. Sharpe, *J. Electrochem. Soc.* **122**, 845 (1975).
79. T. F. Sharpe, *J. Electrochem. Soc.* **124**, 168 (1977).
80. A. C. Simon *in* "Batteries 2" (D. H. Collins, ed.), p. 63. Pergamon, New York (1965).
81. A. C. Simon, S. M. Caulder and J. T. Stemmlie, *J. Electrochem. Soc.* **122**, 461 (1975).
82. A. C. Simon and E. L. Jones, *J. Electrochem. Soc.* **109**, 760 (1962).
83. G. Sterr, *Electrochim. Acta* **15**, 1221 (1970).
84. J. Stange, *Electrochim. Acta* **19**, 111 (1974).
85. A. Tolkacev, *Vestn. Leningrad. Univ., Ser. Fiz. Khim.* **1**, 152 (1958).
86. E. M. L. Valeriote and D. Gallop, *J. Electrochem. Soc.* **124**, 370 (1977).
87. E. Voss and J. Freundlich *in* "Batteries" (D. H. Collins, ed.), p. 173. Pergamon, New York (1963).
88. R. Weiss, Thesis, Nancy, France (1959).
89. W. B. White and R. Roy, *J. Amer. Ceram. Soc.* **47**, 242 (1964).

90. K. Wiesener, W. Hoffman and O. R. Rademacher, *Electrochim. Acta* **18**, 913 (1973).
91. R. W. Wyckoff *in* "Crystal Structures", Vol. 1, 2nd Edition, p. 259. Interscience, New York (1963).
92. E. G. Yampol'skaya, M. I. Ershova, I. I. Astakhov and B. N. Kabanov, *Elektrokhimiya* **2**, 1327 (1966).
93. A. I. Zaslavsky, D. Kondrashov and S. S. Tolkachev, *Dokl. Akad. Nauk SSSR* **75**, 559 (1950).

4 The Lead/Lead Dioxide Couple in Lead-dissolving Electrolytes

F. BECK

Gesamthochschule Duisburg, Lotharstrasse, Germany

Introduction

Both lead and lead dioxide are low-cost materials with interesting electrochemical properties. In most applications the electrodes are used as inert electrodes in electrolysis and electrosynthesis or as porous electrodes of the second kind in batteries. In some other cases, a lead ion electrode

$$Pb(PbO_2)/Pb^{2+}(H_2O)$$

is realized. Examples of this are the electrowinning and electrorefining of lead, the galvanic deposition of lead or lead dioxide to yield composite electrodes and primary batteries of high power density.

In the following, we shall focus on the application of these dissolving lead electrodes in batteries. The possibilities of an extension to secondary batteries will be discussed. As it is pointed out on p. 92, some advantages of technical interest should be expected from such an accumulator. The behaviour of the single electrodes will be described in the next section. It is well recognized that galvanic cells cannot be regarded alone in terms of single electrodes, quite contrary to driven cells.[1] Thus, the real behaviour as a galvanic couple in primary and secondary cells will be outlined on pp. 85–97.

Fundamental Electrochemical Aspects of the System

The electrolyte/solvent system

A hindrance for technical application is due to the fact that the usual acids with low equivalent weight like sulphuric or hydrofluoric acid yield only sparingly soluble lead salts. Others like nitric, formic or acetic acid are electrochemically unstable. These candidates are mentioned in Table I in parentheses. Thus, rather exotic acids like perchloric or tetrafluoboric acid must be used. Table I shows some properties, the most important of which is the high solubility of the lead salts. Concentrated aqueous solutions contain up to 40 wt % lead. In Table II, some densities of these solutions are compiled. The solutions contain free acid to maintain a low pH as a favourable condition for the PbO_2-electrode.[37]

Table I. Solubilities of lead salts in water at 25°C.[2]

Lead salt		c_s(wt %) (25°C)	Equivalent weight of the anion	wt % Pb^{2+} in the saturated solution
Pb $(ClO_4)_2$	Perchlorate	81·5	99·5	41·5
[Pb $(ClO_3)_2$]	Chlorate	81·5	83·5	45·1
$PbSiF_6$	Hexafluosilicate	~70	71·0	~42
$Pb(BF_4)_2$	Tetrafluoborate	67·3[a]	86·8	36·5
$Pb(SO_3NH_2)_2$	Amidosulphonate	~50	96·1	~26
[$PbC_{10}H_5NH_2(SO_3)_2$]	2-naphthylamine-5,7-disulphonate	48·1	150·6	19·6
[$Pb(NH_4)_4(CH_3COO)_6$]	Ammonacetate	~40	213	~13
[$Pb(NO_3)_2$]	Nitrate	37·0	62·0	23·2
[$Pb(CH_3COO)_2$]	Acetate	35·6	59·0	22·6
Pb $CH_2S_2O_8$	Methane disulphonate	7·75	87·1	4·2
[$Pb(HCOO)_2$]	Formiate	2·6	45·0	1·8
[$PbCl_2$]	Chloride	1·08 1·9[b]	35·5	0·81
PbF_2	Fluoride	0·067	19·0	0·057
$PbSO_4$	Sulphate	0·0044 0·00012[c]	48·0	0·0030
$PbO.PbSO_4$	Basic sulphate	0·00134[d]	28·0	0·0011

[a] Our own determination.
[b] In 26% NaCl at 13°C.
[c] In 20 wt % H_2SO_4.
[d] 18°C.

Perchlorates are not considered for practical use due to the costs and the fire hazard, concentrated solutions of $Pb(ClO_4)_2$ being able to ignite organic materials. On the other hand, BF_4^- and SiF_6^{2-} ions are

Table II. Density of aqueous lead(II) solutions at 25°C.

Formula	Corresponding acid c(wt %) of common solution of the acid	Molecular weight of lead salt	Density [g cm^{-3}]				
			1 M Pb salt 1 M acid	2 M Pb salt 1 M acid	3 M Pb salt 1 M acid	4 M Pb salt 1 M acid	saturated Pb salt
$HClO_4$	72	406	1·36				2·78
HBF_4	50	381	1·34	1·65	1·95	2·24	2·20
H_2SiF_6	30	349	1·42	1·73			
HSO_3NH_2	solid	399	1·36				

subject to hydrolysis. The first step, which is slow, is the following:

$$BF_4^- + H_2O \rightleftharpoons BF_3(OH)^- + HF \tag{1}$$

The equilibria involved are extremely complex and have only been recently investigated in more detail.[31,32] HF and hydroxofluoborate anions are generated, without precipitation of insoluble lead salts.

Water is the only solvent of practical interest. The PbO_2-electrode will presumably not work reversibly in other solvents. One pre-condition for a non-aqueous solvent would be the capability to act as an O^{2-} donor and acceptor. Low costs and beneficial chemical as well as electrochemical properties are other unique properties of water. Decomposition may be quantitatively reversed. Both electrodes are metastable with reference to water due to its high hydrogen and oxygen overvoltages.

Our own results for the specific conductivities in the HBF_4 system are shown in Fig. 1. The $HClO_4$ curve is drawn for comparison.[3]

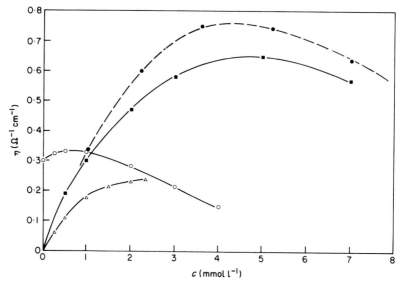

Fig. 1. Specific conductivities of aqueous HBF_4 and $Pb(BF_4)_2$ at 25°C: ■ HBF_4; △ $Pb(BF_4)_2$; ○ c [M] $Pb(BF_4)_2 + 1$ M HBF_4; ● $HClO_4$.[3]

Especially with mixtures rich in free acid, high conductivities up to 0.7 mho cm^{-1}, which resemble those in the lead sulphate battery, are to be expected. In systems with zinc-dissolving electrodes, the conductivity is lower, namely about 0.4–0.1 mho cm^{-1} for $K_2Zn(OH)_4$ and 0.1 mho cm^{-1} for $ZnCl_2$.

The lead-dissolving electrode

The lead-dissolving electrode,

$$Pb \rightleftharpoons Pb^{2+} + 2e \quad (^{\circ}U_H = -0.13 \text{ V}) \tag{2}$$

is of the first kind, the equilibrium potential of which is shifted by $+29.5$ mV/log $c_{Pb^{2+}}$ at 25°C.[4] In sulphuric acid solution, an electrode of the second kind,

$$Pb/PbSO_4/SO_4^{2-}(^{\circ}U_H = -0.35 \text{ V})$$

with a more negative normal potential, is established.

In both directions, the Pb/Pb^{2+} electrode is only slightly polarizable. The exchange current density is $j_0 = 0.3$ A cm^{-2} in 1 N $HClO_4$.[120] The small overvoltages of about 10 mV (cf. Fig. 6) depend among other factors on the crystal plane and on the anion.[5,6] Only with impedance measurements, are any irreversible characteristics detectable.[33] Anodic passivation in perchloric acid occurs at high c.d.[116]

Current efficiencies for electrochemical deposition or dissolution are virtually quantitative. This led formerly to the proposal of a $Pb(BF_4)_2$ coulometer.[7] The quality of the deposits are excellent, especially in the tetrafluoborate electrolyte.[8,9] Excess HBF_4 promotes finer-grained deposits and inhibits dendrite formation. Such electrolytes are technically used in lead-refining baths.[30,98,117] Recycling of lead in batteries via anodic dissolution has been proposed recently[118] (in lead-dissolving electrolytes). Dendrites are avoided, if convection is adapted to local current density. The latter may become very high at peaks, where plane diffusion changes to spherical diffusion:[10]

$$j_{lim} = \frac{zFDc}{r} \tag{3}$$

Thus, a rather smooth substrate is necessary for good results. Generally the lead system is much less subject to dendrite formation than, for example, the zinc system. The addition of inhibitors[8,28,29] is not necessary in the normal case.

The hydrogen overvoltage on lead is very high ($\eta_{H_2} \sim 1$ V). The Pb/Pb^{2+} electrode is at least 1 V more positive than the Pb/H_2 electrode. The electrochemical deposition and dissolution of lead becomes a quantitative process by this way. Many impurities like Fe^{2+}, Co^{2+} or Ni^{2+} are harmless due to the positive lead potential. Other examples like Cu^{2+}, Bi^{3+} or SbO are deposited. However, in the course of lead deposition, these codeposits are "buried" continuously, and no accumulation at the surface is possible.

The lead-dioxide-dissolving electrode

General behaviour

Three physical properties of lead dioxide should be taken into consideration:

1. The electronic *conductivity* is as high as that of mercury, namely about 10^4 mhos cm^{-1} at 25°C. This is believed to be correlated with the non-stoichiometry of the oxide.

2. In PbO_x, an *oxygen deficit* ($x < 2$) was found, independently of the method of preparation. The upper limit is 1·99, lower limits of $x = 1·90,$[13] $1·93,$[14] $1·94,$[15-17] $1·95,$[18] $1·96$[19] and $1·98$[20] having been reported. Below this limit, which seems to depend on the sensitivity of the X-ray method, an α-PbO_x phase is detectable. O^{2-} vacancies according to equation (4) are highly probably. Thus, a fraction of Pb^{4+} in the PbO_2-lattice is substituted by Pb^{2+} and eventually Pb^{3+}. This fraction is increased by interstitial protons, which are bound to O^{2-} to yield OH^-, and is reduced by free electrons, which can be generated via

$$(2\ O^{2-})_{latt} \rightarrow (2\square°_{latt} + 4e) + (O_2)_g \qquad (4)$$

The removal of oxygen proceeds under relatively mild conditions and leads finally to Pb_3O_4.[26,27]

3. PbO_2 electrocrystallizes in two modifications, namely the tetragonal β-PbO_2 in acid solution and the rhombic α-PbO_2 in alkaline solution. The steric influence of the anions is described.[13] In the following, some differences in the electrochemical properties will be considered.

The lead dioxide electrode in lead-dissolving electrolytes is an electrode of the first kind. According to the overall reaction

$$Pb^{2+} + 2\,H_2O \;\rightleftharpoons\; PbO_2 + 4\,H^+ + 2e \;\; (^\circ U_H = 1\cdot46\;V) \qquad (5)$$

the equilibrium potential is shifted at 25°C by $+118$ mV/log c_{H^+} and by $-29\cdot5$ mV/log $c_{Pb^{2+}}$. In sulphuric acid solution, an electrode of the second kind is established with a more positive standard potential:

$$PbO_2/PbSO_4/SO_4^{2-} \,(^\circ U_H = 1\cdot69\;V)$$

The reversible O_2 electrode has a standard potential of $1\cdot23$ V. Thus, the PbO_2 electrode is only metastable in acid solutions, but becomes stable above pH $4\cdot3$ due to the fact that the oxygen electrode shifts with 59 mV/log c_{H^+}. The rest potential has been interpreted in terms of a corrosion potential with oxygen evolution as the anodic process. In alkaline solution, the potential of the PbO_2 electrode is too negative to be of practical interest. Freshly prepared PbO_2 electrodes show a rather unstable rest potential. In alkaline solution, the negative shift may last for months. In acid solutions, a thickness-dependent potential decay over 200 mV, lasting for hours, was observed.[25] As an interpretation, the out-diffusion of oxygen was assumed.

Usually the kinetics of the PbO_2 electrode are discussed in terms of a redox electrode[21] according to

$$Pb^{2+} \;\rightleftharpoons\; Pb^{4+} + 2e \qquad (6)$$

In non-aqueous solutions, e.g. glacial acetic acid, lead (IV) salts can be synthesized with inert anodes.[22] However, in aqueous solutions, kinetics are complicated by rapid chemical follow-up reactions with H_2O or OH^-, due to the extremely high acidity of the Pb^{4+} ion. The situation is quite similar to that in organic electrochemistry. The neutral reaction product PbO_2 electrocrystallizes to form the lattice. In the steady state, the flux-ratio O:Pb crossing the phase boundary must be smaller than $2\cdot0$. The activity of Pb^{4+} or PbO_2 in the PbO_2 lattice is set equal to unity, analogous to the Pb atoms in the metal electrode. Another analogy is due to the fact that the surface of the electrode is subject to changes by electrocrystallization and surface layers. Likewise, a continuous change from $Pb^{2+} \rightarrow Pb^{4+}$ can be assumed. The exchange current density of PbO_2 when measured was found to be about $0\cdot1$ mA cm^{-2}.[119]

It has been proposed that the PbO_2 electrode be regarded as a simultaneous lead *and* oxygen electrode, which is possible from a thermodynamic point of view.[23,24]

Anodic Deposition of Lead Dioxide

In the initial stages of anodic electrodeposition of PbO_2 onto inert substrates, nucleation phenomena play an important role.[34,35] In the steady state, electrocrystallization of PbO_2 leads to a layer which can grow to a thickness of several centimetres. In battery application, thicknesses of some $0 \cdot 1$ mm are used. For each $0 \cdot 1$ mm, capacity is about $0 \cdot 02$ A h dm^{-2}. The electrode process comprises electrochemical and chemical steps.[21,25,34,36–41] The following mechanism,[37] which is in agreement with the careful kinetic measurements of Hampson,[38,39] is a "c.e.e.c." sequence, each step involving the formation of one proton:

$$Pb^{2+}(aq) \underset{}{\overset{K}{\rightleftharpoons}} Pb(OH)^+(aq') + H^+$$
$$\overset{slow}{\longrightarrow} Pb(OH)_2^+ + H^+ + e$$
$$\longrightarrow (PbOOH)^+ + H^+ + e$$
$$\longrightarrow PbO_2 + H^+ \tag{7}$$

The slow step leads to the Pb(III) intermediate $Pb(OH)_2^+$. As a consequence, the reaction order for protons is found to be about -1.[25,34,37,38] Even in strong acid solutions, no soluble Pb(III) or Pb(IV) species are detectable,[25,35] indicating that the follow-up reactions are very fast. The intermediates in equation (7) are assumed to react in the adsorbed state. In strongly acid perchlorate solution, the anodic reaction is first order with respect to Pb^{2+} ions.[38] However, negative reaction orders have been observed as well.[37]

Besides deprotonation steps, condensation and coordination with anions proceed, leading finally to the PbO_2 lattice. Condensation is promoted by protons, analogous to the formation of polyacids of Si, Sn, Cr, Mo and other metals:

$$2\, Pb(OH)_4 \overset{H^+}{\longrightarrow} \quad -Pb-O-Pb- \tag{8}$$

Current efficiencies for the deposition process are as high as 95–98%.[40] Only at low concentrations of Pb^{2+}, especially in the presence of high acid concentrations, may this value decrease appreciably,[41,42] cf. Table III. With some electrolytes, e.g. the organophosphates,

Table III. Current efficiencies and incorporation of additives for the anodic electrodeposition of $PbO_2(10 \text{ mA cm}^{-2}, 2 \text{ h},$ graphite-filled polypropylene, 25°C, stirring).[42]

2 : 1   = 2 M Pb $(BF_4)_2$/1 M HBF_4
0.2 : 4 = 0·2 M Pb $(BF_4)_2$/4 M HBF_4.

10 mmol l⁻¹ Additive	Current efficiency (%)ᵃ		MeOₓ in PbO₂ (%)	
	2 : 1	0·2 : 4	2 : 1	0·2 : 4
—	100	85	—	—
Sb^{3+}	40	49	3·5	13
Bi^{3+}	100	67	5·5	10·7
Tl^+	76	28	7·0	11·5
Mn^{2+}	78	54	2·7	5·3
Co^{2+}	97	89	1·2	1·8
Ni^{2+}	100	89	0·24	0·05
Cu^{2+}	100	90	0·04	0·05

ᵃBased on deposited mass.

PbO_2 deposition becomes impossible. This is shown with the cyclic current voltage curve in Fig. 2. No anodic deposit of PbO_2 can be seen on the Pb electrode, and no cathodic peak occurs. Lead deposition and dissolution are possible under these conditions.

The anodic overvoltage η_a is as high as 300–400 mV. This was explained in terms of a high concentration polarization by the intermediate species.[25] The anions exhibit a marked influence on η_a,[40] cf. Fig. 7. Nitrate ions lower η_a distinctly. The virtual low η_a in the presence of acetate ions is due to the anodic oxidation of the anion itself. The anionic influence on the formation of different modifications[13] has been mentioned already. As it is expected, the influence of the substrate is pronounced in the early stages of the electrodeposition.

An interesting reverse stirring effect was detected by Spahrbier[25]. According to the potentiostatic experiment in Fig. 3, the current dropped dramatically when the stirrer was switched on. This unexpected behaviour was only observed at pH 0, but not at pH 1 and

2. Without forced convection, the current rises slowly to a rather high steady-state value, which is presumably influenced by natural convection at the cylindrical electrode. This effect is regarded as a type of

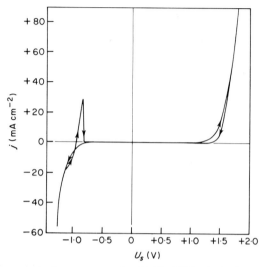

Fig. 2. Cyclic current/voltage curves, 1 mV s⁻¹, 25°C, electrolyte: 30% 1-hydroxy-ethane-1, 1-diphosphonic acid:

$$
\begin{array}{c}
\quad\;\; O \;\; CH_3 \;\; O \\
\quad\;\; \| \quad\; | \quad\;\; \| \\
(HO-P-C\!\!-\!\!-\!\!-\!\!P-OH) \\
\quad\;\; | \quad\; | \quad\;\; | \\
\quad\;\; OH \;\; OH \;\; OH
\end{array}
$$

saturated with PbO ($\sim$3 M H⁺, $\sim$0·1 M Pb²⁺). Base electrode Pt.

autocatalysis. The intermediate, which is produced in the rate-determining step (equation 7), is also generated in a parallel conproportion reaction:

$$PbO_2 + Pb(OH)^+ + H^+ + H_2O \rightarrow 2\,Pb(OH)_2^+ \qquad (9)$$

This reaction *consumes* H⁺; therefore a high surface concentration of H⁺ (low pH, no stirring) favours equation (9).

In the presence of additives, the properties of the electrodeposited layers may be influenced. Oxidizable species like Fe²⁺ [42] or NO₂⁻ [43] lower the current efficiency for the plating process, as shown in Fig. 4. At a low concentration level, however, only a slight loss is to be expected. Other additives like those mentioned in Table III[42] are partially incorporated into the PbO_2. Current efficiencies, especially at low Pb^{2+} concentrations, may decline in parallel. The reason for this is

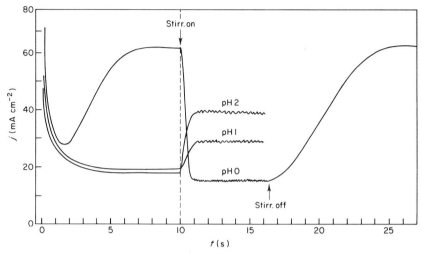

Fig. 3. Potentiostatic deposition of PbO_2 onto Pt after Spahrbier.[25] Cylinder electrode, vertical; $\eta_a = 400 \text{ mV}$. Electrolyte: 1 N $Pb(ClO_4)_2/1$ N $HClO_4$. No stirring between 0 and 10 s and from 16·5–27 s, 25°C.

a depression of the oxygen overvoltage in the presence of the co-deposits. However, Co^{2+}, Ni^{2+} and Cu^{2+} show the opposite effect, leading to quantitative current efficiencies even in dilute solution. Our finding with Co^{2+} does not agree with other results, where traces of Co^{2+} (down to $0·1$ mmol l^{-1}) were found to lower η_{O_2} at a Pb/PbO_2[44,45] or Pt/PbO_2[46,47] anode down to 50% of the initial value. In the lead sulphate battery, concentrations as low as $0·1$ p.p.m. Co^{2+} cause an increase in gassing.[48] The improvement in texture and stability of PbO_2 by traces of Co^{2+} has been reported.[51] SbO^+, Cu^{2+} and Ag^+ [49,50] show similar effects.

Cathodic Dissolution of Lead Dioxide

Analogous to the anodic mechanism, the step in equation (7), leading to "Pb(III)" intermediates, is rate-determining:

$$(PbOOH)^+ + H^+ + e \xrightarrow{\text{slow}} Pb(OH)_2^+ \qquad (10)$$

Reaction orders for H^+ are found to be lower than 2, which can be explained by adsorption effects. The reaction order for Pb^{2+} turns out to be zero.

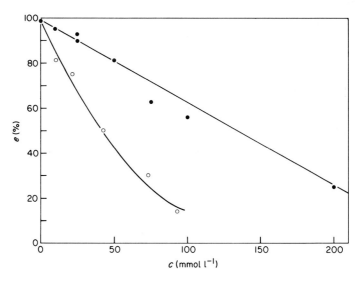

Fig. 4. Current efficiencies (e) for the anodic deposition of PbO_2 onto Pt in the presence of oxidizable additives of concentration c in stirred solution at 25°C. ●——●, Fe^{2+} in the lead tetrafluoborate bath[42]; ○——○, NO_2^- in the lead nitrate bath.[43]

A deficiency of protons may lead to reaction-limited currents, which are due to slow chemical steps, e.g. the dissolution of lower lead oxides. A geleous surface layer has been identified by SEM techniques.[37,40] These currents increase appreciably with temperature.[37] Finally, passivation may occur. In this case, the potential may drop to that of lead, depositing lead eventually as dendrites on top of the oxide, which is not reducible even at these negative potentials.[40,52,67] The passivation is prevented by high proton concentrations,[37,52] low Pb^{2+} concentrations[53] and low current densities.[52,53,62] The last mentioned influence is clearly demonstrated in Fig. 5 according to our own results. At the end of discharge of a cell with rather thick (0·35 mm) layers of HBF_4, the voltage drops sharply. If the current is halved, when the voltage reaches 0·5 (O)V, the curve in Fig. 5 is obtained. In this way, a further 0·43 A h are recovered, which is about 60% of the total charge deficit (7·4–6·7) A h.

According to Hampson,[52] passivation time τ and current density j are correlated by modifying Sand's equation:

$$(j-j_1) \sqrt{\tau} = K, \tag{11}$$

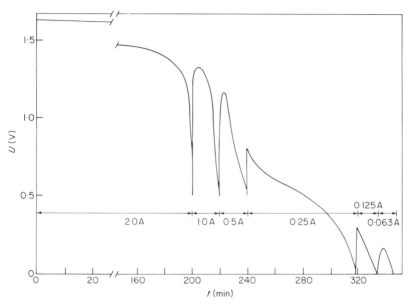

Fig. 5. The 38th discharge in the course of a cyclization experiment at room temperature. Electrolyte: $2 \text{ M Pb(BF}_4)_2/1 \text{ M HBF}_4$. Conversion of electrolyte after charging with 20 mA cm^{-2}: 66%. Charge density: $0\cdot065 \text{ A h cm}^{-2}$. Base electrodes: natural graphite-filled polypropylene. Area = 113 cm^2.
After deep discharge with 20 mA cm^{-2} to $0\cdot5$ V, the current was halved, and after the recovery period halved again. In this way, a total of $0\cdot43$ A h is gained in addition to the $6\cdot7$ A h at $2\cdot0$ A.

where j_1 is a lower limit of c.d., below which no passivation occurs. In the course of cathodic dissolution of PbO_2 in 30% H_2SiF_6,[121] a rapid passivation of the layer can be observed in the presence of $>3\%$ HF or $>0\cdot6\%$ H_2SO_4 or HCl.[122] In *alkaline* solution, PbO_2 is cathodically dissolved to plumbite. Even at very low current densities ($\approx 1 \text{ mA cm}^{-2}$), passivation may occur under these conditions. The electroreduction was shown to proceed in a homogeneous system down to about $PbO_{1\cdot35}$.[54,55] This means, that PbO_x "dissolves" continuously in PbO_2.[61] It was possible to reduce ferricyanide or H_2O_2 at these semiconducting oxides.[56] The concentration of lower oxides of

PbO_x, which will cause additional voltage drops in the electrode is also raised in the presence of reducing agents like formaldehyde.[57] In powdered PbO_2 samples, the bad conductivities of compressed powders[12] and of filled plastic materials,[58] the deviation of X-ray patterns[39] and the disturbance in potential transfer mechanism as a slurry electrode[59] is indicative of the presence of lower oxides at the surface of the PbO_2 particles.

In acid solutions, cathodic dissolution of PbO_2 must be essentially a heterogeneous process. This means that the lattice is dissolved electrochemically layer by layer. Homogeneous mechanisms, discussed in alkaline solutions for PbO_2 or for MnO_2[60]

$$MnO_2(s) + H^+ + e \rightarrow MnOOH \text{ (dissolved in } MnO_2\text{)} \qquad (12)$$

cannot be valid in acid solutions due to the solubility of the reduction products in the liquid phase. Thus, the discharge curve of PbO_2 at low pH values exhibits a nearly constant potential (cf. Figs 10, 14 and 15), depending only on the species in solution, which can be held at a nearly constant level, while in the case of homogeneous reduction the potential varies much more. Normally, the reaction zone is not really two-dimensional, but it is represented by a layer of final thickness of about 1 μm,[40] where the Pb(II) states of the solution vary towards the Pb(IV) states in the bulk of the electrode.[37,40] The assumption of a disturbed surface layer was also necessary in the interpretation of isotope exchange experiments.[64] The reduction proceeds predominantly along the grain boundaries of the polycrystalline oxide, at least in the case of β-PbO_2.[25] This may cause a seeding effect. α-PbO_2 is reduced at higher overvoltages.[25]

It should be mentioned briefly that in the case of the $Pb/PbO_2/PbSO_4$ electrode of the second kind, the potential dependent composition of the multiphase layer at the interphase is even more complicated, involving α- and β-PbO_2, PbO_x, PbO and (basic) lead sulphates.[65,66]

The overvoltages η_c for the reduction process in lead-dissolving electrolytes are lower than for the deposition. η_c is strongly influenced by the proton concentration and by anions, NO_3^- (and SO_4^{2-}) promoting low overvoltages.[40,41] Current efficiencies are not quantitative due to "cathodically stimulated O_2 evolution". However, an overall two-electron reaction must be anticipated, in spite of the fact that this has been occasionally doubted.[63]

Cyclic behaviour of the electrodes

In this section, results for cyclic (alternate charging and dis-charging) polarization of single electrodes will be discussed in so far as they complete the findings already presented in the preceding sections.

Overvoltage/time curves for the galvanostatic deposition and redissolution of lead are given in Fig. 6. With lead as a substrate, a

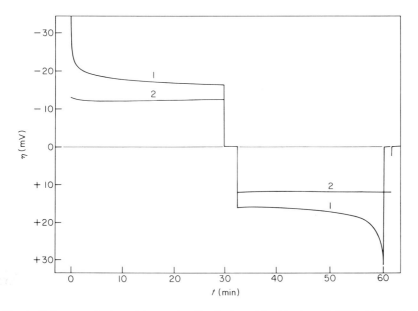

Fig. 6. Galvanostatic deposition and redissolution of lead in 1 M Pb(BF$_4$)$_2$/1 M HBF$_4$ (stirred) at 25°C. $j = 20$ mA cm^{-2}, $\tau_{ch} = 0.5$ h.
1. Base electrode: graphite-filled polypropylene; 2. Base electrode: pure lead.

constant η_c of 12 mV is established from the beginning. The anodic curve is symmetric to the cathodic one. With natural graphite-filled polypropylene as a base electrode, higher overvoltages are measured in the initial stages due to the fact that deposition starts on isolated graphite flakes. In the steady state, a smooth deposit is built up. The overvoltage then slightly exceeds the lead curve due to ohmic losses in the material. The anodic curve is nearly symmetrical.

The cyclic behaviour of the lead dioxide electrode was studied by slow cyclic voltammetry.[37,40] In Fig. 7, new results on the influence of the anion (ClO_4^-, BF_4^-, SiF_6^{2-}, $NH_2SO_3^-$, CH_3COO^- and NO_3^-) are presented. The measurements were performed with Pt electrodes at sweep rates of $1\ mV\ s^{-1}$. The proton concentration was kept constant at 1 M. The experiment starts at the rest potential $U_s = 800\ mV$ in a positive direction. After an anodic current rise to $100\ mA\ cm^{-2}$, the potential sweep is reversed. The cathodic redissolution shows interesting features. The overvoltage at the initial rise increases in the order

$$NO_3^- < ClO_4^- < BF_4^-,\ SiF_6^{2-} < NH_2SO_3^- < CH_3COO^-.$$

Due to the limited capacity and due to passivation effects, the cathodic current goes through a maximum. With $HClO_4$, two maxima can be seen. At $U_s = 200\ mV$, the potential sweep is reversed again. In most cases, a residual, brass-yellow PbO_x remains on the substrate, which cannot be further reduced. On anodizing the electrode again, PbO_2 is deposited on top of this layer at potentials which are more negative than those for the bulk deposition. The capacity of this favoured deposition is limited, leading to a peak at $U_s = 1000$–$1100\ mV$. This peak is indicative of the formation of a residual oxide layer. The strong decrease of the anodic peak in cycle 3 is due to a splitting off of the residual oxide.

In the presence of reducing additives like Fe^{2+}, the formation of this passivating layer is prevented, and no anodic peak appears.[42,110] This is clearly shown in the cyclic galvanostatic experiment in Fig. 8. Without Fe^{2+}, the second and following charging curves show a plateau at lower overpotentials, which corresponds to the anodic peak in Fig. 7. In the presence of Fe^{2+}, this step disappears, and the charging curves resemble the initial one. The current efficiencies are diminished due to an electrochemical oxidation of Fe^{2+}, running in parallel with the diffusion-limited current density j_{lim}. If the current density for charging is j_{ch}, for discharging j_d, the current efficiency α is diminished from virtually one to:

$$\alpha = \frac{j_d t_d}{j_{ch} t_{ch}} = \frac{1 - j_{lim}/j_{ch}}{1 + j_{lim}/j_d} \tag{13}$$

α approaches zero, if $j_{ch} \rightarrow j_{lim}$ or $j_d \ll j_{lim}$. The formation of an inert residual oxide, which can only be reduced by a soluble reductant, can be understood if an insulating layer is assumed between Pt and the

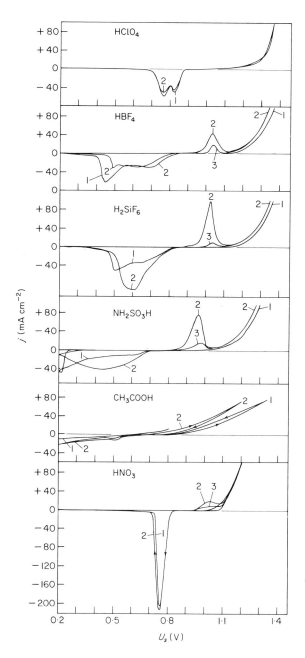

Fig. 7. Slow cyclic voltammetry for deposition and redissolution of lead dioxide on Pt at 25°C in stirred solution. Two $\frac{1}{2}$ cycles, $1\,\mathrm{mV\,s^{-1}}$, starting at rest potential $U_s =$ 0·8 V. Electrolytes: 1. $1\,\mathrm{M}$ $Pb(ClO_4)_2/1\,\mathrm{M}$ $HClO_4$; 2. $1\,\mathrm{M}$ $Pb(BF_4)_2/1\,\mathrm{M}$ HBF_4; 3. $1\,\mathrm{M}$ $PbSiF_6/0·5\,\mathrm{M}$ H_2SiF_6; 4. $1\,\mathrm{M}$ $Pb(NH_2SO_3)_2/1\,\mathrm{M}$ NH_2SO_3H; 5. $1\,\mathrm{M}$ Pb $(CH_3COO)_2/1\,\mathrm{M}$ CH_3COOH; 6. $1\,\mathrm{M}$ $Pb(NO_3)_2/1\,\mathrm{M}$ HNO_3.

residual oxide, preventing the electron injection from the base electrode, but not from the ion. Reducing agents with a similar standard
potential to Fe^{2+}, e.g. NO_2^- or hydroquinone, behave similarly.

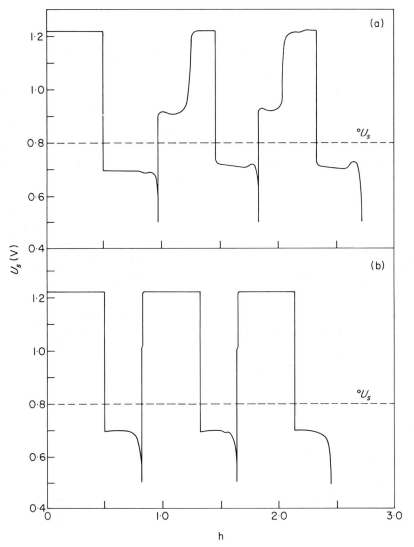

Fig. 8. Galvanostatic deposition and redissolution of lead dioxide in
1 M $Pb(BF_4)_2$/1 M HBF_4 at 25°C in stirred solution. $j = 20$ mA cm^{-2}, $\tau_{ch} = 0.5$ h, conversion of electrolyte < 1%, Pt base electrode. (a) Pure solution; (b)
25 mmol l^{-1} $Fe(BF_4)_2$ as an additive.

However, Ti^{3+} or Sn^{2+}, in spite of the negative potential, exhibit a poor performance. This can be understood in terms of redox reactions at semiconductor electrodes.[68]

Corrosion

Most of corrosion research with lead has been done in sulphuric acid solution, where the formation of passivating layers of $PbSO_4 . PBO_x$, $PbSO_4$ and PbO_2 is possible.[4,65,66,69,70] In lead-dissolving acids, few data are available. In 0.01 N H_2SiF_6, the corrosion rate was found to be as low as $1 \mu A cm^{-2}$. Pb was reported to be essentially stable in HBF_4.[72] In 1 M $HClO_4$, a corrosion rate of $80 \mu A cm^{-2}$ was found at 30°C.[115] According to our own measurements, Pb corrodes very slowly in aqueous HBF_4 or $HClO_4$ due to its high hydrogen overvoltage (cf. Table IV). These data show that corrosion rate is somewhat accelerated in the presence of the lead salt. The experiment with N_2 indicates that O_2 corrosion is a part of the total process. Localized cells, e.g. with graphite-filled polypropylene "CPP", promotes this part, in accordance with results in 20% HCl. At 66°C, a corrosion rate of $40 \mu A cm^{-2}$ was reported in 10% H_2SiF_6.[83]

Table IV. Corrosion of lead in lead-dissolving acids.[73a] The aqueous electrolytes were air-saturated and stirred at 25°C. Lead purity 99·985%, area = 20 cm².

Corrosion medium	j_{corr} ($\mu A cm^{-2}$)
1 M HBF_4	19·5
50 wt % $HClO_4$	5·2
2 : 1[a]	35
2 : 1, N_2	9·0
2 : 1, O_2	131
2 : 1, Pb/CPP localized cell 1 : 1	75
2 : 1, Pb/CPP localized cell, N_2	15
2 : 1, 2 mmol l^{-1} Bi^{3+}	139[b]
2 : 1, 10 mmol l^{-1} Bi^{3+}	360[b]
2 : 1, 10 mmol l^{-1} Fe^{3+}	780[b]
2 : 1, 10 mmol l^{-1} quinone	550[b]
2 : 1, 5 mmol l^{-1} Fe^{2+}	39
2 : 1, 40 mmol l^{-1} Co^{2+}	21
2 : 1, 10 mmol l^{-1} HNO_3	49

[a] $2 : 1 = 2$ M $Pb(BF_4)_2/1$ M HBF_4
[b] Initial corrosion rates (20 h), decreasing tendency.

Impurities which are more noble than lead (Bi^{3+}, Cu^{2+}, H_2TeO_4), accelerate the corrosion of lead. Soluble oxidants like Fe^{3+} or quinone accelerate it even more, as shown in Table IV. The inactivity of Fe^{2+} or Co^{2+} shows clearly that these ions cannot act as mediators for oxygen in the HBF_4 system. HNO_3 is inactive as well due to kinetic hindrance.[77]

Lead dioxide is subject to corrosion in acid solution according to the overall equation:

$$PbO_2 + 2\,HClO_4 \rightarrow Pb(ClO_4)_2 + H_2O + \tfrac{1}{2}O_2 \qquad (14)$$

The stability of this material is high due to the high η_{O_2}, which is 0·9 V for β-PbO_2 and 0·7 V for α-PbO_2 at 1 mA cm^{-2} in 4·4 M H_2SO_4.[78] The value of η_{O_2} is lower with Pb/PbO_2 electrodes of great surface roughness than with smooth Pt/PbO_2 electrodes.[42] Pure β-PbO_2 powders exhibit corrosion rates in 4 M H_2SO_4 as low as 10^{-4} μA cm^{-2}.[79] In 1 M HBF_4/1 M Pb(BF_4)$_2$ ("1:1-electrolyte"), we have measured with dense layers on graphite-filled polypropylene corrosion rates scattering between 1 and 10 μA cm^{-2}.[73] In Fig. 9, the curve without additives

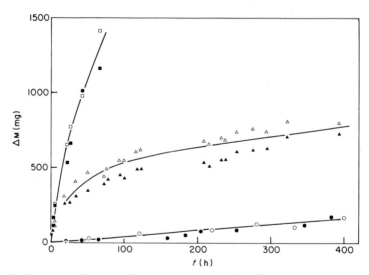

Fig. 9. Corrosion of dense PbO_2 layers on graphite-filled polypropylene in 1 M Pb(BF_4)$_2$/1 M HBF_4 at 20°C. Stirred, air-saturated solution. Area = 26 cm^2: $\bigcirc$, $\bullet$ without additive; $\triangle$, $\blacktriangle$ in the presence of 25 mmol l^{-1} Fe(BF_4)$_2$; $\square$, $\blacksquare$ in the presence of 25 mmol l^{-1} Fe(BF_4)$_2$ and with a "counter-electrode" of Pb, arranged in parallel at a distance of 1·5 cm.

corresponds to $4\,\mu A\,cm^{-2}$. In the perchlorate system, values of $50\,\mu A\,cm^{-2}$ [25] and $0.2\,\mu A\,cm^{-2}$ [41] have been reported. The stability in $HClO_4$ seems to be inferior to that in HBF_4.[97] The much higher corrosion rates with Pb/PbO_2 samples ($\sim 1\,mA\,cm^{-2}$)[80] are presumably due to localized cell action with the lead substrate.

In the presence of $10\,mmol\,l^{-1}$ Fe^{2+} in the 1:1-electrolyte, the corrosion rate is appreciably raised to about $500\,\mu A\,cm^{-2}$, which is lower than the corresponding experiment with Pb/Fe^{3+} (cf. Table IV). With V^{3+}, similar results are obtained. With hydroquinone or Tl^+, higher initial values of about $1.2\,mA\,cm^{-2}$ have been observed.[73b] As demonstrated in Fig. 9, the high rate in the presence of Fe^{2+} decreases with time and finally attains a value which is equivalent to the normal acid corrosion. Of course this is due to the consumption of the reductant. If the sample is arranged in parallel with an isolated lead sheet, the rate decreases after some time to about 50% of the initial value. This is due to the reductive regeneration of the Fe^{2+} ions at the Pb surface.

Oxygen atoms are able to migrate through PbO_2 layers and attack the substrate such as Pb or C.[81] Co^{2+} is reported to inhibit this migration nearly quantitatively.[46,62]

The Pb/PbO₂ Couple as a Primary Battery

The Pb/PbO_2 couple in lead-dissolving acids, predominantly 50–60% $HClO_4$, 40–50% HBF_4 or 30% H_2SiF_6, is practically used as a primary battery of the reserve type. The power density of this battery is very high $(300–500\,W\,kg^{-1})$ due to the fact that current densities of $0.5\,A\,cm^{-2}$ for the perchlorate system[84] and $0.3\,A\,cm^{-2}$ for the fluoborate system[85] are technically feasible. This property is important for military applications such as a high rate "single shot" battery, e.g. for torpedo propulsion or for fuses.[86] Another important feature of the battery is the excellent low temperature behaviour (down to $-60°C$ at lower current densities). This leads to applications in arctic zones and in radio sondes.[87] The lead system is much less costly than silver-containing couples, which are frequently used for the applications mentioned above.

The negatives are usually manufactured from thin lead sheets, containing 0.08% Ca as hardener.[93] The positive is either a thin PbO_2

layer, electrochemically plated onto thin sheets or expanded sheets of nickel or nickel-plated steel,[84,87,88,93] or a porous, plastic-bonded layer of PbO_2 powder hot-pressed onto a nickel sheet base.[85,89,90] The capacities range between 0.002[85] and 0.07 A h cm^{-2}.[87,84] The acid is allowed to fill the battery immediately before current drain with vacuum, membrane-punching or rotating cell-techniques.[86,87,90,91] The time of activated stand is not limited by the active masses (their corrosion rates have been shown in the section on corrosion to be very low, at least in HBF_4), but by the acid attack on the substrate material of the positive through the pores of the PbO_2 layer. A decrease of the initial capacity by about 1% min^{-1} has been reported.[85]

The open circuit voltage of the Pb/PbO_2 couple in contact with concentrated acids is given in Table V. The initial voltages in $HClO_4$

Table V. Open circuit voltages at 25°C for the Pb/PbO_2 couple in concentrated acids.[87]

wt % acid	$HClO_4$ $°U$(V)	HBF_4 $°U$(V)
42	1·86	1·82
50	1·92	1·86
55	1·96	1·92
60	2·05	—
70	2·18	—

are somewhat higher. In HBF_4, α-PbO_2 is reported to show higher voltages (differences of 0.1 V at 25°C and 0.21 V at $-40°$C) than β-PbO_2.[92] According to the general equation for the thermodynamic voltage of the cell for 25°C

$$°U(V) = 1·59 + 0·118 \log c_{H^+} - 0·059 \log c_{Pb^{2+}}, \qquad (15)$$

these voltages, which resemble the standard voltage of the Pb/PbO_2/H_2SO_4 system, are only understandable on the assumption that very low Pb^{2+} concentrations (<1 mmol l^{-1}) are present under these conditions.

In the course of discharge of the Pb/PbO_2 couple in $HClO_4$ or HBF_4, the voltage decreases slightly due to hydrogen ions being consumed and lead ions generated. The slope of the curve depends on the conversion of the electrolyte. The examples presented in Fig. 10 show once again the already-discussed influence of temperature and PbO_2 structure. $HClO_4$ generally gives better results than the other acids of practical interest,[84,87,93] but HBF_4 is preferred for its lower

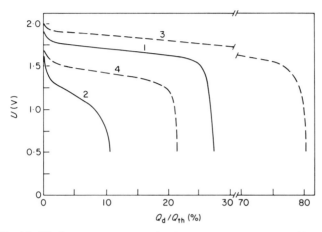

Fig. 10. Discharge curves in Pb/PbO_2 primary batteries, $j = 10$ mA cm^{-2}:

1. 48% HBF_4, 23°C ⎫
2. 48% HBF_4, −60°C ⎭ Bonded PbO_2, after Weissman[85,89]

3. 60% $HClO_4$, 25°C ⎫
4. 60% $HClO_4$, −40°C ⎭ Plated PbO_2, after White *et al.*[84]

costs and more secure handling. At the end of discharge, the voltage drops sharply due to a passivation process at the positive which depends on the current density (cf. section on the lead-dioxide-dissolving electrode). The formation of basic lead perchlorates at $^\circ U \sim 1$ V has been mentioned.[84] Thus the capacity of the battery is limited by the passivation at the positive. In Fig. 11, the capacity, referred to as the theoretical value, is plotted *vs* the current density for some systems in $HClO_4$ and HBF_4. As is clearly demonstrated, the capacity depends on discharge c.d., but not on charge c.d.[95] However, the decline in the case of H_2SO_4 electrolyte with porous electrodes is much more pronounced. In this case, Peukert's rule is valid.[94] Even with very thin (0·1 μm) layers of PbO_2 on Pt, active mass utilization does not exceed 50% in H_2SO_4.[63,96] It drops down to 10% at high c.d. On the other hand, we have found degrees of active mass utilization as high as 98% in HBF$_4$. Figure 11 also shows that forced convection of the electrolyte leads to higher values of this important parameter than free convection, the latter being dependent on the distance of the electrodes.[53]

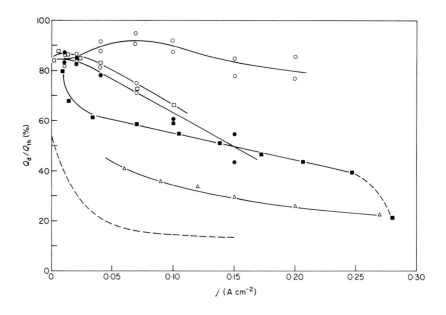

Fig. 11. Active mass utilization in Pb/PbO$_2$ primary batteries as a function of current density at 25°C. Deep discharge at constant *j*.

Initial electro-lyte composition	Q_A(Ah cm^{-2})	Base electrode	Author and further conditions
○,● ~4 M HBF$_4$	0·056	CPP	Beck[95] Forced convection ○j_{ch} plotted, $j_d = 20$ mA cm^{-2} ●j_d plotted, $j_{ch} = 20$ mA cm^{-2}
△ 4 M HBF$_4$	0·059	Ni·	Hampson and Bushrod[53] 15°C, free convection
■ 61% HClO$_4$	0·07	Ni	White *et al.*[84] Free convection
□ ~20% HClO$_4$	0·06	CPP	Beck[101] Forced convection
--- 36% H$_2$SO$_4$	~0·06	Pb(Sb)	Conventional system

The Pb/PbO$_2$ Couple as a Secondary Battery

Introduction and principal statements

The rechargeability of the Pb/PbO$_2$ couple in lead-dissoving electrolytes according to

$$2 Pb(BF_4)_2 + 2 H_2O \underset{\text{discharge}}{\overset{\text{charge}}{\rightleftharpoons}} Pb + PbO_2 + 4 HBF_4 \qquad (16)$$

should be possible in principle. Both corresponding electrode processes are reasonably well understood. Moreover, the cathodic deposition of lead in electrolytic refining and winning of the metal[29,30,98,117] and the anodic deposition of PbO$_2$ in the manufacture of PbO$_2$/C and PbO$_2$/Ti composite electrodes[99] are known to be performed in practice. Some experience can be transferred for the reverse electrode reactions from primary batteries with this system. However, it will be pointed out in the next section that a working secondary battery on this basis demands much more than the total of these electrode processes.

The conventional lead accumulator

$$2 PbSO_4(s) + 2 H_2O \underset{\text{discharge}}{\overset{\text{charge}}{\rightleftharpoons}} Pb + PbO_2 + 2 H_2SO_4 \qquad (17)$$

has been known since the fundamental polarization studies of Gaston Planté in 1859. The early statement[100] that this system seems to be the most economical and practical secondary battery is valid up to present time. It continues to share over 90% of the total production volume of accumulators. A common feature of the lead battery and other commercialized systems is the application of porous electrodes of the second kind. Frequently this has been declared as a necessary precondition for a practical working secondary battery. However, nearly all systems under development comprise at least one electrode of the first kind. An example is the zinc electrode, which is combined with positive electrodes such as O$_2$, Cl$_2$, Br$_2$, PbO$_2$, NiOOH, MnO$_2$, AgO and CuO. The oxide electrodes are normally electrodes of the second kind.

The secondary battery under discussion would contain an oxide electrode of the first kind in combination with a soluble negative. As it has been pointed out by the author,[101] some major disadvantages of the conventional system such as poor active mass utilization, limitation of power density (at least at low temperatures), and high manufacturing costs could be circumvented by such a combination, which is called a *Lösungsakkumulator* (dissolution battery). From a practical point of view, four couples should be of some interest (see Table VI).

Table VI. Practical dissolved state $Me/Me^{()}O_x$ secondary batteries.

No.	System	$°U$ (V)	$E_{s,theor}$ (W h kg^{-1})
1	$Pb/PbO_2/HBF_4$	1·59	107
2	$Pb/MnO_2/HBF_4$	1·36	113
3	$Zn/MnO_2/H_2SO_4$	1·99	306
4	$Mn/MnO_2/H_2SO_4$	2·41	382

Due to unique kinetic properties and corrosion resistance, the Pb/PbO_2 system seems to be the only practical candidate at the moment. Leuchs was the first to propose this cell with HBF_4 or H_2SiF_6 as a secondary battery.[102] Upon charging, he observed the deposition of dense, black and good adhering layers of PbO_2 at the anode and smooth layers of Pb at the cathode. The battery is also mentioned by Betts.[117] Further proposals in regard to electrolyte composition,[103] base electrodes,[104,105] and bipolar layout of the cell[106] were made later on.* Based on the identity of active masses, a comparison with the conventional lead battery is made. In spite of the fact that the theoretical energy density of the lead tetrafluoborate accumulator is rather low (107 W h kg^{-1}, cf. Table VI) and inferior to that of the lead sulphate accumulator ($E_{s,theor} = 170$ W h kg^{-1}), the practical values, which should be attainable, are relatively high. Under the assumption of a bipolar battery design with thin graphite-filled polypropylene electrodes, an active mass utilization of β and neglecting the minor influence of cell walls and end-plates, the following equation for E_s can be easily derived:

$$E_s = \frac{Q_A \cdot \bar{U}_d}{\frac{Q_A}{\beta c F} \cdot s_E + d' s_B} \qquad (18)$$

Q_A = charge density of the electrode layers
$\bar{U}_d$ = average discharge voltage
c = lead concentration

* The reversibility of the lead amidosulphonate system was reported in the literature,[123] in disagreement with our results.

F = Faraday constant
s_E, s_B = densities of electrolyte and base electrodes
d' = thickness of bipolar base electrodes

In Fig. 12, equation (18) is plotted for $\beta = 95\%$ and some electrolyte

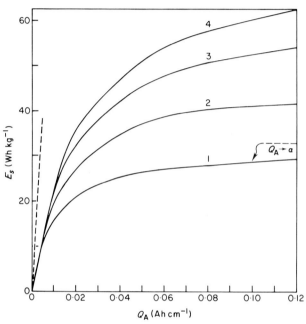

Fig. 12. Lead tetrafluoborate accumulator, energy density E_s vs charge density Q_A
for four different electrolytes.
Electrolyte conversion $\beta = 95\%$; $\bar{U}_d = 1\cdot65$ V, $d' = 3$ mm, $s_B = 1\cdot74$ g cm^{-3}, s_E, *see*
Table II. Electrolytes:

1. 1 M Pb(BF$_4$)$_2$/1 M HBF$_4$
2. 2 M Pb(BF$_4$)$_2$/1 M HBF$_4$
3. 3 M Pb(BF$_4$)$_2$/1 M HBF$_4$
4. 4 M Pb(BF$_4$)$_2$/1 M HBF$_4$

compositions. For small Q_A values, E_s increases proportionally to Q_A,
and only the base electrodes themselves determine E_s:

$$E_s = \frac{\bar{U}_d}{d' s_B} \cdot Q_A \qquad (19)$$

With 1-mm thick bipolar electrodes, which seem technically feasible,
even a steeper initial rise (dashed line in Fig. 12) is possible. With high

Q_A, E_s approaches a limiting value:

$$E_{s,\ lim} = \frac{1}{s_E}\ \bar{U}_d\beta cF,\qquad(20)$$

which is indicated for the lower curve. Equation (20) also shows that E_s increases linearly with β at low values of β.

  These calculations show clearly, that interesting E_s values should be realizable, which exceed those of the lead sulphate battery at a 2 h rate. Moreover, power density and low temperature behaviour of the battery under discussion are improved, as discussed in an earlier section. A detailed analysis shows that lead requirement is only 20–30% per unit of stored energy. Advantages in reference to lead recycling and manufacture costs can be foreseen. Thus, the starting point to build a new lead secondary battery looks promising.

Cyclic behaviour of the battery

A schematical representation of the situation of a cycled battery is given in Fig. 13. A new cycle is started with the clean base electrodes,

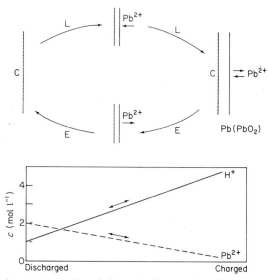

Fig. 13. Schematic representation of the periodic growth and redissolution of active mass layers in a dissolved lead secondary battery upon cyclization. Change in Pb^{2+} and H^+ concentration is indicated.

i.e. all lead is in the dissolved state. Upon charging, layers of Pb and PbO_2 are smoothly deposited. At the end of this stage, lead ions are converted up to 98%. The acid concentration increases appreciably. In the course of discharge, the deposits of active masses are redissolved layer by layer, annihilating all faults created during deposition and ending finally in a state identical to that at the beginning of the cycle. In this way, a very high cycle life should be possible, depending only on the long-term behaviour of the base electrodes and the electrolyte. All limitations of conventional batteries with porous electrodes due to irreversible changes of the solid state, the crystal size and structure, as well as concentration polarization in the pores, are in principle avoided.

It is to be expected that deviations of this ideal behaviour may occur in any real system. The following general discussion is based on our practical experience with the system. Three important points will be considered in the following sections.

The Role of Convection

Convection turns out to be much more important than in systems with porous electrodes, where low concentrations of metal ions run in parallel with short distances for their transport ($\delta \approx$ radius of the pores).[107] In the case of a dissolved state secondary battery, these distances are larger by some orders of magnitude, and convection is needed to overcome this. The difference between free and forced convection has been already discussed in connection with Fig. 11. Inadequate convection leads to disarranged deposits, namely localized ones like lead dendrites or lead oxide nodules on the one hand and an uneven distribution of active masses over the total electrode area on the other hand. The latter, known as shape change, is primarily a problem of convection[108] rather than a problem of current distribution[109] as was pointed out in the case of zinc/zincate electrodes. With the lead system, the accumulation of a dense electrolyte layer at the bottom of the cell after discharge is important, and an even redistribution of electrolyte preceding the following charging process, e.g. by pulsation stirring,[110] is an essential feature. It has been mentioned already that a single distortion during charging is not critical due to the levelling nature of the discharging phase, but the accumulation of local distortions may lead rapidly to the total failure of the battery.

One-sided Accumulation of Active Masses

A one-sided accumulation of active masses at the negative *or* positive is possible upon cyclization. If λ is the current efficiency for the charging process (the deposited masses are related to the given charge Q_{ch}) and ε is the c.e. for discharging (the charge Q_d is related to the previously generated deposits), then the equilibrium condition is given by

$$\lambda_+ \cdot \varepsilon_+ = \lambda_- \cdot \varepsilon_-, \tag{21}$$

if $(1 - \varepsilon)$ is due to corrosion processes, consuming the layers without providing electrons to an external circuit. If this condition is not maintained, a deficit

$$\delta = \lambda_+ \cdot \varepsilon_+ - \lambda_- \cdot \varepsilon_- \tag{22}$$

is established, leading to an unbalanced accumulation of lead on the negative if $\delta < 0$, or an unbalanced accumulation of lead dioxide on the positive if $\delta > 0$. Thus, only a limited number of cycles can be expected. As an approximation, the following simple relationship can be derived:[111]

$$Z_{lim} = \frac{2}{\delta} \tag{23}$$

In this case, a continuous decrease of Q_d will occur during cyclization.

The high rate of transport processes, on which the advantages of the battery are based, turns out to be the reason for this serious, system-specific problem as well. In principle, two ways are open to reach a high cycle life, namely:

(i) minimizing δ, predominantly by improving current efficiencies at the positive, and

(ii) regeneration of the battery after one or a few cycles, which is greatly facilitated by the high solubility of the lead electrolyte, leading to dissolution of the layers via corrosion.

In both cases, the information which comes from measurements with single electrodes are extremely helpful.

Accumulation of Active Masses on Both Electrodes

A third problem is correlated with the accumulation of active masses on both electrodes. The reason for this is the passivation of the oxide electrode. An amount of lead equivalent to the residual PbO_x remains at the negative. The active masses immobilized by this mechanism tend to become steady upon cyclization.

This minor problem can be handled by several measures. In the presence of small concentrations of redox couples like Fe^{2+}/Fe^{3+}, the equalization of the layers is possible by an "electrochemical short circuit", the intensity of which is controlled by the convection. The presence of Fe ions leads to an increased self discharge, but much more in the case of the conventional lead battery,[48,112,113] as we found with smooth electrodes. Other possibilities consist of a residual discharge across a low resistance and of corroding both layers.

Cyclization experiments have been performed with monopolar cells, containing natural graphite-filled polypropylene electrodes with an area of 100 cm². Slow circulation was applied. The battery was deep discharged. With partial charging, corresponding to a 10–30% conversion of the electrolyte, it was possible to yield more than 2000 cycles with the lead tetrafluoborate accumulator.

In Fig. 14, some examples of an experiment with 20% conversion of the electrolyte are presented. After an initial formation at the positive, steady curves are obtained from approximately the 30th cycle. Charging is then onto a residual oxide rather than onto a clean positive, leading to quantitative current efficiencies at *both* electrodes. Thus, equation (21) becomes valid.

However, high conversion of the electrolyte is necessary to reach high energy densities (cf. Fig. 12). Under special conditions, we have obtained until now about 200 cycles with full charge and at constant Q_d, representing a high active mass utilization. In Fig. 15, a cyclization experiment with full charge and deep discharge is represented. A part of the charging process is performed at lower voltages, i.e. under the influence of the residual oxide. The charging process was interrupted when the voltage approached 2·6 V. The rest potential drops after a short time from 1·7 V to 0·8 V, indicating the early disappearance of residual lead, but not of residual lead oxide.

A very thorough further optimization is needed for an extension to bipolar secondary batteries, otherwise cells would quickly get out of step, giving rise to the failure of the system.[114]

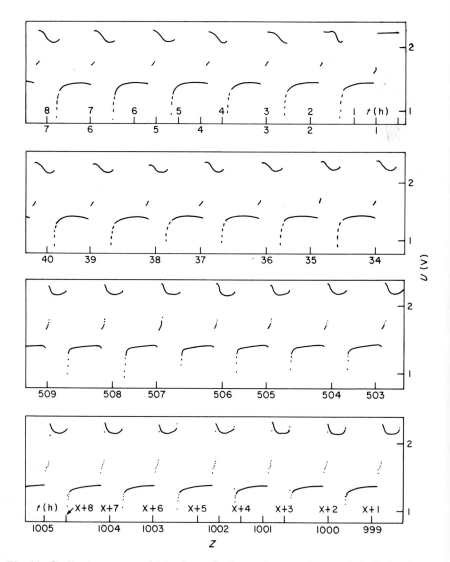

Fig. 14. Cyclization curves of a lead tetrafluoborate battery with vertical filled polypropylene electrodes. Area = 100 cm², sandblasted, at ambient temperature. Electrolyte: 1 M Pb(BF$_4$)$_2$/1 M HBF$_4$; Q_{th} = 0·017 A h cm⁻², j_{ch} = 40 mA cm⁻², j_d = 20 mA cm⁻², 5-min rest between charge and discharge step. β = 20% (partial charging), deep discharging, no *iR* correction, forced convection.

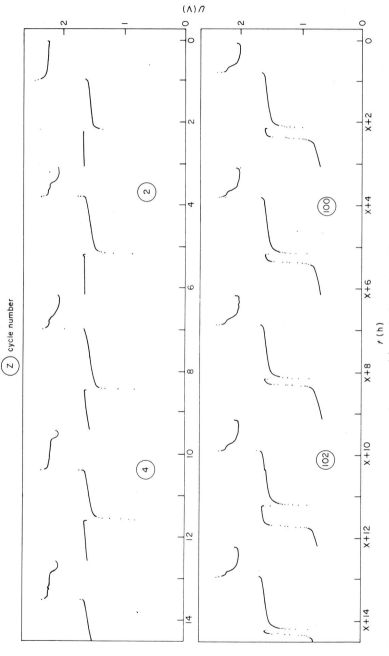

Fig. 15. Cyclization curves of a lead tetrafluoborate battery with vertical graphite-filled polypropylene electrodes containing additives.

Area = 100 cm², sandblasted, at ambient temperature.

Electrolyte: 1·5 M Pb(BF₄)₂/1 M HBF₄; $Q_{th} = 0.020$ A h cm⁻², $j_{ch} = 20$ mA cm², $j_d = 10$ mA cm². 1-h rest after every cycle. β → 100% (full charging), deep discharging, forced convection.

Concluding Remarks

It has been the aim of this chapter to review the electrochemistry of the lead/lead dioxide couple in lead-dissolving electrolytes. For most practical applications, these acids are HBF_4 and H_2SiF_6. The kinetic behaviour of the PbO_2 electrode is much more complicated than that of the Pb electrode. Thus, the limitation in a primary battery is due to passivation and other side reactions of the positive electrode. The possibility of extending the system to secondary batteries, which is promising for several reasons, is impeded predominantly by the problems at the PbO_2 electrode as well. The cathodic discharging is a critical step once again. An analysis of this special problem has been given, and some results have been reported.

Acknowledgments

This review includes hitherto unpublished results of our research work, which was sponsored by BMFT (NT 4450 A).

I thank M. Müller and H. Vogel for valuable experimental assistance.

Symbols

U Cell voltage
$°U$ Standard potential (voltage)
U_H Potential versus standard hydrogen electrode
U_S Potential versus $Hg/Hg_2SO_4/1$ M H_2SO_4-electrode
j_{ch} charging current density
j_d discharging current density
η_a anodic overvoltage
η_c cathodic overvoltage
Q charge
z number of electrons
Z cycle number
α current efficiency
β active mass utilization
CPP   graphite-filled polypropylene, 80:20

References

1. J. O'M. Bockris and S. Srinivasan, *C.R.J. Int. d'Etude Piles Comb.*, *Brux.* II/68 (1965).
2. Gmelin's Handbuch der Anorganischen Chemie, 8. "Lead". Verlag Chemie, Weinheim (1969/1971); cf. A. Seidell and W. F. Linke, "Solubilities". American Chemical Society, Washington (1965).
3. L. H. Brickwedde, *J. Res. Nat. Bur. Stand* **42**, 309 (1949).
4. T. F. Sharpe, *in* "Encyclopedia of the Electrochemistry of the Elements" (A. J. Bard, ed.), Vol. 1, p. 235. Dekker, New York (1973).
5. R. Piontelli and G. Poli, *Z. Elektrochem.* **62**, 320 (1958).
6. R. Piontelli, *Z. Elektrochem.* **55**, 128 (1951).
7. F. Fischer, K. Thiele and E. B. Maxted, *Z. Anorg. Chem.* **67**, 339 (1910).
8. A. H. Du Rose and W. Blum *in* "Modern Electroplating" (F. A. Lowenheim, ed.). Wiley, New York (1967).
9. J. W. Dini and J. R. Helm, *Metal Finish.* **67**, 53 (8/1969).
10. J. O'M. Bockris and D. Drazic, "Electrochemical Science". Taylor and Francis, London (1972).
11. W. Mindt, *J. Electrochem. Soc.* **116**, 1077 (1969).
12. U. B. Thomas, *J. Electrochem. Soc.* **94**, 42 (1948).
13. N. E. Bagshaw, R. L. Clarke and B. Halliwell, *J. Appl. Chem.* **16**, 180 (1960).
14. J. A. Duisman and W. F. Giauque, *J. Phys. Chem.* **72**, 562 (1968).
15. R. P. Ruetschi and B. D. Cahan, *J. Electrochem. Soc.* **105**, 369 (1959).
16. G. Butler and I. L. Copp, *J. Chem. Soc.* 725 (1956).
17. G. L. Clarke and R. Rowan, *J. Amer. Chem. Soc.* **63**, 1305 (1941).
18. A. Byström, *Arkiv. Kemi. Min. Geol.* **A25**, 13 (1947).
19. J. S. Anderson and M. Sterns, *J. Inorg. Nucl. Chem.* **11**, 272 (1959).
20. W. R. Fischer, Doctoral Thesis, University Dortmund (1971).
21. K. Elbs, *Z. Elektrochem.* **3**, 70 (1896).
22. C. Schall and W. Melzer, *Z. Elektrochem.* **28,** 474 (1922); *Chem. Ztg.* **51**, 344 (1927).
23. W. Fischer and H. Rickert, *Ber. Bunsenges. Phys. Chem.* **77**, 975 (1973).
24. H. Rickert, *Z. Phys. Chem.* **95**, 47 (1975); J. P. Pohl and H. Rickert, *Z. Phys. Chem.* **95**, 59 (1975).
25. D. Spahrbier, Doctoral Thesis, University Stuttgart (1960).
26. E. M. Otto, *J. Electrochem. Soc.* **113**, 525 (1966).
27. M. Le Blanc and E. Eberius, *Z. Phys. Chem.* **A160**, 69 (1932).
28. G. Z. Kyriakow *et al.* 1955, cited in ref. 2, p. 296 and p. 306.
29. R. R. Vandervoort *et al.*, *Plating* **57**, 362 (1970).
30. K. Emicke, G. Holzapfel and E. Kniprath, *Erzmetall* **24**, 205 (1971).
31. C. A. Wamser, *J. Amer. Chem. Soc.* **70**, 1209 (1948).
32. R. E. Mesmer, K. M. Palen and C. F. Baes, Jr., *Inorg. Chem.* **12**, 89 (1973).

33. A. N. Fleming and J. A. Harrison, *Electrochim. Acta* **21**, 905 (1976).
34. M. Fleischmann and M. Liler, *Trans. Far. Soc.* **54**, 1370 (1958).
35. H. A. Laitinen and N. H. Watkins, *J. Electrochem. Soc.* **123**, 804 (1976).
36. J. Lander, *J. Amer. Electrochem. Soc.* **98**, 220 (1951).
37. F. Beck, *J. Electroanal. Chem.* **65**, 231 (1975); cf. Extended Abstracts 25th ISE-Meeting, Brighton 9 (1974).
38. N. A. Hampson, P. A. Jones and R. F. Phillips, *Can. J. Them.* **46**, 1325 (1968); **47**, 2171 (1969).
39. J. P. Carr and N. A. Hampson, *Chem. Rev.* **72**, 679 (1972).
40. F. Beck and H. Böhn, *Ber. Bunsenges Phys. Chem.* **79**, 233 (1975).
41. H. B. Mark and W. C. Vosburgh, *J. Electrochem. Soc.* **108**, 615 (1961).
42. F. Beck, unpublished results.
43. C. I. Bushrod and N. A. Hampson, *Trans. Inst. Metal Finish.* **48**, 131 (1970).
44. D. F. A. Koch, *Electrochim. Acta* **1**, 32 (1959).
45. E. V. Krivolapowa and B. N. Kabanow, *Tr. Sowesh. po Elektrokhim. Akad. Nauk SSSR, Otd. Khim. Nauk* 539 (1950).
46. M. Rey, P. Coheur and H. Herbiet, *Trans. Electrochem. Soc.* **73**, 315 (1938).
47. N. F. Razina and G. Z. Kiryakow, *Isv. Akad. Nawk Kaz. SSR. Ser. Khim Nauk* 26 (1959).
48. J. R. Pierson, C. E. Weinlein and C. E. Wright *in* "Power Sources 5" (D. H. Collins, ed.), pp. 97–108. Academic Press, London and New York (1975).
49. Z. A. Nysanbaewa and Yu. D. Dunaev, *Trudy Inst. Organ. Katal. Elektrokhim.* **T3** (1972).
50. M. P. J. Brennan, B. N. Stirrup and N. A. Hampson, *J. Appl. Electrochem.* **4**, 49 (1974).
51. Y. Shibasaki, *J. Electrochem. Soc.* **105**, 624 (1958).
52. N. A. Hampson, P. A. Jones and R. F. Philips, *Can. J. Chem.* **45**, 2039 (1967).
53. N. A. Hampson and C. J. Bushrod, *J. Appl. Electrochem.* **4**, 1 (1974).
54. L. I. Lyamina, N. I. Korolkowa and K. M. Gorbunova, *Elektrokhimiya* **6**, 389 (1970).
55. L. I. Lyamina, N. I. Korolkowa and K. M. Gorbunowa, *Elektrokhimiya* **8**, 651 (1972).
56. L. I. Lyamina, N. I. Tarasowa and K. M. Gorbunowa, *Elektrokhimiya* **11**, 260 (1975).
57. I. M. Issa, M. S. Abdelaal and A. A. El Miligy, *J. Appl. Electrochem.* **5**, 271 (1975).
58. F. Beck and J. Haufe, unpublished work (1969).
59. K. Wiesener, *Exten. Abstr. 27th ISE Meet.* 266 (1976).
60. A. Kozawa *in* "Batteries" (K. Kordesch, ed.), Vol. 1, p. 385. Dekker, New York (1974).
61. S. Ikari and S. Yoshizawa, *J. Electrochem. Soc. Jap.* **28**, E 244, N. 10–12 (1960).

62. L. M. Borzunowa, V. F. Lazarev and A. I. Levin, *J. Appl. Chem. USSR* **45**, 818 (1972).
63. K. Arndt, *Elektrotech. Z.* **57**, 1344 (1936).
64. J. J. Bone, M. Fleischmann and W. F. K. Wynne-Jones, *Trans. Far. Soc.* **55**, 1783 (1959).
65. J. Burbank, *J. Electrochem. Soc.* **104**, 693 (1957); **106**, 369 (1959).
66. P. Ruetschi, *J. Electrochem. Soc.* **120**, 331 (1973).
67. K. M. Gorbunowa, L. J. Lyamina and N. J. Korolkowa, *Electrochim. Acta* **15**, 1597 (1970).
68. F. Beck and H. Gerischer, *Z. Elektrochem. Ber. Bunsenges* **65**, 943 (1959).
69. Gmelin's Handbuch dor Anorganischen Chemie, 8. "Lead" Vol. B 1, pp. 342–349. Verlag Chemie, Weinheim.
70. W. Hofmann, "Lead and Lead Alloys", Springer, New York (1970).
71. W. Katz, *Metalloberfläche* **7A**, 161 (1953).
72. A. F. Clifford, H. C. Beachell and W. M. Jack, *J. Inorg. Nucl. Chem.* **5**, 63 (1957).
73. (a) F. Beck, *Werkst. Korros.* **28**, 688 (1977); (b) F. Beck, *Z. Naturforsch* **32a**, 1042 (1977).
74. J. W. Dini and J. R. Helms, *J. Electrochem. Soc.* **117**, 269 (1970).
75. O. Heckler and H. Hanemann, *Z. Metallkunde* **30**, 410 (1938).
76. V. A. Titow and E. N. Semkowa, *Zh. Prikl. Khim.* **32**, 2492 (1959).
77. J. A. Shropshire and B. L. Tarmy, ACS-Series No. 47 "Fuel Cell Systems", Washington (1965).
78. P. Ruetschi, R. T. Angstadt and B. D. Cahan, *J. Electrochem. Soc.* **106**, 547 (1959).
79. P. Ruetschi, J. Sklarchuk and R. T. Angstadt, *Electrochim. Acta* **8**, 333 (1963).
80. W. H. Beck, R. Jones and W. F. K. Wynne-Jones, *Trans. Far. Soc.* **50**, 1249 (1954).
81. B. N. Kabanow *et al.*, *Electrochim. Acta* **9**, 1197 (1964).
82. J. J. Lander, *J. Electrochem. Soc.* **105**, 289 (1958).
83. DECHEMA-Werkstofftabelle, Frankfurt 1959, data sheet "H_2SiF_6".
84. J. C. White, W. H. Powers, R. L. McMurtric and R. T. Pierce Jr., *Trans. Electrochem. Soc.* **91**, 73 (1947).
85. G. D. McDonald, E. Y. Weissman and T. S. Roemer, *J. Electrochem. Soc.* **119**, 660 (1972).
86. F. G. Turrill and W. C. Kirchberger, *Proc. 24th Power Sources Conf.*, Atlantic City, PSC Publication Committee (1970).
87. J. P. Schrodt, W. J. Otting, J. O. Schoegler and D. N. Craig, *Trans. Elektrochem. Soc.* **90**, 405 (1946).
88. M. Barak, Lecture given at the Symposium "Electrochemistry of Lead", Salford, England (1975).
89. U.S. Patent 3,770,507 (Globe-Union, 1972).
90. T. J. Kilduff and E. F. Horsey, *Proc. 24th Power Sources Conf.*, Atlantic City, PSC Publication Committee (1970).
91. U.S. Patent 3,507,707 (A. M. Biggar, 2/1967).

92. U.S. Patent 3,033,908 (Union Carbide Corp., 6/1959).
93. C. J. Bushrod, private communication (1975).
94. W. Peukert, *Elektrotech. Z.* **18**, 287 (1897); cf. H. Niklas and D. Berndt, *Elektrotech. Z.* A, **94**, 694 (1973).
95. F. Beck, unpublished results.
96. S. Ikari, S. Yoshizawa and S. Okada, *J. Electrochem. Soc. Jap.* **27**, 552 (1959).
97. L. Greenspan, *Trans. Electrochem. Soc.* **90**, 416 (1946).
98. A. Schmidt, "Angewandte Elektrochemie". Verlag Chemie, Weinheim (1976).
99. F. Beck, "Elektroorganische Chemie". Verlag Chemie, Weinheim (1974).
100. E. J. Wade, "Secondary Batteries", London (1902).
101. F. Beck, *Chem.-Ing.-Techn.* **46**, 127 (1974).
102. German Patent 38,193 (G. Leuchs, 5/1886), *see also Ber. Dtsch. Chem. Ges.* **20**, III, 152 (1887).
103. U.S. Patent 1,425,163 (H. Bardt, 9/1921).
104. British Patent 449,893 (Philips, 10/1934).
105. German Patent 2,144,198 (BASF, 9/1971); cf. U.S. Patent 4,001,037.
106. German Patent 636,830 (Philips, 9/1936).
107. K. J. Vetter, *Chem.-Ing.-Techn.* **45**, 213 (1973).
108. K. W. Choi, D. N. Bennion and J. Newman, *J. Electrochem. Soc.* **123**, 1616, 1628 (1976).
109. J. McBreen, *J. Electrochem. Soc.* **119**, 1620 (1972).
110. F. Beck, *Exten. Abstr. 27th ISE Meet.* 9, 30 (1976).
111. F. Beck, *Chem.-Ing.-Techn.* **49**, (1977).
112. K. Elbs, *Z. Elektrochem.* **7**, 261 (1900).
113. E. Willihnganz, *J. Electrochem. Soc.* **100**, 527 (1953).
114. M. Barak, personal communication (1975).
115. S. A. Awad, Kh. H. Kamel and Z. A. Elhady, *J. Electroanal. Chem.* **34**, 431 (1972).
116. C. J. Bushrod and N. A. Hampson, *Brit. Corros. J.* **6**, 87 (1971).
117. A. G. Betts, "Electrolytic Refining of Lead", (1902), German Translation, Halle (1910).
118. M. V. Ginatta and U. Ginatta, Symposium Electrochemistry of Lead, Salford (1975).
119. N. A. Hampson, P. C. Jones and R. F. Phillips, *Can. J. Chem.* **45**, 2045 (1967).
120. S. Haruyama, *J. Electrochem. Soc. Jap.* **35**, 62 (1967) (*Chem. Abstr.* **68**, 26230 (1968)).
121. A. I. Levin, A. F. Kondratew, V. F. Lazarow and A. A. Proskurnikow, *Zh. Prikl. Khim.* **41**, 101 (1968) (*Chem. Abstr.* **68**, 101 082 (1968)).
122. A. F. Kondratew, V. F. Lazarow and A. I. Levin, *Trans. Ural Polithek. Inst.* 1968, No. 170, p. 83 (*Chem. Abstr.* **71**, 18145b (1969)).
123. S. Sarangapani, V. K. Venkatesan and M. A. V. Devanathan, *Trans. SAEST* **5** IV, 146 (1970).

5 Electrochemistry of Lead in Molten Salts

W. D. DAVIS

Research Laboratories, Associated Lead Manufacturers Ltd., Greenford, Middlesex

Introduction

According to Delimarskii and Markov,[1] the electrochemistry of lead in molten salts may be dated from Faraday's experiments in 1833 on the electrolytic decomposition of fused lead halides. From the mid-nineteenth century onwards, electrochemical studies have been reported which are concerned with lead as a constituent either of a fused electrolyte or of an electrode. A broad distinction can be drawn between research whose purpose is first and foremost scientific, and research which is aimed at a potential industrial application, and this distinction is maintained as far as possible in this chapter. An exception is made occasionally when a scientific technique of particular and unusual interest has been developed with the direct purpose of investigating some effect observed in more empirical work.

In much of the purely scientific material, the lead is one solute or one electrode metal among several and the emphasis of the research is not specific to the element. For this reason, and because nowadays the science of molten salts is already copiously documented and reviewed, the emphasis of this chapter is laid on the applied, particularly the electrometallurgical, side of the subject. A little of the justification for this emphasis vanished recently when a monograph[2] on the electrochemical refining of heavy metals in melts appeared, but as this is in Russian and not likely to circulate widely in the West, the original concept has been adhered to.

Scientific Aspects

General

Sources of Information

During the sixth and seventh decades of the present century the science of molten salts developed very rapidly, with increasing precision and sophistication in the purification of melts and in the techniques of measurement. The monographs of Blander[3] and Sundheim[4] represent an important step in the documentation of the fully developed science. Both of these include accounts[5,6] on the progress of electrochemistry. Bloom[7] has contributed a useful introduction to the subject of molten salt chemistry, including its electrochemical aspects, which also formed the subject of an important monograph.[1] The thermodynamics of molten salt mixtures has been exhaustively treated[8] and, more recently, much valuable information with particular relevance to extraction metallurgy has been assembled.[9] Reference electrodes in fused salts have been discussed by Laity[10] and by Alabyshev et al.[11]

A Molten Salts Data Centre has been established at the Rensselaer Polytechnic Institute, Troy, New York, and under its auspices, a "Molten Salts Handbook" has been compiled[12] which tabulates, with references, much of the data published up to that time on the physical, thermodynamic and electrochemical properties of melts and on their spectroscopy and structure, together with information on practical and experimental matters such as the purification of salts and the choice of containers. The U.S. National Bureau of Standards has contributed reference works on the Gibbs free energies and excess free energies for equilibrium type cells[13] and on electrical conductivities.[14] Tables of the conductivities, with other properties, of nitrates, nitrites and their binary mixtures[15] and of fluoride melts[16] have appeared, and similar compilations for chloride, bromide, iodide and mixed melts are in preparation.

In addition, a series of reviews entitled "Advances in Molten Salt Chemistry" edited by Braunstein et al.[17-19] has provided an ongoing account of further progress, and Inman et al.[20-22] have contributed a series of specialist reports on the electrochemistry of molten salts which attempt to maintain an encyclopaedic survey of progress in the subject. These draw a helpful distinction between transport properties, including electrical conductivities; processes at equilibrium, including electrode potentials; and non-equilibrium processes, including polarography, chronopotentiometry and all manner of electrode processes. This subdivision will be adopted in this chapter, after some general remarks on lead and its fusible compounds.

Molten Electrolytes Based on Lead Compounds

Stable melts in which lead salts are the only or major components include the dihalides, the sulphide, PbS, and the oxide, PbO. The halides (fluoride excepted) are low-melting ($PbCl_2$ m.p. 498°C, $PbBr_2$ m.p. 370°C, PbI_2 m.p. 412°C), and form eutectics with other compounds, mainly other halides, to give melts with lower melting points still, extreme examples being the ternary eutectics formed by lead, zinc and potassium chlorides, with a minimum melting point of 211°C.[23] Lead carbonate and lead nitrate decompose on heating, without melting. Lead sulphide melts at 1112°C and forms molten mixtures with alkali-metal and other sulphides and with lead chloride, the last examples presenting considerable interest in the electrometallurgical field. Lead fluoride (m.p. 822°C) and lead monoxide (m.p. 886°C) also enter as major components into lower melting systems, among which the lead-containing glasses are of particular technical importance. However, the most studied lead-based melts are the halides, in particular lead chloride and its binary mixtures with the alkali-metal chlorides. In this series, interactions between the Pb^{2+} and Cl^- ions to form complexes increase with increasing atomic number of the alkali-metal.[7] Thus, compounds are present in the $PbCl_2$–KCl, $PbCl_2$–RbCl and $PbCl_2$–CsCl phase diagrams, and conductance minima are present in the corresponding melts. Lithium and sodium chlorides form no compounds with lead chloride and raise the conductivity of the melt when added to molten lead chloride.

As compared with the compounds of, for example, the alkali-metals, the free energies of formation are low, and this limits the voltage range within which lead-containing melts are stable. Lead chloride is decomposed into its elements by an applied voltage of 1·23 V at 550°C;[24] in contrast to this, the LiCl–KCl eutectic, widely used as a solvent melt in e.m.f. and electrode kinetic studies, has a potential range of about 3·6 V.[25] It follows that the lead-based melts are of restricted value as solvent melts, and apart from conductivity determinations and thermodynamic studies carried out to provide necessary data, or for the sake of the light they shed on bonding and structure in melts, much research involving lead-based melts is of an applied character.

Interactions Between Lead and Molten Salts

Metallic lead is only slightly soluble in lead halides; Corbett and van Wimbush[26] determined the solubility in lead chloride by a gravimetric method as 0·020 mol % at 600°C, 0·052 mol % at 700°C and 0·123 mol % at 800°C. Corbett et al.[27] found a somewhat higher solubility in molten lead iodide (rising to 0·4 mol % at 700°C). The ionic species present in the solution was identified as Pb_2^{2+}, rather than Pb^+, by equilibrating lead chloride with lead–gold alloys in which the activity of the lead was known, and plotting the concentration of the sub-ion against the activity of lead in the alloy phase.[28] The equilibrium:

$$Pb_{alloy} + Pb_{melt}^{2+} \rightleftharpoons Pb_{2_{melt}}^{2+},$$

requires the concentration of the sub-ion to depend linearly on the activity of lead in the alloy, whereas the alternative:

$$Pb + Pb^{2+} \rightleftharpoons 2 Pb^+,$$

implies a dependence of the concentration of the sub-ion on $a_{Pb}^{1/2}$. The reduced ion was determined by anodic polarography on a platinum microelectrode in the melt. The diffusion currents, taken as a measure of the concentration of the sub-ion, were proportional to a_{Pb}.

Van Norman et al.[29] found the solubility to be 0·0057 mol % at 518°C by a coulometric method, and found the same dependence of the concentration of the reduced species, as determined by anodic chronopotentiometry, on a_{Pb}. Topol,[30] using a potentiometric method, concluded that the dissolved lead in lead iodide was also present as Pb_2^{2+}. The identification thus appears fairly certain and doubt must attach to an earlier conclusion[31] that the reduced ion is Pb^+. Delimarskii and Markov[1] point out that the solubility is reduced to a negligible level by the dilution of the lead halides with others, and the dissolution of lead is of no practical consequence in the mixed chloride electrolytes used in applied electrorefining studies, although because it makes back-reaction possible it has a bearing on the current efficiency when lead is electrowon from pure molten lead chloride.[32]

Metallic lead is in equilibrium with halide melts in which lead ions are those most easily reduced, and such a system makes an excellent reference electrode.[11] Lead is, however, oxidized in molten nitrates (which are frequently used as solvent melts) with the formation of a layer of oxide; the sulphate ion is also reduced by metallic lead.[33] The Pb/Pb^{2+} couple is reversible in silicate, borate and phosphate melts.

Transport properties

Electrical Conductivity

Janz[12] lists the conductivities of pure molten lead compounds and gives references to binary systems. The most important compound, lead chloride, has a conductivity (10^3 Hz) of $1 \cdot 45$ ohm^{-1} cm^{-1} at 500°C, $1 \cdot 93$ at 600°C and $2 \cdot 32$ at 700°C.

Smithells,[34] in addition, gives values for a number of binary mixtures in which lead compounds occur. The correlation between conductivity and complex ion formation in binary mixtures of lead chloride and alkali-metal chlorides has already been referred to; more details are given by Delimarskii and Markov[1] and Bloom.[7]

Only Winterhager and Werner[35] have carried out determinations of the conductivity of lead fluoride. Their values vary from $5 \cdot 3$ ohm^{-1} cm^{-1} at 877°C to $5 \cdot 7$ ohm^{-1} cm^{-1} at 977°C.

Bell and Flengas[36] studied conductivity in the technically important $PbCl_2$–PbS binary system. Addition of the sulphide to the chloride initially reduces the conductivity but at about 40 mol % PbS, electronic conductivity sets in and there is a marked increase. The pure sulphide has a specific conductivity of 120 ohm^{-1} cm^{-1} and the temperature dependence is positive, indicating semiconducting behaviour.

Although lead nitrate itself cannot be melted, the conductivities of molten binary mixtures of lead nitrate with potassium, sodium, silver and thallium nitrates at temperatures up to 415°C have been determined.[37] Increasing the amount of lead nitrate added to a monovalent metal nitrate decreased its conductivity; this was attributed to hindered migration of ions due to higher coulombic interactions, rather than to structural changes in the melts.

Conductivity determinations on various melts continue to be reported. Umetsu and Ishii[38] and also Lyubimov et al.[39] have studied the lead chloride–zinc chloride system, while molten glasses containing lead oxide have been the subject of conductivity studies.[40–42] Redeterminations are still sometimes made on compounds and systems for which data adequate for most purposes already exist; lead bromide, for example, has been studied recently[43] and also the $PbCl_2$–KCl system.[44]

Organic melts have received hardly any attention, but Ekwunife et al.[45] have broken new ground by determining the conductivities of the lead salts of the alternate members of the series of carboxylic acids with C_{2n} ($n = 3$–9) carbon atoms in the chain. Melting points varied from 69°C to 114°C and decomposition set in at 197°C. The plots of $\log_e K$ against $1/T$ were curved, but followed a similar course for all the salts, regardless of chain length, leading to the conclusion that the lead ion is the main current-carrying species.

Processes at equilibrium

The interaction of lead with the oxyanions precludes the determination of the standard electrode potentials of lead in nitrate melts and most of the values available have been determined in halide media, chlorides in particular. As no universally agreed reference couple for molten salts exists, different authors have selected different reference points.

Plambeck[25] has presented a useful review of e.m.f. series compiled up to that date. Cells of the type

$$M_1/M_1^{n+} \text{ (solvent melt)} \,/\, \text{(solvent melt)} \ M_2^{n+}/M_2$$

were favoured, with a porous plug or frit dividing the two half-cells. Where the solute concentration is below 0·1 M and the solvent melt is uniform, the solution junction potential is negligible. The series for the LiCl–KCl eutectic (59 mol % of LiCl) at 450°C is most extensive; many of the values in it were determined using the Pt/Pt^{2+} couple as reference.[46] Nernstian behaviour for the Pb/Pb^{2+} couple up to 5 mol % $PbCl_2$ was observed.

In halide media, and in oxidic electrolytes such as silicates in which no interaction between the metal and the oxyanion takes place, the Pb/Pb^{2+} couple is highly reversible and reproducible. This makes it a widely used reference electrode; Laity[10] and Alabyshev et al.[11] both cite numerous examples of its use, with pride of place given to the halide melts. Cells of the type

$$Pb_{liq}/PbCl_2-MCl_n/Cl_{2(g)}, C$$

where MCl_n is an alkali-metal or other diluent chloride, have been used to determine the thermodynamics of lead chloride-based melts, and the activities of lead in alloys with metals noble to it may be determined from the e.m.f. of cells of the type

$$Pb_{alloy}/PbCl_2-MCl_n/Pb_{liq}$$

Oxide melts may replace halides in such uses. Sridhar and Jeffes[47] studied the thermodynamics of the lead oxide–silica system using the formation cell:

$$Fe, Pb/PbO_{(x)}, SiO_{2(1-x)}/O_{2(g)}Pt,$$

which proved highly reversible. The same authors[48] and also Caley and Masson[49] have employed the Pb/Pb^{2+} couple in determining the thermodynamics of lead phosphate melts. Hager and Walker[50] and Hager and Zambrano[51] offer examples of the use of lead silicate melts in determining the thermodynamic properties of, respectively, lead–gold and lead–gold–silver alloys.

Outside the halides and oxides, a reversible couple was said[52,53] to exist between metallic lead and lead ions in molten potassium thiocyanate up to its decomposition temperature of 276°C.

Delimarskii et al.[54] used the Pb/Pb^{2+} couple to determine the stability constants of the complex ions PbF^+, PbI^+, $Pb(CN)_2$ and $Pb(CN)_3^-$ in equimolar KCl–NaCl at 706°C, by analysing the e.m.f.'s of cells of the type

$$Pb/KCl-NaCl+2\cdot5 \text{ mol } \% \text{ } PbCl_2, KCl-NaCl+x \text{ } PbCl_2+yKX/Pb,$$

where KX is potassium fluoride, iodide or cyanide.

Non-equilibrium processes

Electrode Kinetics of Reactions Involving Lead Ions

Piontelli and Sternheim[55] investigated both the anodic and cathodic processes on molten lead in lead chloride at 530°C and in $PbCl_2$–KCl melts at 560°C and at 670°C. Up to at least 2 A cm^{-2}, there was no appreciable activation overpotential. This points to a very rapid charge-transfer process and to a high exchange current density. A number of authors have estimated this quantity, applying various techniques to molten lead electrodes in chloride solvent melts containing small concentrations of lead chloride. Laitinen et al.[56] used a double-pulse method in LiCl–KCl at 450°C, and Ukshe and Bukun[57] used an a.c. impedance technique to measure the faradaic impedance in equimolar KCl–NaCl at 720°C. By extrapolation of these and other results to pure lead chloride, Hart[58] arrived at values between 20 and 1400 A cm^{-2}. Faradaic impedance measurements have also been applied in the KCl–NaCl melt at 800°C,[59] in the KCl–LiCl eutectic at 450°C,[60] and in a KCl–ZnCl$_2$ melt at 500°C.[61]

Very little work of any kind has been reported in melts based on lead fluoride, but Pizzini and Agace[62] found that molten lead in PbF$_2$–NaF melts between 600°C and 800°C functions as anode or cathode without detectable activation overpotential.

Polarography and Related Techniques

Laitinen and Osteryoung[6] pointed out that the general application of polarography in molten salts has been the study of the solute species dissolved in a particular solvent melt. With lead as the solute, polarographic methods have been employed to determine its diffusion coefficient and also to study its complexation in solution. Laitinen and Osteryoung's review, together with those of Liu et al.[5] and of Inman et al.[20–22] provide between them a fairly complete assembly of the available information, and this section need only make some generalizations.

Reversible reduction of the lead ion is normally observed, exceptions being encountered sometimes in L.S.V. studies at high sweep rates. For example, Francini et al.[63] found some irreversibility at 10 V s^{-1} in the NaNO$_3$–KNO$_3$ eutectic melts at 270°C. A similar effect occurred above 1·5 V s^{-1} in molten KSCN–NaSCN[64] and in LiOAc–NaOAc–KOAc[65] at 6 V s^{-1}. All these effects were detected using the dropping mercury electrode.

The dropping mercury electrode is usable up to about 250°C and a number of workers have observed the reduction of lead from low-melting nitrate melts.[66–68] Lead, once deposited, diffuses rapidly into mercury and thus the current–voltage curves, if reduction is reversible, should obey the Heyrovsky–Ilkovic equation:

$$E = E_{1/2} + \frac{RT}{nF} \ln \frac{i_d - i}{i}$$

Most authors agree that this happens, although diffusion coefficients calculated from the Ilkovic equation sometimes differ appreciably. Swofford and Holifield,[69] however, found sharp maxima in the polarographic waves and marked departure from Heyrovsky–Ilkovic behaviour. This was tentatively attributed to the oxidation of lead by the NO_3^- ion, giving an insoluble product. Tridot et al.,[70] on the other hand, found reversible two-electron reduction of lead ions at 240°C. The dropping mercury electrode has been used at 140°C in an $AlCl_3$–KCl–NaCl eutectic,[71] and at 160°C in an $AlCl_3$–LiCl–KCl melt.[72] Reversible reduction was observed.

The widely used LiCl–KCl eutectic has too high a melting point for the dropping mercury electrode. Egan and Heus[73] used a dropping bismuth electrode in this melt at 450°C. Diffusion currents proportional to concentration were obtained, and a diffusion coefficient could be calculated from the Ilkovic equation. Schwabe and Ross[74] also confirmed that the Ilkovic equation applies to the reduction of Pb^{2+} at the dropping bismuth electrode in this electrolyte.

Laitinen et al.[75] used platinum, carbon and tungsten electrodes in the same electrolyte. Above the melting point of lead, polarograms conforming to the Kolthoff–Lingane equation:

$$E = E_{1/2} + \frac{RT}{nF} \ln (i_d - i)$$

were obtained on tungsten, appropriate to the absence of alloy formation.

Naryshkin et al.,[76] using linear-sweep voltammetry in LiCl–KCl at 450°C, found reversible reduction of Pb^{2+} on platinum. Peak current was proportional to the logarithm of the concentration, indicating an insoluble product. Gaur et al.,[77] on the other hand, used a platinum microelectrode in $MgCl_2$–KCl at 975°C and observed Heyrovsky–Ilkovic behaviour. In their case it appeared that enough time was allowed for the deposited lead to form an alloy. Kolthoff–Lingane behaviour was observed on tungsten. Further evidence that rapid-sweep L.S.V. on platinum leaves insufficient time for alloy formation to take place was found[78] in a $NaNO_3$–$Ba(NO_3)_2$ melt at 350°C. Evidence of interaction of the lead with NO_3^- was obtained.

Christie and Osteryoung[79] used polarography on the dropping mercury electrode in $LiNO_3$–KNO_3 at 180°C to determine the stability constants of the chloro-complexes of Pb^{2+}. The stepwise constants were 42 and 3 for $PbCl^+$ and $PbCl_2$. In the presence of bromide, maxima were obtained, but Inman et al.[80] succeeded in obtaining stability constants for the corresponding bromo-complexes at 145°C in the $LiNO_3$–$NaNO_3$–KNO_3 eutectic. Susic et al.[81] observed a shift in the half-wave potential of lead in $LiNO_3$–KNO_3 at 170°C when the di-sodium salt of EDTA was added, and also the diffusion currents were reduced because the complex has a smaller diffusion coefficient.

Polarographic studies in electrolytes other than nitrates and halides are few, but Delimarskii and Tumanova[82] have observed the reduction of Pb^{2+} (introduced as the oxide) from the lithium carbonate–potassium carbonate eutectic at 640°C. Limiting current was proportional to concentration.

Chronopotentiometry has also been applied to the study of the reduction of lead ions. The reviews mentioned earlier give more details but typically the plots of potential vs log $\tau_{1/2}$–$t_{1/2}$ indicate reversible two-electron reduction.

The Dropping Lead Electrode

Interactions between lead and many solute metals would seem to limit the use of this electrode for polarography, but Naryshkin et al.[83] have studied the reduction of zinc ions at 450°C in molten LiCl–KCl and found Heyrovsky–Ilkovic behaviour.

Heus et al.[84] describe the use of dropping lead or bismuth electrodes to study the double-layer capacity of these metals in LiCl–KCl. Careful removal of oxide by filtration is essential before the metal enters the capillary. Assiduous purification and pre-electrolysis are applied to the melt. The measured capacitance varied only by 3% between 1 and 20 kHz, and on lead, temperature had little effect in the 390–480°C range.

Applied Aspects

General

The low melting point and the high electrochemical equivalent of lead, together with the availability of stable low-melting electrolytes based on lead compounds, make the element a particularly favourable subject for attempts to develop electrometallurgical processes using fused salt electrolytes. Most of these fall into three categories: (i) conventional anode-to-cathode electrorefining of lead, leaving impurities in the anode, (ii) electrowinning lead from sulphide concentrates dissolved or suspended in a chloride melt, with evolution of elemental sulphur at the anode, and (iii) the removal of certain impurities, particularly bismuth, from molten lead by making it the *cathode* in suitable electrolytes. These three branches of the subject are discussed in turn in the next three sections.

Rather less attention has been given to the removal, partial or complete, of certain impurities by making lead the anode in a suitable electrolyte, and this possibility, together with some miscellaneous electrometallurgical researches, are referred to in a fourth section. Although, to the author's knowledge, no fully commercial process has yet been commissioned, very recent papers in all three of the main categories testify to continuing interest in them. However, all these potential processes have to compete with established practice, and must either offer some economic or other advantage over existing processes or do something that these will not. In anode-to-cathode electrorefining, the advantages of working with molten metal electrodes and a molten electrolyte have been stated by Amstein *et al.*[85] Very high current densities ($2–3$ A cm^{-2}) can be employed, with negligible polarization; many complications attached to aqueous electrorefining, such as the accumulation of anode slime, the formation of dendritic or uneven deposits and the drag-out of electrolyte, are eliminated. In the particular cases of antimony and bismuth alloyed with lead, the concentration of the alloying element in the anode can be taken virtually to 100% with little contamination of the lead at the cathode. In other words, alloys of lead with antimony or bismuth may be parted into the elements. However, fused-salt refining cells work at $2–3$ V compared with $0·35–0·7$ V for modern aqueous refining cells[86] and the extra power is an additional cost to be compensated for elsewhere.

Renewed interest in the direct electrolysis of galena concentrates has been justified by the avoidance of environmental pollution by sulphur dioxide in situations where a sulphuric acid plant cannot be operated economically in line with a roaster.[32,87,88] "Cathodic" refining to remove bismuth from lead is cleaner than the established Betterton–Kroll process, but its capital cost is higher in its present state of development.[89]

Anode-to-cathode electrorefining

Borchers[90] demonstrated the separation, on a pilot-plant scale, of lead–bismuth alloy into "very pure" lead and an anode residue containing 90–95% bismuth in an electrolyte consisting of lead oxychloride dissolved in equimolar KCl–NaCl. Since then, the subject has been taken up in the USSR, first by Alabyshev and Gel'man,[91] who also obtained very pure lead from a lead–bismuth anode, and subsequently by Delimarskii and his collaborators. This work has covered chloride and oxide electrolytes and mixtures of the two; chlorides, notably the ternary eutectic $PbCl_2$–KCl–NaCl, have received most attention and have been used in the only scaled-up experiments reported.

In the chlorides, a series of direct trials in small externally heated cells consisting essentially of one crucible inside another[92–95] have established an order of efficacy of the separation of different elements. Gold and bismuth separate most cleanly, antimony and arsenic almost as well, silver concentrates in the anode but larger amounts appear in the cathode, and copper and tin are hardly separated. The behaviour of tin is hardly surprising because the standard electrode potentials of tin and lead lie very close together in fused chlorides, and the fact that the two elements electrolyse more or less together has been adapted to the refining of lead–tin solder,[96] with special reference to the removal of gold.[97] As examples of typical results, lead–bismuth was parted into 96% Bi–4% Pb and lead containing 0·0015% Bi, and the electrolysis of solder containing gold gave an anode residue containing 7·5% Au and a refined product with less than 0·0003%. A refining ratio (anode content divided by cathode content) of about 20 was found for silver.

Silicates and mixed silicate–chloride electrolytes have been investigated as refining media by Delimarskii et al.[98] Compared with the chloride media, the separation of bismuth was indifferent, but that of tin and silver was improved. An electrolyte having the empirical composition $PbSiO_3$–1·27PbO–2PbCl_2 was favoured by Delimarskii and Kosmatyi[99,100] and evidence of dissolution of lead as a sub-ion was obtained. A range of sodium silicate–lead silicate electrolytes has been patented[101] for parting lead and silver at 880–1200°C, and described in more detail by the same authors.[102] None of these electrolytes appear to have been employed in scaled-up experiments.

As regards the theory of the process, some very clean separations rest fairly obviously on the stability, or lack of it, of compounds of the impurity elements. The complete retention in the anode of gold in chloride electrolytes and of silver in silicates must reflect the fact that gold chloride and silver oxide are thermodynamically unstable at cell-operating temperatures. The clean separation of antimony in chloride melts forms another special case; here the high volatility of antimony trichloride prevents much transfer of the element across the cell. As for the other elements, Delimarskii and his colleagues have sought to correlate their refining behaviour with their positions in e.m.f. series drawn up according to the potentials of cells of the type:

$$Pb/ZnCl_2-PbCl_2-KCl,\!/ZnCl_2-PbCl_2-KCl + 5 \text{ mol } \% \text{ MeCl}_n/Me$$

with asbestos dividing the half-cells. Delimarskii and Zarubitskii[2] report the following values at 450°C for the two electrolytes they favour:

$PbCl_2-KCl-NaCl : Pb/Pb^{2+}0\cdot00, Cu/Cu^+0\cdot12,$
    $Ag/Ag^+0\cdot28, Bi/Bi^{3+}0\cdot36 \text{ V}$
$ZnCl_2-PbCl_2-KCl : Pb/Pb^{2+}0\cdot00, Cu/Cu^+0\cdot18,$
    $Ag/Ag^+0\cdot36, Bi/Bi^{3+}0\cdot40 \text{ V}$

On this basis the poor performance of copper is partly accounted for by its proximity to lead in the series but silver in practice is separated from lead less cleanly than bismuth in spite of the wide e.m.f. margin between silver and lead. A more sophisticated analysis of the refining process has been attempted by Hart et al.[103] and elaborated in more detail by Hart.[58] They assumed that thermodynamic equilibrium exists at both electrodes and applied the friction coefficient formalism[104] to the relative rates of transport of the major and impurity metal ions across the cell. The treatment leads to two major conclusions. First, for each impurity element in a given major metal a "refinability parameter" can be defined. When this exceeds unity, the impurity is likely to concentrate in the cathode rather than the anode, and vice versa. Elements with the cleanest retention in the anode should thus have very small parameters. Secondly, when the parameter indicates that an impurity should concentrate in the anode, the refining ratio should increase with current density to a maximum value which, taking reasonable values for the factors in the equation, is of the order of $0\cdot5 \text{ A cm}^{-2}$, which is in the lower end of the range in which molten salt cells operate. Figure 1 illustrates the effect predicted. It was necessary to make a number of simplifying assumptions, to estimate the unlike-ion friction coefficients from the conductivities of the pure chlorides, and to ignore the like-ion friction coefficients. However, at 800 K (527°C) the refinability parameters for copper, silver and bismuth were estimated as 2·7, 0·17 and 10^{-8}, and the separation of these elements can be classified as poor, fair and excellent.

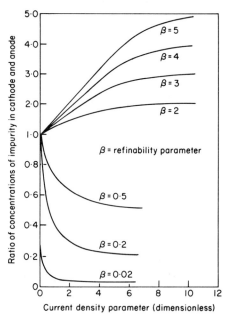

Fig. 1. The effect of current density on refining ratio for different values of the refinability parameter (reproduced by permission of the Institution of Mining and Metallurgy).

Inman[105] criticized the assumption that friction coefficients calculated from the equivalent conductivities of the pure chlorides of the impurity elements could be applied to the very different conditions obtaining in the refining cell where the chlorides are present in low concentration. While this criticism must have some force, a valuable contribution has still been made in that the importance of the transport of impurity ions, apart from their equilibrium concentration, has been formally taken into account.

Bartlett *et al.*[106] took into account the thermodynamic equilibrium concentrations of impurity ions at both electrodes and the rates of diffusion of the impurity and major metal ions, and arrived at the conclusion that the refining effect increases with cell voltage up to a limiting value. Where the voltage drop is entirely ohmic, this means that a limiting current density exists above that at which the refining effect levels off, so that this conclusion agrees with that of Hart et al.[103] The effect of convection was also considered and it was shown that when it is present a higher current density is needed to achieve the maximum refining effect. No estimates of relative refinability were made, but the predicted effect of current density was tested experimentally for the separation of copper impurity from tin, in a convective cell, with satisfactory results.

It is thus possible to sum up the findings of theoretical approaches as confirming that, qualitatively, the transport properties of the metal ions must be considered and that, for impurities which concentrate in the anode, increased current density does not impair refining. Current density can thus be selected for optimum yield or optimum temperature, i.e. on purely engineering grounds. As for the refining effect to be expected with different impurities, empirical studies seem as good a guide as theory.

In attempting to scale up a process using two molten metal electrodes, there is an obvious difficulty in opposing them while keeping the distance between them short and, if possible, uniform. The arrangement using two pools of molten metal lying side by side with an insulating wall between is poor because the current path is shortest near the wall and thus the current concentrates there. This gets worse as the size of cell increases. One solution, with almost ideal geometry, is to separate the two electrodes by a porous diaphragm permeable by electrolyte but not by molten metal. In the USSR, Pavlenko and Grinyuk[107] have described rigid porous diaphragms in the form of hollow containers set in cavities in hollow graphite blocks or sleeves which constituted the cathode (Fig. 2). Anode metal was located inside these porous liners, the walls of which constituted vertical diaphragms. The porous materials developed cracks or were otherwise unstable. Recent work in the U.K.[85,108] has been concerned with horizontally disposed diaphragms made from the "blankets" of ceramic fibres sold for hot-face thermal insulation (Fig. 3). Cells rated at 800 A and 3200 A functioned for over a month, and the scale has been increased to 8000 A. The lead chloride-based electrolytes had to be kept scrupulously free from dissolved oxide to prevent attack on the diaphragm. The emphasis of this work was on the parting of lead and antimony and the cells had to be run above the melting point of antimony (630°C), but the complete retention of bismuth in the anode was confirmed. Copper, again, was hardly removed from the lead at all.

Soviet work on pilot-plant cells has been directed mostly to separating lead from bismuth, and apart from the experiments of Pavlenko and

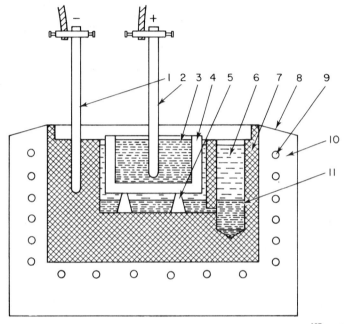

Fig. 2. Diagram of refining cell with porous silicon carbide diaphragm.[107] 1. Cathode conductor; 2. Anode conductor; 3. Anodic alloy; 4. Diaphragm; 5. Heat-resisting glass supports; 6. Molten electrolyte; 7. Graphite cathode block; 8. Steel body; 9. Nichrome heaters; 10. Thermally insulating packing; 11. Cathodic lead.

Grinyuk,[107] diaphragms have been avoided. Roms *et al.*[109,110] described a cell in which the anode metal formed a pool on the bottom and the cathode lead was deposited on vertical graphite plates suspended from the lid (Fig. 4). It was then collected in horizontal troughs underneath the cathodes but at a higher level than the anode surface. By dividing the anode compartment into three by low walls and passing the anode alloy from one compartment to another as its bismuth content rose, part of the lead could be obtained with a very low bismuth content from an anode containing, in the first compartment, mostly lead.

In the particular case of the separation of lead and bismuth, another way of overcoming the basic geometrical difficulty is to use a very low-melting electrolyte containing zinc chloride and to operate at a

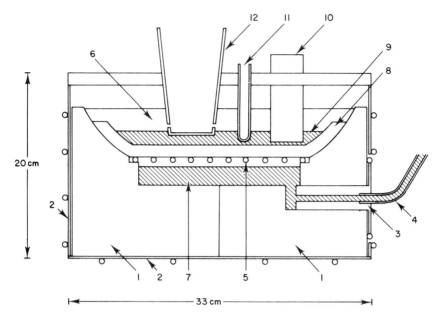

Fig. 3. Diagram of 800-A cell with ceramic-fibre "blanket" diaphragm (reproduced by permission of the Institution of Mining and Metallurgy). 1. Ceramic block; 2. Water-cooled plate; 3. Cathode conductor; 4. Siphon tube; 5. Support rods; 6. Molten electrolyte; 7. Cathode metal; 8. Diaphragm; 9. Anode metal; 10. Anode conductor; 11. Thermocouple sheath; 12. Baffle crucible.

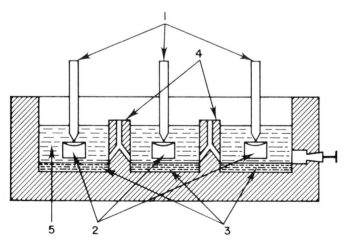

Fig. 4. Diagram of refining cell with vertical cathodes.[109,110] 1. Graphite cathodes; 2. Collectors for cathodically deposited lead; 3. Molten anodes; 4. Ceramic partitions; 5. Molten electrolyte.

temperature at which the lead–bismuth alloy anode and the electrolyte are molten but the lead cathode is solid. The method has been described and patented by the same authors.[111,112] The temperature range of operation was 230–280°C and the cathode current density had to be kept below 0.4 A cm^{-2} to avoid dendrite formation. Zarubitskii and Budnik,[113] in a later patent, extended the method to provide for forced cooling of the solid cathode on which the lead was deposited. In this way a temperature gradient was maintained in the cell; the anode was at 360°C, so that all compositions in the lead–bismuth diagram could be treated when molten, and the cathode was at 260°C. The cathode current density was kept at 0.3 A cm^{-2}.

Electrowinning lead from sulphide ore

Lead sulphide dissolves readily in molten lead chloride and forms a eutectic melting at 441°C[114] containing 22% by weight (24.7% M) of the sulphide. Because the sulphide has a lower decomposition potential than the chloride (0.466 V at 550°C compared to 1.24 V), electrolysis of the solution produces metallic lead and gaseous sulphur and, in principle, lead can be produced directly from mineral galena by this method. Townsend[115] was granted a patent on the process in 1906. Rideal,[116] writing in 1918, refers (giving, unfortunately, a false reference) to good electrical efficiency having been obtained in the electrolysis of finely crushed galena dissolved in an electrolyte of molten lead chloride and sodium chloride. The accumulation in the electrolyte of impurities from the ore raised its melting point and the initial good performance was not maintained.

A pilot-plant was operated at the Halkyn mine in Wales for some years up to the outbreak of war in 1939 and the process is still often referred to as the Halkyn process. It was based on a patent granted to Gibson and Robson,[117] and Richardson[87] and Newall[118] gave a limited amount of additional information. Bipolar electrodes were employed, at a voltage of 1.2–1.4 V per stage and an energy consumption of 740–790 W h kg^{-1}, which implies a current efficiency of the order of 50%. The concentration of dissolved lead sulphide was 3–5% (polysulphide formation took place at higher concentration), the temperature 550°C, and the current density 0.6–0.75 A cm^{-2}. The bath was agitated, and addition of feed and removal of gangue took place in a special compartment. Delimarskii and Zarubitskii[2] stated that a pilot-plant was operated at one time in Australia, and that the adoption of the process foundered once more on the accumulation of gangue.

On a small scale, modifications of the process have been studied.[24,119–123] Some of these authors favoured the addition to the melt of alkali chlorides, which reduce the tendency to form metal fog and also lower the maximum temperature at which electrolysis can be conducted. This in turn improves the current efficiency. Izgaryshev and Grigor'ev[124] advocated the suspension of the galena in a melt of mixed sodium and potassium chlorides, and Gul'din and Buzhinskaya[125] favour a similar electrolyte with a small quantity of dissolved sodium sulphide, and justify the choice on the grounds of its safety and low cost as compared to lead chloride. They describe a pilot cell working at 1500 A and the very high voltage of 7–11 V, at 630–730°C. Apart from silver, gold and bismuth, which reported in the lead, other impurities were retained in a sludge which had to be removed regularly. Furukawa[126] has patented the use of a pressed anode of fused or partly fused galena, suspended in a lead chloride electrolyte, as an alternative to dissolving crushed galena. The advantages are not clear.

Welch et al.[88] initiated a revival of research into the process and established once more that high current efficiency is favoured by a low concentration of lead sulphide, high current density and a low temperature. Because Bell and Flengas[36] have shown that lead dissolves hardly at all in lead chloride–lead sulphide melts, back-reaction would seem to be attributable either to the dissolution of sulphur in the melt or to the formation of subsulphide or polysulphide ions at the anode. Reactions at the anode have thus taken pride of place in attempts to understand the process. Linear sweep voltammetry studies on the anode reactions in melts saturated with lead sulphide demonstrated that a polysulphide was formed at a lower potential than that required for the evolution of sulphur, and the reaction of this species with the lead was held responsible for the low current efficiencies. Prolonged electrolysis of melts containing high concentrations of sulphide (530°C, 25% M PbS) produced deposits on the anode which at low current density (20 mA cm^{-2}) were smooth and insulating. Chemical analyses were inconclusive but an excess of sulphur was present. At higher current densities (100 mA cm^{-2}) the potential was sufficiently anodic for sulphur evolution to take place as well and a honeycombed structure was formed.

Further studies showed that for several metals (zinc, silver, copper and tin) a chloride of the same cation must be present in the electrolyte to promote appreciable solubility of the sulphide. Thus, other sulphides contaminating the galena will not dissolve in lead chloride or $PbCl_2$–NaCl and will form an insoluble gangue, and, in consequence, low-grade ores might be amenable to treatment if the problem of the separation of gangue could be solved. To elucidate further what happens at the anode, King and Welch[127] developed an ingenious optical cell (Fig. 5) consisting of two microscope slides held horizontally 0·8 mm apart by two strips of graphite aligned with the two long sides and serving both as electrodes and spacers. An electrolyte containing dissolved lead sulphide could be introduced into the intervening space and its electrolysis observed and filmed. The furnace arrangement is shown in Fig. 6. In particular the total volume of gaseous sulphur which had been produced at any moment could be determined by summing the volumes of the (cylindrical) gas bubbles. The rate of increase in total volume at 530°C was lower than that calculated from the coulombs passed, indicating that the sulphur was dissolving, and at 445°C the rate of sulphur dissolution was an order of magnitude less than that at 530°C. Strong evidence was thus adduced that back-reaction with

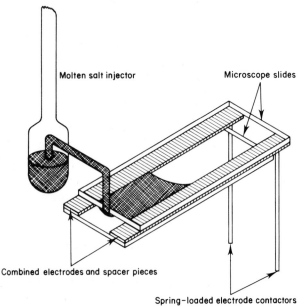

Fig. 5. Molten salt optical cell (reproduced by permission of the authors).

dissolved sulphur contributes to the reduced current efficiencies at higher temperatures observed in practical electrowinning studies. The most recent of these have been reported,[128] and Welch *et al.*[129] have summarized the findings of the current revival of interest in the process.

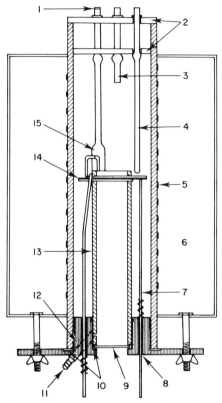

Fig. 6. Molten salt optical cell, furnace arrangements (reproduced by permission of the authors). 1. Silicone rubber septum; 2. Armourglass; 3. Gas outlet; 4. Thermocouple well; 5. Kanthal resistance wire furnace winding; 6. Kieselguhr furnace insulation; 7. Spring-loaded electrode contactor; 8. PTFE bushes; 9. Window; 10. Silicone rubber seals; 11. Inert gas inlet; 12. Brass endpiece; 13. Alumina tube; 14. Optical cell; 15. Molten salt injector.

The removal of bismuth and other impurities from lead cathodes

Dittmer[130] first patented the removal of bismuth from lead by stirring in elemental sodium under a cover of molten sodium hydroxide. Rapid transfer of bismuth to the melt took place when the lead contained about 4% of sodium. In a later patent,[131] Dittmer introduced a variant in which the necessary concentration of sodium in the molten lead was generated electrolytically by making the lead the cathode. The anode was a steel pan of molten lead in which the bismuth was concentrated by oxidation of the Na_3Bi species present in the melt, e.g.

$$Na_3Bi \rightarrow 3\,Na^+ + Bi + 3e$$

Delimarskii *et al.*[132] point out that when lead is present at the anode, the further oxidation of bismuth to the oxide is prevented, as lead is oxidized more readily than bismuth. Accounts of practical studies have been contributed,[133–136] the latest of which describes a 600-A cell in which a 50-kg charge of lead containing initially 0·29% Bi was refined to 0·0036% Bi with the consumption of 350 W h kg^{-1} lead. The reactions associated with the transfer of bismuth account for only a small fraction of the cell current and it is meaningless to refer to current efficiency, but some perspective can be gained by noting that the alternative possibility of electrolysing the lead away from the bismuth in a lead chloride-based melt would consume 390 W h kg^{-1} at 1·5 V and 520 W h kg^{-1} at 2.0 V. An unpublished account[137] revealed that the process is now operated on a large, though unstated, scale in the USSR. The precise mechanism of the process is in some doubt, but the sodium bismuthide is probably present in the melt as a stable sol containing neutral and charged particles represented respectively as $(Na_3Bi)_x(NaOH)_y$ and $(Na_3Bi)_x(NaOH)_y–OH^-$, the magnitude of x and y being unknown. Although the process is normally carried out at a temperature just above the melting point of lead, in the 340–400°C range, a patent[138] claims that a cleaner separation can be achieved by working in the temperature range 450–500°C. All the compounds present in the lead–sodium system are then molten and, presumably, less likely to enter the electrolyte phase. They are in any case less readily formed than Na_3Bi. Another patent[139] claims some benefit from the addition of 2% of zinc oxide to the melt. This may be a result of an enhanced stability of the suspended Na_3Bi due to the increased viscosity of the melt.

    The most informative account of recent developments has been given by Gardner and Denholm.[89] Lead containing bismuth (about 0·03% by weight) was made the cathode in molten sodium hydroxide and charged with 3–5% sodium, most of the bismuth being extracted into the melt. A further reduction of the bismuth content could then be effected, if desired, by equilibrating the metal once more with clean molten sodium hydroxide. The metal was then freed from sodium by air and steam and the Na_3Bi and Na suspended in the sodium hydroxide were oxidized non-electrolytically to globules of alloy which could be filtered out. Figure 7 represents a flow diagram. The process was taken to the 4000 A

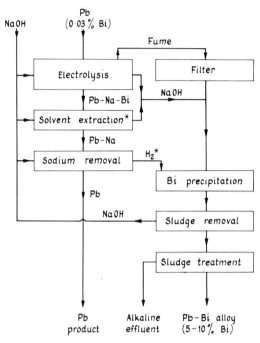

Fig. 7. Flowsheet for the Dittmer–CSIRO process for debismuthizing lead. The asterisked stages may prove to be unnecessary upon scaling up (reproduced by permission of the Australasian Institute of Mining and Metallurgy and the authors).

scale of operation, and at best the bismuth content of the lead was reduced to $0·0002\%$, with an energy consumption of 230 W h kg^{-1}. The process would be varied so that either, or both, of the two phases, lead and sodium hydroxide electrolyte, could be used continuously or batchwise. In a brief discussion of the mechanism, the new finding was announced that pure bismuth oxide fume was deposited in one part of the apparatus, and this was tentatively attributed to the oxidation of bismuthine produced by the reactions:

$$Na + NaOH \rightarrow Na_2O + \tfrac{1}{2} H_2$$

$$3 H_2 + 2 Na_3Bi \rightarrow 2 H_3Bi + 6 Na$$

Dorin *et al.*[140] have plotted the potential–pO^{2-} diagram for bismuth in molten sodium hydroxide at 623 K (350°C) and pointed out that areas for hydrogen stability and for sodium bismuthide formation are adjacent. Participation of hydrides as intermediates in fume formation is thus likely, and Gardner and Denholm[89] also suggest that hydrides may take part in the transfer of Na_3Bi from the metal to the melt. Although most of the published material on the subject is devoted to the removal of bismuth from lead, Delimarskii *et al.*[141] have reported the removal of tellurium from lead by an analogous process (at a cathode in sodium hydroxide).

Moreover, although no pilot-scale work has been reported, analogous processes involving the formation and extraction into the melt of intermetallic compounds occur in halide melts as well as sodium hydroxide. Heymann and Weber,[142] for example, found that although bismuth, antimony and gold are insoluble in molten sodium bromide, they were transferred into the melt from lead in the presence of sodium. Delimarskii *et al.*[143] reported briefly on the removal of arsenic, antimony and bismuth from a lead cathode in a melt of calcium, sodium and potassium chloride at 600°C. No tin was removed and little copper. Silver could be removed only if zinc was introduced into the lead, when Ag_2Zn_5 (m.p. 636°C) was extracted.

Miscellaneous applied studies

Some Soviet work has dealt with the removal of certain impurities from lead by taking advantage of their stronger tendency to ionize when the lead is made the anode in a melt containing no lead. Irgaliev *et al.*[144] described small-scale work in which arsenic, antimony and tin were removed from a stirred molten lead anode in sodium hydroxide. The noble impurities, bismuth, silver and copper, were unaffected. The impurities removed are precisely those which are removed in the well-known Harris process in which molten lead is treated with oxidizing agents under a sodium hydroxide cover. Delimarskii and Zarubitskii[2] discuss anodization in halide electrolytes and the possibility of combining an anodic step with the cathodic treatments referred to in the previous section.

Tran Kim Hoa and Welch[33] have considered the electrowinning of lead from lead sulphate dissolved in the LiCl–KCl and the Li_2SO_4–

Na_2SO_4–K_2SO_4 eutectics. Chronopotentiometric measurements, determinations of the dependence of cathodic current efficiency on current density, and direct observation of the interaction of molten lead with the melts, confirmed that initially the lead ions are reversibly discharged at the cathode but that the lead then reacts with dissolved lead sulphate. Monobasic lead sulphate was found in all the used melts, and this prompted the suggestion that the secondary reaction was:

$$4\,Pb + 5\,PbSO_4 \rightarrow PbS + 4\,(PbO.PbSO_4)$$

Jenkins and Welch[145] have complemented this work by investigating the anode reactions (in the LiCl–KCl eutectic) when a consumable carbon anode is employed.

Murphy et al.[32] have described preliminary investigations on what may be called a two-stage Halkyn process in which the galena is leached with aqueous ferric chloride solution to produce lead chloride which, after purification, is fed to a molten chloride electrolyte bath for electrowinning. Lead chloride and its binary eutectics with lithium, sodium and potassium chlorides were employed. Lead chloride itself gave poor current efficiency, presumably because of back-reaction of chlorine with dissolved lead. The lithium chloride eutectic at 450°C gave 99% current efficiency. This work was again justified by the increasing control being exercised on the sulphur dioxide emissions associated with the conventional roasting of galena, and while it is at a very early stage, it would seem in principle to be free from the difficulties caused in the Halkyn process by the accumulation of gangue in the cell.

Shelley and Charles[146] studied the electrolysis of lead silicate melts at 700°C between a molten lead (or tin) electrode and an arc struck between a pair of carbon electrodes above the melt surface. Lead oxide fume was produced in approximately faradaic yield whether the arc acted as cathode or as anode, and it was suggested that at the cathodic arc the deposited lead was immediately oxidized and at the anodic arc a negative lead-bearing ion such as PbO_2^{2-} is discharged. It was suggested that the work could lead to new extraction processes, perhaps with two arc electrodes giving total recovery of lead as fume, but apart from some similar work on tin by the same authors, there appear to have been no further developments.

References

1. Yu. K. Delimarskii and B. F. Markov, "Electrochemistry of Fused Salts" (English translation by A. Peiperl). Sigma Press, Washington (1961).
2. Yu. K. Delimarskii and O. G. Zarubitskii, "Electrolytic Refining of Heavy Metals in Ionic Melts". Metallurgiya, Moscow (1975).
3. M. Blander (ed.), "Molten Salt Chemistry". Interscience, New York (1964).
4. B. R. Sundheim, (ed.), "Fused Salts". McGraw-Hill, New York (1964).
5. C. H. Liu, K. E. Johnson and H. A. Laitinen in "Molten Salt Chemistry" (M. Blander, ed.). Interscience, New York (1964).
6. H. A. Laitinen and R. A. Osteryoung in "Fused Salts" (B. R. Sundheim, ed.). McGraw-Hill, New York (1964).
7. H. Bloom, "The Chemistry of Molten Salts". Benjamin, New York (1967).
8. J. Lumsden, "Thermodynamics of Molten Salt Mixtures". Academic Press, London and New York (1966).
9. F. D. Richardson, "Physical Chemistry of Melts in Metallurgy". Academic Press, London and New York (1975).
10. R. W. Laity in "Reference Electrodes: Theory and Practice" (D. J. G. Ives and G. J. Janz, eds). Academic Press, New York and London (1961).
11. A. F. Alabyshev, M. F. Lantratov and A. G. Morachevskii, "Reference Electrodes for Fused Salts" (English translation by A. Peiperl). Sigma Press, Washington (1965).
12. G. J. Janz, "Molten Salts Handbook". Academic Press, New York and London (1967).
13. G. J. Janz and C. G. M. Dijkhuis, "Molten Salts, Volume 2". National Bureau of Standards, Washington (1969).
14. G. J. Janz, F. W. Dampier, G. R. Lakshminarayanan, P. K. Lorenz and R. P. T. Tompkins, "Molten Salts, Volume 1". National Bureau of Standards, Washington (1968).
15. G. J. Janz, U. Krebs, H. G. Siegenthaler and R. P. T. Tompkins, *J. Phys. Chem. Ref. Data* 1, 581–746 (1972).
16. G. J. Janz, G. L. Gardner, U. Krebs and R. P. T. Tompkins, *J. Phys. Chem. Ref. Data* 3, 1–115 (1974).
17. J. Braunstein, G. Mamantov and G. P. Smith (eds), "Advances in Molten Salt Chemistry, Volume 1". Plenum Press, New York (1971).
18. J. Braunstein, G. Mamantov and G. P. Smith (eds), "Advances in Molten Salt Chemistry, Volume 2". Plenum Press, New York (1973).
19. J. Braunstein, G. Mamantov and G. P. Smith (eds), "Advances in Molten Salt Chemistry, Volume 3". Plenum Press, New York (1975).
20. D. Inman, A. D. Graves and R. S. Sethi in "Electrochemistry, Volume 1". Specialist Periodical Reports, Chemical Society, London (1970).
21. D. Inman, A. D. Graves and A. A. Nobile in "Electrochemistry, Volume 2". Specialist Periodical Reports, Chemical Society, London (1972).

22. D. Inman, J. E. Bowling, D. G. Lovering and S. H. White *in* "Electrochemistry, Volume 4". Specialist Periodical Reports, Chemical Society, London (1974).
23. Ya. V. Ugai and V. A. Shatillo, *Zh. Fiz. Khim.* **23**, 744–754 (1949).
24. H. Winterhager, *Forschungsber. Wirtsch-u. Verkehrsministeriums Nordrhein-Westfalen* No. 134 (1955).
25. J. A. Plambeck, *J. Chem. Eng. Data* **12**, 77–82 (1967).
26. J. D. Corbett and S. van Wimbush, *J. Amer. Chem. Soc.* **77**, 3964–3966 (1955).
27. J. D. Corbett, S. van Wimbush and F. C. Albers, *J. Amer. Chem. Soc.* **79**, 3020–3024 (1957).
28. J. J. Egan, *J. Phys. Chem.* **65**, 2222–2224 (1961).
29. J. D. van Norman, S. J. Bookless and J. J. Egan *J. Phys. Chem.* **70**, 1276–1281 (1966).
30. C. E. Topol. *J. Phys. Chem.* **67**, 2222–2225 (1963).
31. S. V. Karpachev, A. G. Stromberg and E. Iordan. *Zh. Fiz. Khim.* **18**, 47–52 (1944).
32. J. E. Murphy, F. P. Haver and M. M. Wong, *U.S. Bur. Mines.*, *Report of Investigations* 7913 (1974).
33. Tran Kim Hoa and B. J. Welch, *J. Appl. Electrochem.* **3**, 45–52 (1973).
34. C. J. Smithells (ed.), "Metals Reference Book", 5th edn. Butterworth, London (1976).
35. H. Winterhager and L. Werner, *Forschungsber. Wirtsch-u. Verkehrsministeriums Nordrhein-Westfalen* No. 438 (1957).
36. M. C. Bell and S. N. Flengas, *J. Electrochem. Soc.* **113**, 27–37 (1966).
37. H. C. Gaur and J. K. Jain, *J. Chem. Eng. Data* **17**, 200–203 (1972).
38. Y. Umetsu and Y. Ishii, *Nippon Kinzoku Gakkaishi* **37**, 997–1004 (1973).
39. V. N. Lyubimov, V. A. Fedotov and A. P. Kuznetsov, *Izv. Sib. Otd. Akad. Nauk SSSR, Ser. Khim. Nauk* No. 5, 6–12 (1973).
40. H. Saito, K. Goto and M. Someno, *Tetsu to Hagane* **55**, 539–549 (1969).
41. M. Ashizuka and M. Ohtani, *Nippon Kinzoku Gakkaishi* **33**, 498–503 (1969).
42. E. A. Erznkyan and K. A. Kostanyan, *Arm. Khim. Zh.* **22**, 103–110 (1969).
43. J. J. McNicholas, *J. Electrochem. Soc.* **115**, 936–937 (1968).
44. A. J. Easteal and I. M. Hodge, *J. Phys. Chem.* **74**, 730–735 (1970).
45. M. E. Ekwunife, M. U. Nwachaku, F. P. Rinehart and S. J. Sime, *J. Chem. Soc., Far. Trans. I* **71**, 1432–1440 (1975).
46. H. A. Laitinen and C. H. Liu, *J. Amer. Chem. Soc.* **80**, 1015–1020 (1958).
47. R. Sridhar and J. H. E. Jeffes, *Trans. Inst. Min. Metall.* **76**, C44–C50 (1967).
48. R. Sridhar and J. H. E. Jeffes, *Trans. Inst. Min. Metall.* **78**, C14–C20 (1969).
49. W. F. Caley and C. R. Masson, *Metal–Slag–Gas Reactions and Processes*, 140–154 (1975).

50. J. P. Hager and R. A. Walker, *Trans. Met. Soc. AIME* **245**, 2307–2312 (1969).
51. J. P. Hager and A. R. Zambrano, *Trans. Met. Soc. AIME* **245**, 2313–2318 (1969).
52. G. Metzger, *Comm. Energie At. (France) Rappt.* No. R-2566 (1964).
53. W. G. Bennett and W. E. Jensen, *J. Inorg. Nucl. Chem.* **28**, 1829–1835 (1966).
54. Yu. K. Delimarskii, V. F. Grishchenko and L. I. Zarubitskaya, *Russ. J. Inorg. Chem.* **20**, 921–925 (1975).
55. R. Piontelli and G. Sternheim, *J. Chem. Phys.* **23**, 1358 and 1971–1972 (1955).
56. H. A. Laitinen, R. P. Tischer and D. K. Roe, *J. Electrochem. Soc.* **107**, 546–555 (1960).
57. E. A. Ukshe and N. G. Bukun, *Russ. J. Phys. Chem.* **35**, 1330–1333 (1961).
58. P. F. Hart, Thesis, London (1969).
59. R. Yu. Bek and A. S. Lifshits, *Izv. Sib. Otdel. Akad. Nauk SSSR, Ser. Khim. Nauk* **14**(6), 70–78 (1967).
60. A. S. Lifshits and R. Yu. Bek, *Izv. Sib. Otd. Aka. Nauk SSSR, Ser. Khim. Nauk* **12**(5), 8–12 (1968).
61. O. G. Zarubitskii and V. S. Kikhno, *Izv. Vyssh. Ucheb. Zaved. Tsvetn Metal.* **1**, 50–52 (1973).
62. S. Pizzini and L. Agace, *Corrosion Sci.* **5**, 193–198 (1965).
63. M. Francini, S. Martini and C. Monfrini, *Electrochim. Metall.* **2**, 325 (1967).
64. M. Francini, S. Martini and H. Giess, *Electrochim. Metall.* **3**, 355 (1968).
65. H. Giess, M. Francini and S. Martini, *Electrochim. Metall.* **4**, 17–19 (1969).
66. N. H. Nachtrieb and M. Steinberg, *J. Amer. Chem. Soc.* **72**, 3558–3565 (1950).
67. M. Francini and S. Martini, *Z. Naturforsch.* **23a**, 795 (1968).
68. D. Inman, D. G. Lovering and R. Narayan, *Trans. Far. Soc.* **63**, 3017–3025 (1967).
69. H. S. Swoffard and C. L. Holifield, *Anal. Chem.* **37**, 1509–1512 (1965).
70. G. Tridot, G. Nowogrowki, J. Nicole, M. Wozniak and J. Canonne, *C.R. Acad. Sci.* **270**, C204 (1970).
71. R. M. de Fremont, R. Rosset and M. Leroy, *Bull. Soc. Chim. France* No. 4, 706–712 (1964).
72. M. Saito, S. Suzuki and H. Goto, *Nippon Kagaku Zasshi* **83**, 883–886 (1962).
73. J. J. Egan and R. Heus, *J. Electrochem. Soc.* **107**, 824 (1960).
74. K. Schwabe and R. Ross, *Z. Anorg. Allgem. Chem.* **325**, 181–190 (1963).
75. H. A. Laitinen, C. H. Liu and W. Ferguson, *Anal. Chem.* **30**, 1266–1270 (1958).
76. I. I. Naryshkin, V. P. Yurkinskii and P. T. Stangrit, *Elektrokhimiya* **5**, 1476–1479 (1969).
77. H. C. Gaur, H. L. Jindal and R. S. Sethi, *Electrochim. Acta* **15**, 845–851 (1970).

78. W. K. Behl and H. C. Gaur, *J. Electroanal. Chem. Interfacial Electrochem.* **32**, 293–302 (1971).
79. J. H. Christie and R. A. Osteryoung, *J. Amer. Chem. Soc.* **82**, 1841 (1960).
80. D. Inman, D. G. Lovering and R. Narayan, *Trans. Far. Soc.* **64**, 2476–2486 (1968).
81. M. V. Susic, D. A. Markovic and N. N. Hercigonja, *J. Electroanal. Chem. Interfacial Electrochem.* **41**, 119–125 (1973).
82. Yu. K. Delimarskii and N. Kh. Tumanova, *Ukr. Khim. Zh.* **30**, 52–56 (1964).
83. I. I. Naryshkin, V. P. Yurkinskii and B. S. Yavich, *Izv. Vyssh. Ucheb. Zaved., Khim. i Khim. Tekhnol.* **11**, 23–25 (1968).
84. R. J. Heus, T. Tidwell and J. J. Egan *in* "Molten Salts: Characterization and Analysis" (G. Mamantov, ed.), pp. 499–508. Dekker, New York (1969).
85. E. H. Amstein, W. D. Davis and C. Hillyer *in* "Advances in Extraction Metallurgy", pp. 399–412. Institution of Mining and Metallurgy, London (1972).
86. E. R. Freni *in* "Symposium on sulfamic acid and its electrometallurgical applications, Milan 1966", pp. 367–371. Edito della Associazione Italiana di Metallurgia, Milan (1969).
87. J. B. Richardson, *Trans. Inst. Min. Metall.* **46**, 339–461 (1937).
88. B. J. Welch, P. L. King and R. A. Jenkins, *Scand. J. Metall.* **1**, 49–55 (1972).
89. H. J. Gardner and W. T. Denholm, Papers presented at the South Australia Conference, June, pp. 197–201. Australasian Institute of Mining and Metallurgy, Parkville, Victoria (1975).
90. W. Borchers, *Z. Elektrotech. Elektrochem.* **1**, 13–15 (1894).
91. A. F. Alabyshev and E. M. Gel'man, *Tsvetn. Metal.* **2**, 37–43 (1946).
92. Yu. K. Delimarskii, P. P. Turov and E. B. Gitman, *Ukr. Khim. Zh.* **21**, 687–693 (1955).
93. Yu. K. Delimarskii, B. F. Markov, I. D. Panchenko, E. B. Gitman and A. A. Kolotii, *Trudy Chetvertogo Soveshchaniya Elektrokhim.* 1956, pp. 710–714. Moscow (1959).
94. Yu. K. Delimarskii and I. D. Panchenko, *Ukr. Khim. Zh.* **25**, 458–462 (1959).
95. E. B. Gitman and Yu. K. Delimarskii, *J. Appl. Chem. USSR* **32**, 607–610 (1959).
96. Yu. K. Delimarskii, E. S. Grin-Gnatovskii, V. V. Pokrovskii and A. P. Samodelov, *Ukr. Khim. Zh.* **31**, 687–693 (1965).
97. E. K. Kleespies, J. P. Bennetts and T. A. Henrie, *J. Metals, N.Y.* **22**, 42–44 (1970).
98. Yu. K. Delimarskii, I. G. Pavlenko and Yu. E. Kosmatyi, *Zh. Prikl. Khim.* **33**, 1840–1843 (1960).
99. Yu. K. Delimarskii and Yu. E. Kosmatyi, *Ukr. Khim. Zh.* **31**, 179–181 (1965).
100. Yu. K. Delimarskii and Yu. E. Kosmatyi, *Ukr. Khim. Zh.* **34**, 767–773 (1968).

101. Yu. K. Delimarskii, V. G. Budnik and O. G. Zarubitskii, USSR Patent 373226 (1973).
102. Yu. K. Delimarskii, O. G. Zarubitskii and V. G. Budnik, *Zh. Prikl. Khim.* **47**, 1062–1065 (1974).
103. P. F. Hart, A. W. D. Hills and J. W. Tomlinson *in* "Advances in Extractive Metallurgy" pp. 624–651. Institution of Mining and Metallurgy, London (1968).
104. R. W. Laity, *Ann. N.Y. Acad. Sci.* **79**, 997–1022 (1960).
105. D. Inman, *in discussion of paper by* P. F. Hart, A. W. D. Hills and J. W. Tomlinson *in* "Advances in Extractive Metallurgy" pp. 624–651. Institution of Mining and Metallurgy, London (1968).
106. R. W. Bartlett, M. C. van Hecke and C. Q. Hoard, *Met. Trans.* **3**, 2241–2245 (1972).
107. I. G. Pavlenko and A. P. Grinyuk, *Ukr. Khim. Zh.* **29**, 868–873 (1963).
108. W. D. Davis, *Chem. Ind.* No. 9, 386–387 (1975).
109. Yu. G. Roms, B. S. Belen'kii and Yu. K. Delimarskii, *Tsvetn. Metal.* **41**, 34–38 (1968).
110. Yu. G. Roms, B. S. Belen'kii, Yu. E. Kosmatyi and Yu. K. Delimarskii, *Khim. Prom. Ukr.* **4**, 26–27 (1969).
111. Yu. K. Delimarskii, V. G. Budnik, O. A. Salabai and O. G. Zarubitskii, *Zh. Prikl. Khim.* **43**, 2244–2247 (1970).
112. Yu. K. Delimarskii, V. G. Budnik, O. A. Salabai and O. G. Zarubitskii, USSR Patent 291983 (1971).
113. O. G. Zarubitskii and V. G. Budnik, USSR Patent 420699 (1974).
114. W. Truthe, *Z. Anorg. Chem.* **76**, 161–173 (1912).
115. C. P. Townsend, German Patent 182478 (1906).
116. E. K. Rideal, "Industrial Electrometallurgy, including Electrolytic and Electrothermal Processes". Bailliere, Tindall and Cox, London (1918).
117. A. R. Gibson and S. Robson, British Patent 448328 (1934).
118. A. P. Newall *in discussion of paper by* H. J. T. Ellingham, *Trans. Inst. Chem. Eng.* **6**, 77–103 (1938).
119. E. Angel and E. Garnum, *Teknisk Tidsk.* **75**, 279–284 (1945).
120. H. Sawamoto and Ts. Saito, *J. Mining Inst. Japan* **68**, 555–558 (1952).
121. H. Winterhager and R. Kammel, *Z. Erzberg. Metallhutt.* **9**, 97–105 (1956).
122. A. Riesenkampf, *Rudy Metale Niezelazne* **10**, 271–274 (1965).
123. Ts. Yanagase, Y. Suginohara and I. Kyono, *Denki Kagaku* **36**, 129–136 (1968).
124. N. A. Izgaryshev and N. K. Grigor'ev, *Zh. Obshchei Khim.* **6**, 1676–1685 (1936).
125. I. T. Gul'din and A. V. Buzhinskaya, *Zh. Prikl. Khim.* **33**, 2603–2606 (1960).
126. T. Furukawa, Japan Patent 3659('51) (1951).
127. P. L. King and B. J. Welch, *J. Appl. Electrochem.* **2**, 23–29 (1972).
128. P. L. King and B. J. Welch, *Proc. Aust. Inst. Mining Met.* **246**, 7–12 (1973).
129. B. J. Welch, A. J. Cox and P. L. King, "Papers presented at the Richardson Conference on the Physical Chemistry of Process Metallurgy". Institution of Mining and Metallurgy, London (1974).

130. J. C. Dittmer, US Patent 2365177 (1944).
131. J. C. Dittmer, US Patent 2507096 (1950).
132. Yu. K. Delimarskii, O. G. Zarubitskii and I. G. Pavlenko, *Ukr. Khim. Zh.* **31**, 469–473 (1965).
133. V. D. Ponomarev, I. Z. Slutskii and G. A. Kononenko, *Izv. Vyssh. Ucheb. Zaved. Tsvet. Metal.* **7**, 67–72 (1964).
134. Yu. K. Delimarskii, I. G. Pavlenko and O. G. Zarubitskii, *J. Appl. Chem. USSR* **35**, 302–306 (1962).
135. Yu. K. Delimarskii, I. G. Pavlenko, O. G. Zarubitskii and V. A. Oleinik, *J. Appl. Chem. USSR* **38**, 814–819 (1965).
136. Yu. K. Delimarskii, O. G. Zarubitskii, V. G. Budnik and I. V. Aleksanyan, *Tsvetn. Metal.* **43**, 14–17 (1970).
137. Yu. K. Delimarskii, Unpublished account delivered to the Molten Salts Discussion Group, Leatherhead, September (1974).
138. Yu. K. Delimarskii, O. G. Zarubitskii, V. G. Budnik and N. F. Zakharchenko, USSR Patent 393365 (1973).
139. Yu. K. Delimarskii, O. G. Zarubitskii and I. G. Pavlenko, USSR Patent 224092 (1968).
140. R. Dorin, H. J. Gardner and L. J. Rogers, Paper presented at the Electrometallurgy Symposium, Melbourne University, November (1975).
141. Yu. K. Delimarskii, O. G. Zarubitskii and I. I. Penkalo, *Izv. Vyssh. Ucheb. Zaved. Tsvetn Metal.* **4**, 78–81 (1971).
142. E. Heymann and H. P. Weber, *Trans. Far. Soc.* **34**, 1492 (1938).
143. Yu. K. Delimarskii, A. P. Samodelov and E. A. Kontorovich, *J. Appl. Chem. USSR* **44**, 2392–2394 (1971).
144. G. B. Irgaliev, V. D. Ponomarev, I. Z. Slutskii and G. A. Kononenko, *Met. Khim. Prom. Kazakhstana, Nauchno-Tekh. Sb.* **5**, 48–52 (1962).
145. R. A. Jenkins and B. J. Welch, *J. Appl. Electrochem.* **3**, 53–60 (1973).
146. T. R. Shelley and J. A. Charles, *Trans. Inst. Min. Metall.* **78**, C177–C180 (1969).
147. G. J. Janz, R. P. T. Tomkins, C. B. Allen, J. R. Downey, Jr., G. L. Gardner, U. Krebs and S. K. Singer, *J. Phys. Chem. Data* **4**, 871–1178 (1975).
148. G. J. Janz, R. P. T. Tomkins, C. B. Allen, J. R. Downey, Jr. and S. K. Singer, *J. Phys. Chem. Data* **6**, 409–596 (1977).
149. G. J. Janz, C. B. Allen and R. P. T. Tomkins, *J. Electrochem. Soc.* **124**, 51C–54C (1977).
150. D. G. Winter and A. M. Strachan, *in* "Advances in Extractive Metallurgy" pp. 177–184. Institution of Mining and Metallurgy, London (1977).
151. R. K. Jain, H. C. Gaur, E. J. Frazer and B. J. Welch, *J. Electroanal. Chem. Interfacial Electrochem.* **78**, 1–30 (1977).
152. J. H. Cleland and D. J. Fray, *in* "Advances in Extractive Metallurgy" pp. 141–146. Institution of Mining and Metallurgy, London (1977).
153. F. P. Haver, C. H. Elges, D. L. Bixby and M. M. Wong, *U.S. Bur. Mines Rep. Invest.* RI 8166 (1976).
154. M. M. Wong and F. P. Haver, *in* "Molten Salt Electrolysis in Metal Production" pp. 21–29. Institution of Mining and Metallurgy, London (1977).

155. W. T. Denholm, R. Dorin and H. S. Gardner, *in* "Advances in Extractive Metallurgy" pp. 235–244. Institution of Mining and Metallurgy, London (1977).

Note Added in Proof

During the period which has elapsed between the preparation of this chapter and its publication, the Molten Salts Data Centre has produced its promised compilations of data (see p. 104) on molten chlorides[147] and molten bromides and iodides.[148] An article on the work of the centre has also appeared.[149]

Winter and Strachan[150] have published further e.m.f. data, and Jain *et al.*[151] have reviewed the applications of chronopotentiometry in molten salts.

Cleland and Fray[152] have described a novel solution to the problem, referred to on p. 117, of opposing two molten metal electrodes. In their apparatus, a cylindrical container with its axis vertical is divided into two chambers by a vertically disposed diaphragm of the same type as those used by Amstein *et al.*;[85] the molten metal electrodes cascade down each chamber over a packing of conducting particles.

The work mentioned on p. 127 on the recovery of lead from lead chloride produced in its turn by the hydrometallurgical treatment of galena has progressed and more recent accounts are now available.[153,154] A later account of the removal of bismuth from lead by cathodic treatment has been published.[155]

6 Electrodeposition of Lead

J. A. VON FRAUNHOFER

School of Dentistry, University of Louisville Health Sciences Centre, Kentucky, U.S.A.

Introduction

Lead is one of the oldest metals known to man and it has been extensively used over hundreds of years. This usage derives from the somewhat unusual properties of lead, notably good ductility, high malleability and excellent corrosion resistance combined with a relatively low melting point. Consequently, lead may be used in a variety of forms, namely sheet, piping, castings and many different types of coatings. In fact, one of the principal applications of metallic lead is in the provision of corrosion protection for base metals, principally iron and steel but also for zinc, aluminium and copper and its alloys. Radiation protection, soft solders and type metals are other major uses of metallic lead and lead alloys.

The metallurgy and general chemistry of lead have been discussed in detail by a number of workers.[1-9] There have also been a number of authoritative monographs and reviews on the industrial uses of lead and lead compounds: these uses include lead–acid storage batteries, lead-based pigments, the corrosion resistance of lead and its alloys such as terne plate (lead–25% tin), lead cable sheathing, type metals and lead solders.[9-19] These applications of lead are largely outside the scope of this work and the principal interest of this chapter is the electrodeposition of lead and lead alloys, although such electrodeposited coatings are often of major importance in many of the fields indicated above.

Lead Coatings

Lead and lead alloy coatings are applied to metallic substrates (and often non-metallic materials as well) by a wide variety of techniques and for a diversity of industrial purposes. The intended application of the coating determines, in large part, the coating technique adopted.

The principal uses of lead and lead alloy coatings are for corrosion protection,[20–22] bearing surfaces,[23,24] architectural purposes,[25,26] etch resists for printed circuit boards,[27–29] and enhanced solderability.[30,31] These applications derive from either the metallurgical characteristics of lead, for example its use in solders and bearings, or its electrochemical properties, notably its corrosion resistance. Lead and lead alloy coatings may be applied by the following methods: dip coating, fusion, spray coating, hot and cold rolling, cladding and electrodeposition.[2] Whilst only the last named technique is of direct relevance here, it should be noted that the physical and electrochemical properties of all forms of lead coating often differ markedly from those of the massive metal, particularly in the case of electrodeposits.

Electrodeposited lead and lead alloys are used in virtually all of the industrial applications outlined above but the requirements of the electrodeposit and the electrolyte used varies with the precise application.[32–34] Coatings for corrosion resistance and etch resists, for example, are required to be adherent, pore-free and of uniform thickness, but these criteria are often less important for other applications.

Electrodeposition of Lead

The electrodeposition of lead and lead alloys has been discussed in a number of monographs on general electrodeposition of metal finishing.[34–41] The properties and structural characteristics of electrodeposited lead have similarly been discussed in a number of reviews and research reports over the years.[2,22,34,40,41,43–48] It is clear from this that the electrodeposition of lead and lead alloys has commercial and technical importance, although far less than that of other metals such as nickel.

Lead has a high hydrogen overpotential and, consequently, although it is less noble than hydrogen, it may be electrodeposited from aqueous media at high cathode efficiencies. Several electrolytes, notably the fluoborate, silicofluoride and sulphamate baths, have achieved technical importance for commercial lead and lead alloy electrodeposition, although there is increasing interest in the development of novel systems. There is also a growing interest in the deposition of lead from fused salt baths[49-51] and there have been efforts to develop the use of electroless deposition of lead.[52]

Lead may be electrodeposited from the aqueous solutions of its soluble salts such as the acetate and nitrate but the resultant electrodeposits have no commercial or technical usefulness. These lead electrodeposits tend to be acicular and field-oriented textured,[2] that is, there is preferential growth in the direction of the applied electric field. Thus, the deposits are fibrous in texture and possess a degree of porosity. In fact complex salt solutions, that is, those containing complex anions such as the fluoborate, silicofluoride, sulphamate and pyrophosphate ions and many others, are necessary to produce satisfactory, that is, smooth and coherent, electrodeposits. The reason for this behaviour is not altogether clear despite the long-term usage of such electrolytes in the electrodeposition of lead. It should be noted, however, that lead electrodeposits from even complex salt solutions are textured, having a basis-oriented structure, although the presence of organic additives such as gelatin or tannic acid in the electrolyte will obviate the coarse texture of the electrodeposited metal. This chapter will, therefore, be concerned only with complex salt electrolytes used for the electrodeposition of lead.

Lead silicofluoride

The silicofluoride bath is one of the oldest electrolytes used for the electrodeposition of lead. Although it is less important than the fluoborate and sulphamate baths for general electrodeposition of lead and lead alloys (see later), the silicofluoride bath has importance in the Betts process for electrorefining of lead.[53,54]

The lead silicofluoride electrolyte is prepared by the action of silicofluoric acid on lead oxide, the solution containing $75-100 \text{ g l}^{-1}$ of lead as lead oxide and $120-150 \text{ g l}^{-1}$ of silicofluoric acid.[37] In order to achieve dense electrodeposits, additions of organic materials such as glue and sulphonates are made to the solution and, in the case of glue, the addition level is 0.013% of the weight of the electrolyte.[54]

The bath is usually operated at c. 40°C and cathode current densities of $2-8 \text{ A dm}^{-2}$ are possible with solution agitation by means of recirculation.[37,54] Cathode current efficiencies greater than 90% are found in operation and the electrodeposited lead generally has a purity of the order of 99·999%.[54] The quality of the electrodeposit is generally insensitive to small variations in the operating parameters.

Lead fluoborate

The lead fluoborate bath is the oldest[55] and most widely used electrolyte for electrodeposition of lead (and lead alloys).[56] The solution is prepared from basic lead carbonate and fluoboric acid, the bath containing 100 g l^{-1} lead and 42 g l^{-1} free fluoboric acid, with approximately 10 g l^{-1} boric acid as a buffer and at least $0·2 \text{ g l}^{-1}$ of glue to refine the grain structure of the electrodeposit and prevent "treeing". The bath may be operated over the temperature range of 15–40°C and cathode current densities of $1–3 \text{ A dm}^{-2}$.[37] The cathode current efficiency is close to 100%.

Whilst the lead fluoborate electrolyte is normally used as a vat electrodeposition solution, it has been used for brush or localized electrodeposition.[2,57] Since the first investigation and introduction of brush plating techniques, the method has undergone further development[58,59] so that it may be used for rapid electrodeposition procedures rather than merely as a localized or reparative plating method.

The deposition of lead from the fluoborate bath is relatively high, being $3·87 \text{ g A}^{-1} \text{ h}^{-1}$ or $0·57 \text{ μm min}^{-1}$ at 1 A dm^{-2}, but in recent years there has been increased interest in the development of high speed electrodeposition solutions.[61,62]

One approach to increased rates of electrodeposition is by turbulent flow techniques involving either enhanced solution flow rates or accelerated cathode movements.[61] If the solution flow rate is increased to $1·25 \text{ m s}^{-1}$ or greater, current densities in excess of 1 A cm^{-2} and deposition rates greater than 20 μm min^{-1} may be achieved. In a pilot plant, deposition rates of 100 μm min^{-1} were reported to be achieved, although the current efficiency of lead deposition would appear to be reduced at these high deposition rates.

Alternatively, more concentrated solutions, i.e. having lead fluoborate levels of 400 g l^{-1} and 210 g l^{-1} of lead as metal, may be used to achieve greater deposition rates.[62] In these solutions, the free fluoboric acid level is often lower, $c. 2 \text{ g l}^{-1}$, than in the conventional bath but the boric acid level remains approximately the same although electrolytes having much higher fluoboric and boric acid contents have also been studied. Very high limiting current densities, of the order of $200–330$ A dm^{-2}, may be achieved at elevated temperatures (viz. 70°C). Despite the high current densities and raised solution temperature, the deposits were reported to possess refined grain structures provided suitable additives were present in the electrolyte, hydroquinone appearing to be the most satisfactory grain-refining agent.[62]

High speed lead electrodeposition appears to be a commercially viable process, the economics being comparable with those for terne plate, electrotinning and for zinc electrodeposition at similar thicknesses although the solderability of the coating may not be wholly satisfactory.[62] Despite the reported successes obtained with concentrated solutions, high-speed lead electrodeposition does not appear to have received general commercial exploitation except in the field of selective plating procedures.[59]

The standard lead fluoborate electrolyte is somewhat acidic, having a pH of about $1·5$, but there has been some interest in the development of very acidic electrolytes having low metal ion levels.[63] Current densities of up to 7 A dm^{-2} have been reported for a fluoborate electrolyte containing only 10 g l^{-1} of lead with good electrodeposit distribution being obtained. X-ray diffraction studies revealed no differences between these metal-low pH electrolyte deposits and those obtained from the conventional electrolyte.[63]

Low pH fluoborate electrolytes (pH $0·5–2·0$) containing 20 wt % lead fluoborate and operating at 30–40°C and 3 A dm^{-2} have been used for the deposition of screens for fluorescence sensitization purposes.[63a]

Bright lead (and lead–tin) electrodeposits from fluoborate electrolytes, as well as from silicofluoride and sulphamate electrolytes, have been reported.[63b] The brightening additions consist of lower aliphatic and aromatic aldehydes together with a non-ionic polyalkoxylated surface active agent at an addition level of greater than 5 g l^{-1}. Greater levels, e.g. $5–50$ g l^{-1}, of a non-ionic alkoxylated fatty acid alkylol amide surface active agent are used if no brightener additions are present. The bath is operated at 10–60°C, has a low pH (0·5–1·5) and current densities of c. $40–60$ A dm^{-2} are possible with cathode movement.

Despite the widespread usage and commercial importance of the fluoborate electrolyte for lead electrodeposition, other electrolytes have been investigated for lead, notably the sulphamate bath as well as various other complexing media. These alternative electrolytes will be considered below.

Lead sulphamate

The lead sulphamate electrolyte has some commercial importance but less than that of the fluoborate bath. The electrolyte appears to be used mainly in the production of corrosion resistant coatings and for bearing surfaces although it is of growing importance for the electrorefining of lead.[64] It is also being used increasingly in the electrodeposition of lead alloys (see later).

The sulphamate bath is probably the simplest electrolyte for lead electrodeposition, containing 100–150 g l^{-1} of lead as lead sulphamate. It is operated at a temperature of 25–50°C and current densities of 1–3 A dm^{-2}.[37] The solution has a very low pH, 1·5 to 2·5, the pH being adjusted with sulphuric acid.

This electrolyte yields dense, fine-grain lead deposits at cathode current efficiencies close to 100% although it has been shown that binary phenol additions of 0·2–0·5 g l^{-1} increase cathode polarization and facilitate the deposition process.[65]

Other electrolytes

A variety of other electrolytes have been studied over the years and both acid and alkaline media as well as fairly complex solutions have been investigated.

Electrodeposition of lead from pyrophosphate-based electrolytes has been reported in the past[66,67] but there appears to have been little commercial development of the pyrophosphate bath. Pyrophosphate electrolytes have been used, however, for lead alloy electrodeposition, notably lead–tin (see later).

A gluconate bath containing 200 g l^{-1} basic lead carbonate, 100 g l^{-1} sodium gluconate and 100 g l^{-1} sodium hydroxide has also been reported.[68] Additions of 0·2 g l^{-1} glue and c. 0·5 ml l^{-1} of Turkey red oil, however, are made to the electrolyte to increase the throwing power, the same effect being achieved by increasing solution temperatures. The bath operates at 70°C with a current density of 5 A dm^{-2}, the current efficiency being reported to be 99%. Interestingly, the throwing power of this bath was reported to be greater than that of the fluoborate and silicofluoride electrolytes.[68]

Lead acetate has also been used in various complex baths with some success in electrodepositing lead.[47,69-71] A bath based on a mixture of lead acetate, ammonium acetate and sodium sulphate, i.e. one forming an *in situ* lead sulphate complex, has been reported to give homogeneous, fine-grained deposits of a technically useful thickness (10–12 μm).[69] The successful operation of the bath was found to be dependent upon the lead sulphate complex formed in solution. Whilst organic additives such as gelatin and *p*-cresol improved the quality of the electrodeposit, the complexed lead sulphate has a greater grain refining effect.[69] The grain refinement appears to be due to the blockage of active centres on the cathode surface rather than to increased cathodic polarization.

The concept of complexed lead salts formed in solution has been utilized in two other lead acetate-containing solutions. In one, tripolyphosphate complexes are formed by the interaction between sodium tripolyphosphate and lead acetate in the presence of a proprietary form of ethylene diamine tetracetic acid (Trilon B).[70] A practical electrodeposition bath for lead was claimed to have been developed, the lead deposition apparently occurring almost reversibly with little cathodic polarization, although additions of gelatin are required, presumably to yield fine-grained deposits. The effect of the Trilon B additive is not fully understood but studies of its action in a cadmium pyrophosphate electrolyte[72] indicate that it forms a more stable but more soluble "Trilonate" complex of cadmium. At low current densities in this system, there is little effect on the cathodic polarization curve but at greater current densities, there is increased cathodic polarization accompanied by a more uniform current distribution. A similar effect may occur with the lead tripolyphosphate complex.

Mixtures of lead acetate (80 g l^{-1}) and polyethylenepolyamine (110 g l^{-1}) also yield a complexed electrolyte which has been reported to give fine-grained, compact electrodeposits of up to 500-μm thickness.[71] The bath operates at 18–25°C and current densities of up to 2 A dm^{-2} have been obtained.

Whilst at this stage there are no clear indications that complexed lead electrolytes will displace the commercially important fluoborate bath, this type of *in situ* complex formation may result in electrolytes having greater throwing powers and possibly give self-brightening electrodeposits.

There have been reports in the literature on the electrodeposition of lead from perchlorate baths,[73–76] which are characterized by very low pH levels, but apart from their use in coating bearing surfaces, there appears to have been very little commercial or technical development of these electrolytes. An even more acidic solution, $pH < 1$, is used in the little known lead phenolsulphonate bath,[77] but like the even older perchlorate bath, this electrolyte has received little development over the years. The perchlorate electrolyte has, however, been used to study the mechanism of electrodeposition of lead[76] and it has been shown that deposition involves a two-dimensional nucleation mechanism and, during the growth of metal nuclei, charge transfer is the rate-determining step.

Electrodeposition of lead from alkaline media is also possible and there are various references to studies of this process.[78–82] Alkaline lead baths are usually based on sodium plumbite solutions prepared by saturating sodium hydroxide solution with lead dioxide. Studies of the deposition of lead from both acidic and alkaline media indicate that the charged state of the surface determines the initial cathodic polarization.[81] This supports the observations made from the perchlorate electrolytes.[76] It was also found, in alkaline solutions, that the cathodic polarization exhibited a linear dependence on the logarithm of the current density, but at high current densities, electrodeposition appears to be a diffusion controlled process,[81] although the latter observation would appear to be true for the electrodeposition of most metals. There are indications[83] that improved throwing powers are obtained with alkaline electrolytes.

It is clear from the above that lead may be electrodeposited from a wide variety of electrolytes and in virtually all cases, very high cathode current efficiencies are obtained. In commercial practice, the properties and service requirements dictate the precise electrolyte used since the properties of electrodeposited lead vary with the bath.[75] Whilst electrodeposited lead generally contains little hydrogen and the inclusion level is small, the latter usually arising from colloids present in the bath, organic additives are required to obtain satisfactory deposits. Numerous additives can be used with lead and their presence affects the grain structure of the deposit, treeing or dendrite formation, throwing power and brightness.[34,84] The grain size of deposits is an important characteristic since it strongly influences the internal stress.[47] In fact, the internal stress may change in magnitude and even sign depending upon the grain structure of the deposit and the deposition conditions.[34,47] The porosity of lead electrodeposits also may often be a serious consideration. For protective coatings, a minimum coating thickness as high as 75 μm has been quoted as necessary to ensure freedom from porosity,[2,79] although lower limits have been given by others[2] whilst there are reports of commercial processes for producing thin, pore-free deposits.[20]

In general, the electrodeposition of lead has not received detailed investigation although there have been a few reports of studies into the mechanism of nucleation and electrodeposition of lead,[1,34,85,86] such studies usually being performed with rotating disc electrodes. The underpotential electrodeposition of lead monolayers on to gold from a Pb^{2+}-containing perchloric acid solution, however, has been studied by linear sweep voltammetry and reflectance spectroscopy.[86a] The lead was found to deposit at potentials well below the reversible potential for the bulk metal, and although the metal was deposited as ions, the effective ionic charge was thought to be substantially less than +2, this effect being ascribed to strong interactions with the band structure of the gold substrate. At a more negative potential, but one still positive to the reversible potential of bulk lead, there appeared to be a two-dimensional phase transition to form a metallic lead monolayer. The phenomenon of underpotential deposition has received little study in the past but obviously is of great theoretical, and possibly technical, importance in the general field of electrodeposition.

Finally, there has been some investigation[87] of the mechanism of contact deposition of lead from alkaline solution by means of sodium amalgam. Deposition is largely governed by diffusion, the rate being controlled by access of plumbate ions to the amalgam surface and decreases with raised solution pH levels.[87]

Electrodeposition of Lead Alloys

A wide variety of binary and ternary lead alloys may be electrodeposited from aqueous solution, and these alloys are summarized in Table I. There are four principal areas of industrial usage of lead alloy electrodeposited coatings, namely for corrosion resistance, on bearing surfaces, in the production of printed circuit boards and other electronic components, and for solderability, although certain electrodeposited alloys are also used for decorative purposes.

Table I. Electrodeposited binary and ternary alloys of lead.

Binary	Ternary
Antimony	Antimony–tin
Bismuth	Arsenic–tin
Copper	Copper–tin
Indium[a]	Manganese–tin
Iron	Tin–zinc
Nickel	
Silver	
Thallium	
Tin	
Zinc	

[a] Deposited as Pb–In co-deposit. Alloying achieved by heat treatment.

The most important electrodeposited alloy of lead is the lead–tin binary alloy which has two main compositional ranges (*see below*), but other alloys have a growing industrial importance. The principal advantages of the alloys as compared with the pure electrodeposited lead are their increased hardness and greater corrosion resistance combined, in some cases, with superior throwing power for the electrodeposition system.

As with pure lead electrodeposits, there have been many reviews of the mechanical properties, corrosion resistance and the industrial applications of lead alloy electrodeposited coatings[23,24,31,33,34,41,75,87–92] and there is also a growing interest in the hardening of lead alloy deposits by means of particle co-deposition.[93,94] Lead alloys may be electrodeposited by the conventional rack and barrel techniques[41,95–98] as well as by selective plating methods.[41,58,99] There have also been a number of reports on the development of bright lead alloy electrodeposition processes.[97,100–110] The basic electrochemical principles of the electrodeposition of alloys have been discussed in various monographs.[35,37,38,42] The differences in the deposition potentials of the various metals given in Table I and that of lead are sufficiently wide (in some cases such as copper there is a difference in potential of 470 mV) that binary or ternary alloy deposition from simple aqueous solutions is thermodynamically impossible. Consequently, complexing solutions are mandatory for alloy deposition, and electrolytes similar to those used for pure lead are employed.

Lead–tin alloys

The most important electrodeposited binary lead alloys are those with tin, the lower tin content alloys being known loosely as terne plate. There are, in fact, a wide range of binary alloy compositions capable of electrodeposition, but they fall into two broad categories, high lead alloys having 60+% lead contents and high tin alloys (up to 60% tin contents). The high lead alloys are used for coating the running surfaces of bearings and for corrosion resistance whilst the high tin alloys are used for enhanced solderability and as etch resists for printed circuit boards.[91,107–109]

The most widely used electrolyte for the electrodeposition of binary lead–tin alloys is the fluoborate bath. The bath consists essentially of the basic lead fluoborate electrolyte with additions of tin fluoborate, the ratios of the two cations determining in large part the composition of the electrodeposit. Various bath compositions have been proposed for lead–tin alloys and these are summarized in Table II. These baths all contain the fluoborates of lead and tin together with free fluoboric acid,

Table II. Electrolytes for deposition of lead–tin alloys.

Component (g l^{-1})	Bearing overlays[a]			Solderability deposits[b]		
Lead	200	88	88	104	25	25	104
Stannous tin	20	6	6	35	35	55	47
Free fluoboric acid	50	100	40 or 100	30	100	40 or 100	30
Boric acid	25	25	—	12	25	—	12
Peptone	5	0·5	—	—	5	—	—
Glue	5	—	—	0·375	—	—	0·375
Temperature (°C)	18–24	16–38	25–30	—	16–38	25–30	—
Cathode current density (A dm^{-2})	10	3	3	0·7	3	3	0·7
Reference number	24	56	39	113	56	39	113

[a] Bearing overlays: lead–8–12% tin electrodeposits.
[b] Deposits for enhanced solderability: 40% lead–60% tin electrodeposits.

the latter being required to prevent hydrolysis of the stannous ion to an insoluble basic tin salt. The electrolyte solution also contains boric acid to suppress the hydrolysis of fluoboric acid to boric acid and hydrofluoric acid, as the presence of the latter would result in precipitation of the poorly soluble lead fluoride.[24]

Organic addition agents, commonly glue, gelatin or peptone, although resorcinol and hydroquinone are also used, are always present in solution. If such additives are omitted from the solution, then deposits tend to be rough and exhibit "treeing", this effect being observed with both pure lead and lead–tin alloys. These organic additives are necessary, therefore, to ensure the production of smooth, fine-grained deposits, but, more importantly, their presence is essential for co-deposition of tin to actually occur.[24,56,110] The concentration of additives such as peptone in solution has to be markedly increased for the electrodeposition of the tin-rich binary alloys. Thus a level of $0.5 \, g \, l^{-1}$ of peptone is recommended for bearing overlay deposits, i.e. lead–5–10% tin, but $5 \, g \, l^{-1}$ is necessary for solder coatings containing 40% lead–60% tin.[56] The fluoborate bath for lead–tin alloy deposition can be used for both batch plating operations in vats and barrels as well as for the continuous coating of wire,[110a] the electrolyte differing little from those given in Table II.

At present, the precise action of organic additives in the co-deposition of tin with lead is uncertain and the generally advanced explanation that they maintain tin in its stannous oxidation state is hard to accept for substances such as glue and gelatin. It is known, however, that the tin content of deposits is increased by an increase in current density and this phenomenon might account for the effect of organic additives. It is possible that the organic additives form organometallic compounds in solution and thus alter the discharge potentials of the depositing cations. Furthermore, when these organometallic ions or the parent organic additives are deposited at the cathode, there may be a localized poisoning effect which inhibits growth of the lead crystallites such that the local current density is increased and deposition of tin is favoured. Thus finer-grained and more homogeneous deposits are obtained.

The grain-refining and brightening effect of organic additives has in fact resulted in considerable efforts to develop electrolytes for the deposition of bright lead–tin alloy electrodeposits. Furthermore, much of the development effort has been devoted to finding alternatives to glue, gelatin and peptone which tend to co-deposit with the binary alloy and lower its corrosion resistance.

The principal efforts towards the development of bright electro-deposition of binary lead–tin alloys have been for soldering and electronics applications, that is, 40:60 lead–tin deposits. Unfortunately, much of the work on bright solder electrodeposition is incorporated in patent specifications with few details being available in the scientific literature. Thus, whilst there are several reports of bright solder deposits having good corrosion resistance, the actual details are not always available although most of these processes are based on the fluoborate electrolyte and contain somewhat complex organic additives.[31,97,99–109,111–113] One process, the Labolac process, is claimed to give bright, white and easily solderable coatings containing 40–47% lead.[97,107–109] This bath is based on the fluoborates of lead and tin, but the total concentration of the two cations is of the order of $30 \, g \, l^{-1}$. The deposit composition may be altered by selection of the current density for a given solution concentration and/or by variation of the cation concentrations in solutions.[107,108] Whilst the actual organic additives used in this process were not specified, apart from the statement that they were not based on peptone or glue, it was stated that the additives had a four-fold action. Thus, it was reported that they had a reducing action, that they formed organometallic derivatives with lead and tin such that there was a reservoir of stannous tin in solution, that their use resulted in microcrystalline, homogeneous and non-porous deposits and, finally, that they resulted in bright, white electrodeposits having good solderability and corrosion resistance.[97,107–109] Surprisingly, despite the comparatively dilute solutions, high current efficiencies (90–95%) and high cathodic current densities ($0\cdot5$–$10 \, A \, dm^{-2}$) with a deposition rate of $0\cdot5 \, \mu m \, min^{-1} \, A^{-1} \, dm^{-2}$ were claimed for the process.

A wide range of brightening agents has been proposed in the patent literature for lead–tin solder electrodeposits, many of these additives being used in conjunction with surfactants. These agents include the reaction product of acetaldehyde and o-toluidine,[99] formaldehyde and various amines,[101] benzal acetone and triacryloylhexahydro-s-triazine,[105] 2-naphthalene monocarboxaldehyde,[106] chloro-substituted aromatic aldehydes and aliphatic amines,[104] polyethylene oxide–phenol condensation products,[112] with cationic and/or non-ionic surfactants, usually functioning as emulsifying agents, and anti-oxidants such as hydroquinone also being present in solution. A major advantage of surfactants, at least in one case,[113] appears to be the ability to yield essentially gas-free solder deposits such that there is little outgassing during melting of the solder electrodeposit without flux. It also appears that surfactants also tend to facilitate the production of easily solderable coatings whilst the conjoint use of anti-oxidants increases the tin-content of the deposit, particularly for the low-metal, low-current tin–lead deposits.[113] The results of many workers indicate that the corrosion resistance of bright lead–tin solder compares favourably with that of the matt or dull finish deposits.[102,103]

It should be mentioned, however, that despite the great efforts to develop efficient brightening agents for lead–tin alloys and many other electrodeposition processes, the actual brightening action of these organic (and often inorganic) additives is still not fully understood.

There have been attempts, however, to investigate the mechanism by which peptone influences the nature of electrodeposited lead–tin alloys. Chronopotentiometric studies[117] have indicated that the fractional surface coverage of the lead–tin electrode is a function of the peptone concentration. Furthermore, adsorption of peptone on to the electrode appears to be the main factor in the influence of the amino acid on the electrodeposition of lead–tin alloys. These findings indicate that organic additives exert their brightening effect through cathode absorption rather than by formation of organometallic compounds. However, more detailed studies are undoubtedly required before brightening phenomena are fully understood.

Various other electrolytes have been used for the electrodeposition of lead–tin binary alloys. Alternative electrolytes have been sought because fluoborates are toxic and present pollution and effluent problems. One successful approach has been the use of sulphamate baths and both lead-rich and tin-rich electrodeposits may be produced (Table III).[118] The tin content of the deposited alloy is proportional to

Table III. Electrodeposition of lead–tin alloys from sulphamate baths.[118]

Solution composition	Electrodeposited alloy	
	Lead–8–12% tin	Lead–20–60% tin
Lead sulphamate (g l^{-1})	97–138	42–110
Tin sulphamate (g l^{-1})	24	60–120
Sulphamic acid (g l)	51–80	70–120
CTAB[a]	2–10	1·5–10
Temperature (°C)	20–40	18–25
Cathode current density (A dm^{-2})	1–2	1
Current efficiency (%)	100	—

[a] CTAB: Cetyltrimethylammonium bromide.

the tin content of the electrolyte although both the concentration of cetyltrimethylammonium bromide (CTAB) in solution and the current density affect the tin content of the deposit to some degree. The CTAB addition agent exercises both a brightening effect and reduces the porosity of the electrodeposited alloy. Various other brightening additives have been developed for sulphamate baths, typically unsaturated carbonyl compounds such as benzalacetone used in conjunction with acryloyl-containing compound (e.g. 1,3,5-triacryloyl-hexahydro-s-triazine) as a secondary brightener[107] although such baths appear to require the presence of surfactants based on a polyoxyalkylene. This brightening system was primarily developed for solder coatings and tin–3% lead coatings for electrical components but may be used with sulphamate, sulphonate, perchlorate and pyrophosphate baths.[107]

Tin-rich lead–tin electrodeposits (lead–40–99% tin) may also be obtained from solutions based on the phenolsulphonate and/or cresolsulphonate salts of divalent tin and lead.[100] These solutions, which contain 10–60 g l^{-1} of the tin and lead salts of at least one of the organic acids, together with 50–200 g l^{-1} of the free acid in solution, are less toxic than the fluoborate-based electrolytes.[100] Furthermore, the addition of acetaldehyde and the isopropylalcohol extract of the reaction product of acetaldehyde and toluidine enables bright electrodeposits to be produced.[100]

Lead–tin alloys may also be successfully electrodeposited from pyrophosphate electrolytes.[66,119–122] Typically, electrolytes based on stannous pyrophosphate, lead nitrate and sodium or potassium pyrophosphate have been used to electrodeposit a range of lead–tin alloys (Table IV).[120] The sodium or potassium pyrophosphate concentration in solution is always in excess of the stoichiometric requirements of complex formation and, in fact, molar ratios of 3 : 1 or greater are normally used for the pyrophosphate ion to metal in solution.

The tin content of the deposit can be varied from zero to almost 90% by adjusting the ratio of tin to lead in solution, but there is no direct relationship between the tin content in solution and that in the deposit. This is in contradistinction to the behaviour found with the fluoborate electrolyte in which the ratio of the two metals in solution dictates the composition of the deposit.

The pyrophosphate electrolyte yields smooth, adherent, fine-grained deposits, the colour of which varies from grey (high lead-content alloys) to white (high tin-content alloys). The electrolyte requires the presence

Table IV. Pyrophosphate electrolyte for lead–tin alloys.[120,122]

Constituent	Concentration (g l⁻¹)
Stannous pyrophosphate	4·1–61·7
Tin	2·4–35·6
Lead nitrate	3·3–33·1
Lead	2·1–20·7
Pyrophosphate, $(P_2O_7)^{4-}$	39·2–261·8
pH	9·0–9·5
Temperature (°C)	40–80
Cathode current density (A dm⁻²)	0·5–5·0
Tin content of deposit (%)	0–90
Time to deposit 25 μm at 3 A dm⁻² (min)	15

of additives such as gelatin, β-naphthol or sodium β-naphthalene sulphonate to ensure a fine-grained electrodeposit. The grain-refining effect of gelatin is very marked and is comparable with that observed in the deposition of pure lead from the pyrophosphate electrolyte. Furthermore, the alloy deposited from a pyrophosphate electrolyte has a finer-grain structure than that observed with the fluoborate and other electrolytes.

Interestingly, for a given electrolyte composition, the tin content of the electrodeposit is increased by the presence of the three addition agents cited above but other additives have little or no such effect. An increase in the cathode current density also results in an increase in the tin content of the deposit. However, an increase in the pyrophosphate concentration in solution, at higher current densities, yields increased tin contents in the electrodeposit, but decreased tin contents are obtained at higher current densities if the total metal concentration in solution is increased.

Finally, before discussing the electrodeposition of other binary lead alloys, it should be mentioned that there has been some interest in the development of high speed and high throwing power electrodeposition solutions. This work has primarily been concerned with the fluoborate electrolyte but it is feasible that in the future, other electrolytes might receive similar development efforts. One high-speed terne alloy (up to lead–15% tin) deposition electrolyte has been reported to give both anode and cathode efficiencies of approximately 100% at current densities up to 86 A dm⁻².[123] The electrolyte contained 125 or 250 g l⁻¹ of the metal fluoborate, with 5 or 10 g l⁻¹ hydroquinone as the addition agent, and has applicability for strip plating lines. The deposits from this electrolyte have been extensively investigated and the results indicate that they have excellent properties, particularly corrosion resistance.[111] Various workers have described high throwing power fluoborate electrolytes, principally for the production of solder coatings on printed circuit boards.[124,125] The high throwing power solutions are characterized by a low total metal content but high fluoboric acid content together with the presence of a small amount of peptone as an additive.[125,126] The improved throwing power, which yields greater uniformity in the deposit thickness, is associated with a more uniform current distribution,[126] presumably resulting from a combination of concentration and resistance polarization due to the low metal ion concentrations in solution. Rotating disc electrode studies of the microthrowing power within a dilute fluoborate bath indicate that the microthrowing power is dependent upon the agitation rate and the organic additive concentration in solution.[127] In fact, the results may be interpreted in terms of a diffusion model. This again indicates the importance of the organic additive on the characteristics of electrodeposited lead and lead alloys.

Other binary lead alloys

The electrodeposition of various other binary lead alloys has received attention in order to develop coating for particular purposes, notably wear- and corrosion-resistant bearing coatings. The following systems have received the greatest attention: lead–copper,[128-132] lead–antimony,[44,129,133,134] lead–thallium,[135-136] and lead–silver.[128,132]

Lead–copper alloys are usually deposited from a fluoborate electrolyte[128,131] although there have been reports of this binary alloy being electrodeposited from a nitrate bath containing cetyltrimethylammonium bromide as an additive.[130] Deposition of lead–copper alloys from a complex lead acetate–sulphate bath has been investigated in some detail.[129] Using a base electrolyte containing 160 g l^{-1} lead acetate, 250 g l^{-1} ammonium acetate together with 6 g l^{-1} gelatin and a small quantity of p-cresol (80 drops l^{-1}), a range of lead–copper alloys could be deposited by variation of the copper sulphate addition to the electrolyte. Thus electrodeposits containing between 3·2 and 11% copper could be obtained by adding 2–12 g l^{-1} copper sulphate to the base electrolyte. Operating at $1-1 \cdot 2$ A dm^{-2}, pH $6-7 \cdot 5$ and 25–30°C, the cathode current efficiency was found to be 95–100%. An acetate-based electrolyte has also been used[131] to deposit copper and copper-coated lead powder for sintering on to bearing sleeves. This process operated at low current efficiency, *c.* 20%, at current densities of up to 13 A dm^{-2} but yielded fairly coarse grain powders suitable for sintering.[131] This particular example of lead–copper alloy deposition is unique and does not fall into the normal category of electrodeposition processes.

Electrodeposition of lead–copper alloys is a typical example of a less noble/more noble alloy system in that for a given electrolyte composition, the copper content of the deposit is increased with increase in the electrolyte temperature but is decreased with increase in the cathode current density.[129] Whilst bright, dense binary alloys containing up to 12% copper are capable of being electrodeposited, these deposits tend to be brittle and the process is difficult to control although the deposit quality can be improved by periodic current reversal techniques.[132]

The electrodeposition of lead–antimony binary alloys has been investigated by several workers, particularly for the coating of bearing surfaces due to their greater corrosion and wear resistance compared with pure lead.

Lead–antimony alloys may be deposited from a fluoborate bath[44,133] and from a complex acetate–sulphate bath.[129] A successful approach to the electrodeposition of this binary alloy is to use a concentrate of lead fluoborate (503 g l^{-1}) with free fluoboric acid (161 g l^{-1}) and free boric acid (45 g^{-1}), this concentrate being diluted to 460 ml l^{-1} and $0 \cdot 23$ g l^{-1} each of peptone and resorcinol being added to this base electrolyte. Additions of an antimony concentrate, containing 191 g l^{-1} antimony, 36 g l^{-1} free fluoboric acid and 78 g l^{-1} free boric acid are then made to the base electrolyte. At a current density of $2 \cdot 5$ A dm^{-2}, an addition of 2 g l^{-1} of antimony from the concentrate yields a lead–$0 \cdot 8$% antimony electrodeposit, whilst 10 g l^{-1} antimony results in deposits of lead–$13 \cdot 2$% antimony.[44,133] Binary alloys within this range are obtained by variation of the antimony content in solution, the relationship between deposit composition and the antimony level in solution being approximately linear. It was also found, however, that β-naphthol gave deposits that were more resistant to outgassing on heating than electrodeposits from baths containing the peptone/resorcinol additives.[133]

The lead acetate–lead sulphate base electrolyte used for electrodeposition of lead–copper alloys (*see above*) may also be used for lead–antimony alloys.[129] In this case, however, additions of potassium antimonyl tartrate are made to the base electrolyte. An addition of $10 \, g \, l^{-1}$ of potassium antimonyl tartrate yielded a lead–7·2% antimony alloy whilst additions of $20 \, g \, l^{-1}$ and $24 \, g \, l^{-1}$ produced deposits of lead–8·2% antimony and lead–13·2% antimony, respectively.

Lead–antimony alloys have also been electrodeposited from a tartrate-based electrolyte containing tartaric, fluoboric and boric acids.[134,135] Interestingly, the antimony content of the electrodeposit is claimed to be controlled by the free fluoboric acid level in the electrolyte.[135]

The electrodeposition of lead–antimony binary alloys is also a less noble/more noble system and the electrodeposition characteristics parallel those found in the lead–copper system (see above) and similar binary alloy systems. Typically, the antimony content of the deposit may be increased by raising the solution temperature, increased stirring rates, current reversal and a higher antimony content in the electrolyte.[44,129,133,134] In the tartrate electrolyte, both an increase in the antimony concentration and the total metal concentration in solution will increase the antimony content of the electrodeposit.[134] In contradistinction, the antimony content of the electrodeposit is reduced by an increase in the lead concentration in solution and by an increase in the current density.[44,129,133,134] The latter behaviour is characteristic of the electrodeposition of less noble/more noble alloy systems.[35,37,38]

There has also been some investigation of the electrodeposition of lead–thallium and lead–silver binary alloys.[132] The lead–thallium alloy electrodeposits have the great advantage of exhibiting good corrosion resistance at elevated temperatures and pressures, whilst lead–silver electrodeposits possess good wear resistance. Interestingly, lead–thallium alloy electrodeposits are cheaper and possess higher melting points than lead–indium-bearing overlays which are not produced by co-deposition, but instead by alternate deposition of the two metals, the coating then being heat-treated to permit diffusion to occur.[24]

Lead–thallium alloys are deposited from perchlorate solutions and from fluoborate electrolytes.[132] Of the former, a solution containing $39 \, g \, l^{-1}$ lead, $32 \, g \, l^{-1}$ thallium, $30 \, g \, l^{-1}$ free hydrochloric acid and $10 \, g \, l^{-1}$ each of peptone and glue, at a current density of $0·5 \, A \, dm^{-2}$, will deposit $2\bar{0} \, \mu m \, h^{-1}$ of a lead–10% thallium alloy. Using a basic lead fluoborate electrolyte containing $40 \, g \, l^{-1}$ lead and $10 \, g \, l^{-1}$ fluoboric acid together with $10 \, g \, l^{-1}$ each of glue and peptone, lead–thallium alloys containing up to 30% thallium can be electrodeposited. At a current density of $1–1\cdot2$ A dm^{-2}, the thallium content of the solution determines the thallium content of the deposit and for levels of 13–74 g l^{-1} in solution with the above base electrolyte, alloys containing 5–30% thallium may be obtained.

Lead–silver binary alloys have been deposited from an alkaline mixed cyanide–acetate bath.[132] An electrolyte based on 30 g l^{-1} silver cyanide, 45 g l^{-1} potassium cyanide, 40 g l^{-1} Rochelle salt, 3 g l^{-1} potassium hydroxide and 5 g l^{-1} basic lead acetate (although the free cyanide and hydroxyl ion contents can be varied) will deposit alloys in the range lead–$1\cdot5$–3% silver. As stated earlier, this type of binary alloy electrodeposit has the merit of greater wear resistance than pure lead.

Electrodeposited binary lead alloys, other than lead–tin, are principally used as bearing overlays. Consequently, aesthetic considerations such as a bright finish are secondary to the mechanical properties of such deposits. Furthermore, whilst solderability is a very important characteristic for the high tin lead–tin alloys, it is less critical and often of little importance for the electrodeposited alloys for bearing overlays or other corrosion/wear resistance applications.

Ternary lead alloys

A variety of ternary lead alloys are electrodeposited, principally for bearing overlay applications and for corrosion resistance applications.[24,87,89,129,132,137,138] The electrodeposited lead ternary alloys that have been investigated or are in actual use as bearing overlays are lead–tin–copper, lead–antimony–copper and lead–antimony–tin. These alloys are preferred to pure lead, lead–tin and other binary alloys as bearing coatings, due to their greater hardness, increased corrosion resistance, superior anti-friction properties and greater strength. Furthermore, the ternary electrodeposition baths have been claimed to be easier to operate than the traditional lead and lead–tin systems.[87]

Lead–tin–antimony alloys are deposited from fluoborate electrolytes.[87,137,138] One bath, containing lead fluoborate (100 g l^{-1} lead), stannous fluoborate (30 g l^{-1} stannous tin), antimony fluoborate (6 g l^{-1} antimonous antimony), together with 80 g l^{-1} free fluoboric acid and $15\cdot5$ g l^{-1} of hydroquinone and peptone, yields deposits of lead–11% tin–7% antimony at 21 ± 1°C and a current density of

$3 \cdot 9$ A dm^{-2}.[87] A $5°$ rise in the electrolyte temperature increases the tin content of the electrodeposit by $1 \cdot 6\%$ and the antimony content by $0 \cdot 8\%$. This effect parallels the behaviour of the individual binary alloys. Interestingly, a 20% variation of the lead content of the electrolyte from 100 g l^{-1} results in a 2% change in the tin content of the deposit but has no effect on the antimony content. Similarly, a 20% variation in the tin content of the electrolyte effects a 2% change in the lead content of the electrolyte but has no influence on the antimony content. Furthermore, a 30% variation in the free fluoboric acid level in the electrolyte has no effect on the alloy composition. These findings are very interesting but at present there is insufficient knowledge available to adequately account for, or to predict, the effect of variations in solution composition or operating parameters on electrodeposited ternary alloys.

A somewhat similar electrolyte to the above has been reported to yield high tin-containing ternary alloys, namely lead–39% tin–1% antimony, the electrodeposit possessing good corrosion resistance and anti-friction properties. In this case, however, the electrolyte contains fluoboric acid and phosphoric acid together with gluten as an additive.[138] The precise role of the orthophosphoric acid in the electrolyte is uncertain but presumably it functions as some type of modifier. In general, it appears that these ternary lead–tin–antimony systems are operated with lead–tin anodes (90% lead–10% tin) with the antimony level being maintained by solution addition since the dissolution characteristics of the ternary anodes tend to be poor. In fact, it would appear that the antimony in the anode tends to dissolve via its oxide, i.e. it exhibits a limiting dissolution rate and, therefore, there is inadequate maintenance of the antimony level in the electrolyte.

Lead–tin–copper alloys have been used to some extent for bearing overlays[24] although there are difficulties associated with the deposition of the ternary alloy having a reproducible composition.[89,132] A fluoborate electrolyte is used for this system and the electrodeposition of alloys containing 6–10% tin and up to 2% copper has been reported.[89,132] These ternary alloys, particularly lead–6–9% tin–0·75–1% copper,[132] are claimed to exhibit low frictional characteristics and a superior corrosion resistance than the lead–tin–antimony alloy.

A lead–antimony–copper alloy may be electrodeposited from the lead acetate bath used for the binary lead–copper and lead–antimony alloys.[129] The electrodeposited alloy has been stated to be smooth, bright, fine-grained and homogeneous. In this approach, 6 g l^{-1} copper sulphate and 12 g l^{-1} potassium antimonyl tartrate are added to the lead acetate base electrolyte (*see above*) and, at pH 6, lead–7·2% antimony–5·1% copper alloys are obtained. If the potassium antimonyl tartrate level is increased to 24 g l^{-1}, the antimony content of the electrodeposited alloy is increased to 10% but the copper content is decreased to 4·8%. However, if the current density is increased with this composition of electrolyte, there is little change in the electrodeposited alloy. Thus at $1-1\cdot2 \text{ A dm}^{-2}$, a deposit of lead–9·9% antimony–4·6% copper is obtained whilst at $1\cdot5 \text{ A dm}^{-2}$, lead–8·9% antimony–4·4% copper is produced. When the potassium antimonyl tartrate level is held at 24 g l^{-1} but the copper sulphate content is reduced to 4 g l^{-1}, the electrodeposited alloy is markedly different, namely lead–11·2% antimony–3% copper. An increase in electrolyte temperature increases both the antimony and the copper contents of the electrodeposited alloy.[129] It is clear from these results that the interrelationships between electrolyte temperature, current density and other electrodeposition parameters and the composition of electrodeposited ternary alloys are poorly understood. There is, therefore, a real need for further investigation of the electrodeposition of ternary alloys, particularly those of lead.

Finally, there has been work on the deposition of lead-containing quaternary alloys reported in the literature,[137,139] as well as tin-rich lead-containing ternary alloys.[137,140,141] There appear to be few details readily available on the deposition of quaternary lead alloys but an electrodeposited alloy having the composition lead–6–9% tin–1% copper–0·5% antimony has been reported.[139] There have also been reports of the electrodeposition tin-rich alloys containing 10–14% lead and 25–35% manganese.[137,140,141] One electrolyte for the deposition of this type of ternary alloy contains lead oxide, manganese and tin sulphates, ammonium sulphate, the citrate of trisubstituted ammonia, a proprietary form of ethylene diamine tetracetic acid and glue, current densities of $2–20 \text{ A dm}^{-2}$ being possible.[140] These tin-rich ternary tin–lead–manganese alloys exhibit good corrosion resistance and have a microhardness some 20 times greater than that of pure lead.[141] Interestingly, despite the fact that these electrodeposits are essentially tin-based alloys, they have lead-type crystal lattices irrespective of their compositions.[141] Whilst not strictly relevant to the electrodeposition of lead, these alloys are included in this discussion for completeness.

General Comments

Lead is an unusual metal in that it exists in two oxidation states and forms a large number of salts with inorganic and organic acids in both its plumbous and plumbic states. Furthermore, it also forms a large number of organometallic compounds. Thus lead is characterized by its high reactivity and its very wide range of inorganic and organic compounds.

These characteristics of lead are reflected by its electrochemistry, notably the very wide variety of electrolytes from which it may be electrodeposited. However, satisfactory deposits of lead and its alloys are obtained only from electrolytes containing lead salts with large anions such as the fluoborate, silicofluoride, sulphamate, pyrophosphate, perchlorate, gluconate and phenolsulphonate ions. Furthermore, the current efficiency of lead electrodeposition is extraordinarily high, usually in the range 90–100% even from dilute solutions. This high cathode current efficiency in the electrodeposition of lead and lead alloys is a consequence of the high hydrogen overpotential for lead. However, the role of the complex anions given above, in contrast to common complexing anions such as cyanide (although the pyrophosphate anion is regarded by some workers as a complexing agent), in the electrodeposition of lead is not readily understood. The role of organic additives such as gelatin, peptone and glue in effecting grain refinement, however, may be accounted for on the basis of an adsorption mechanism.

A possible explanation of the improved lead electrodeposits from electrolytes containing complex anions such as the silicofluoride and fluoborate ions, as compared with other soluble salts such as lead nitrate and acetate, could be a steric effect. The electrodeposition of lead (and all other metals) is under control by concentration polarization, and plumbous ions can only enter the diffuse double layer, prior to deposition, when the anion ligands are oriented away from the cathode and are lost by the cation. It is at this stage that the complex anions influence the deposition of lead, resulting in technically (and commercially) acceptable electrodeposits. There have been, however, few studies of the mechanism of lead and lead alloy electrodeposition and as yet the process is far from fully understood. There is, therefore, a very real need for detailed mechanistic studies into lead electrodeposition and research efforts should be devoted to the role of complex anions in this process. This would permit a greater understanding of an interesting and technically important electrochemical process.

References

1. "Gmelin's Handbuch der Anorganischen Chemie", System No. 47—Blei, 8th ed. Verlag GmbH, Weinheim (1974).
2. W. Hofman, "Lead and Lead Alloys", 2nd edn (translation). Springer-Verlag, Berlin (1970).
3. J. A. Smythe, "Lead". Longmans Green, London (1923).
4. J. T. Cairns and P. T. Gilbert, "The Technology of Heavy Non-ferrous Metals and Alloys". Newnes, London (1967).
5. D. Greninger, V. Kollonitsch and C. H. Kline, "Lead Chemicals". I.L.Z.R.O, New York (1975).
6. J. W. Mellor, "A Comprehensive Treatise on Inorganic and Theoretical Chemistry", Vol. 7. Longmans Green, London (1929).
7. H. Shapiro and F. W. Frey, "The Organic Compounds of Lead". Interscience, New York (1968).
8. L. C. Willemsens, "Organolead Chemistry". I.L.Z.R.O., New York (1964).
9. J. A. von Fraunhofer, The chemistry and corrosion of lead and its alloys. M.Sc. dissertation, University of London (1967).
10. Lead Development Association, Edited Proceedings of International Conferences on Lead: London (1962), Arnhem (1965), Venice (1968), Hamburg (1971), Paris (1974). Pergamon Press, Oxford.
11. G. Smith, "Storage Batteries", 2nd edn. Pitman, London (1971).
12. W. Garten, "Bleiakkumulatoren", 9th edn. Varta A.G., Frankfurt (1968).
13. T. C. Patton (ed.), "Pigment Handbook", 3 vols. J. Wiley, New York (1973).
14. J. A. von Fraunhofer and J. Boxall, "Protective Paint Coatings for Metals". Portcullis Press, Redhill (1976).
15. H. C. Wesson in "Materials of Construction for Chemical Plant" (I. L. Hepner, ed.), L. Hill, London (1962).
16. J. A. von Fraunhofer, *Anti-Corrosion* **15**(8), 4 (1968).
17. S. A. Hiscock, "Lead and Lead Alloys for Cable Sheathing". E. Benn, London (1961).
18. P. Bergsoe and A. Bergsoe, "Printing Metals", 2nd edn. Glostrup, Denmark (1958).
19. K. H. Manko, "Solders and Soldering". McGraw-Hill, New York (1964).
20. W. Machu in Proceedings of the Symposium for the Protection of Steel Structures by Metallic Coatings, Vienna (1964).

21. T. Biestek, "Selection of Electroplated Coatings in Accordance with their Intended Application". Centrala Handlowa Metali Niezelaznyeh, Poland (1970).
22. V. Q. Kinh, *Corrosion* **20**, 342 (1972).
23. L. F. Spencer, *Metal Finish.* **69**, 69 (1971).
24. J. A. Morris, *Trans. Inst. Metal Finish.* **51**, 56 (1973).
25. A. R. Cook, Lead plated copper. I.L.Z.R.O. Research Report (1965).
26. C. A. MacKay and B. T. K. Barry, *Proc. 5th Int. Lead Conf.* Paris (1974).
27. G. R. Strickland, *Electroplat. Metal Finish.* **24**(7), 5; (8), 5 (1971).
28. Shipley Co. Inc., U.S. Patent 3,582,415 (1969).
29. G. C. Wilson, *Electroplat. Metal Finish.* **25**(8), 32 (1972).
30. L. Missel, *Electroplat. Metal Finish.* **25**(4), 20 (1972).
31. J. Korpius, *Metalloberfläche* **27**, 357 (1973).
32. R. L. Seth, *Electroplat. Metal Finish.* **25**(2), 5 (1972).
33. W. H. Safranek, *Draht* **24**, 420 (1973).
34. R. Walker, *Int. Metall. Rev.* **19**, 1 (1974).
35. F. Lowenheim (ed.), "Modern Electroplating", 3rd edn. Interscience, New York (1974).
36. A. H. DuRose "Blei" *in* "Handbuch der Galvanotechnik", Band 2. Carl Hanser Verlag, Munich (1966).
37. J. A. von Fraunhofer, "Basic Metal Finishing". Elek Science, London (1976).
38. D. R. Gabe, "Principles of Metal Surface Treatment and Protection". Pergamon Press, Oxford (1972).
39. W. Canning & Co., "Canning Handbook on Electroplating", 21st edn. W. Canning, Birmingham (1970).
40. D. G. Foulke and F. D. Crane, "Electroplaters' Process Control Handbook". Reinhold, New York (1963).
41. H. W. Dettner and J. Elze, "Handbook of Electroplating Technology". Carl Hanser Verlag, Munich (1967).
42. A. Brenner, "Electrodeposition of Alloys", Vols I and II. Academic Press, New York and London (1963).
43. H. K. Vengberg, *Norsk Galvano-Teknisk* **6**, 44 (1963).
44. J. W. Dini and J. W. Helms, *Metal Finish.* **67**(8), 53 (1969).
45. N. A. Pangarov, *J. Electroanal. Chem.* **9**, 70 (1965).
46. L. P. Bicelli and G. Poli, *Electrochim. Acta* **11**, 289 (1966).
47. M. Ya. Popereka, *Zh. Fiz. Kkim.* **39**, 1321 (1965).
48. R. Walker, *Metallurgia* **78**, 131 (1968).
49. G. L. Schnable and J. G. Javes, *Electrochem. Technol.* **2**(7/8), 201 (1964).
50. E. A. Ukshe and V. N. Devyatkin, *Zh. Prikl. Khim.* **38**, 1153 (1965).
51. R. Y. Bek and A. S. Lifshits, *Izv. Sibirsk. Old. Akad. Nauk SSSR, Ser. Kkim. Nauk* (14), 70 (1967).
52. Texas Instruments, U.S. Patent 3,409,466 (1965).
53. A. G. Betts, U.S. Patents 713,277 and 713,278 (1902); 891,395 and 891,396 (1908); 918, 647 (1909).

54. C. L. Mantell, "Electrochemical Engineering", 4th edn. McGraw-Hill, New York (1960).
55. G. Leuchs, D.R. Patent 38,193 (1886).
56. H. Benninghoff, *Galvanotechnik* **57**, 168 (1966).
57. N. P. Fedotieff, B. P. Artamonoff and N. J. Rasmerova, *Metal Abstr.* **4**, 448 (1937).
58. Metachemical Processes Ltd., "The Dalic Process". Company publication (1963).
59. M. Rubinstein, *Plating* **59**, 540 (1972).
60. R. Draper, "Tables and Operating Data for Electroplaters". Portcullis Press, Redhill (1975).
61. H. Safranek and C. H. Layer, *Trans. Inst. Metal Finish.* **53**, 121 (1975).
62. A. K. Graham and H. L. Pinkerton, *Trans. Inst. Metal Finish.* **40**, 249 (1963); and 50th Ann. Tech. Prog. Am. Electroplaters' Soc., Atlantic City, pp. 139–146 (1963).
63. F. Haskó, S. Farkas, P. Bujtas and R. Fath, Proceedings 8th Congress of the International Union for Electrodeposition and Surface Finishing, Basel (1972); Forster Verlag A.G., Zurich (1973).
63a. Fuji Photo Film Co. Ltd., British Patent 1,372,526 (1974).
63b. Kenvert International Corporation, British Patent 1,368,318 (1974).
64. B. A. Shenoi, R. Subramanian and K. S. Indira, *Electroplat. Metal Finish.* **21**(10,12), 336, 399 (1968); **22**(1), 25 (1969).
65. A. I. Levin, V. F. Lazorev and V. A. Mukhin, *Zh. Prikl. Khim.* **38**, 1569 (1965).
66. V. Sree, J. Vaid and T. L. Rama Char, *J Electrochem. Soc. Japan* **26**, 224 (1958).
67. V. Sree, *J. Sci. Ind. Res.* **18A**, 478 (1959).
68. A. Uenc and K. Okubo, *Kinzoku Hyomen Gijutsu* **15**, 216 (1964).
69. W. Machu and M. Lofty Ali el Sayed Badawy, *Werkst. Korros.* **14**, 939 (1963); **15**, 8 (1964).
70. O. V. Izbekova, O. K. Kudra and V. I. Suprunchuk, *Proc. Metals* **6**, 555 (1970).
71. T. K. Zotova and V. P. Artamonov, *Zash. Metall.* **11**, 113 (1975).
72. N. T. Kudryavtsev and G. A. Selivanova, *Izv. VUZ, Khimya Khim. Tekhnol* **13**, 932 (1970).
73. F. C. Mathers, *Trans. Electrochem. Soc.* **17**, 261 (1910).
74. M. Schlötter *in* "Die Korrosion metallischer Werkstoffe", Vol. 3, p. 393. Hirzel, Leipzig (1940).
75. R. Walker and S. Thorley, *Metal. Finish.* **74**, 30 (1976).
76. L. Peraldo Bicelli and C. Ramagnani, *Electrochim. Metal.* **4**, 140 (1969).
77. M. Schlötter, *Korros. Metallsch.* **3**, 30 (1927).
78. F. C. Mathers, *Trans. Electrochem. Soc.* **38**, 121 (1930).
79. W. Machu, "Moderne Galvanotechnik". Verlag Chemie, Weinheim (1954).
80. G. Hansel, *Elektrie* **17**, 264 (1962).
81. G. A. Emelyanenko, E. Y. Bairbarova and T. A. Koretskaya, *Ukr. Khim. Zh.* **35**, 22 (1969).

82. V. N. Flerov and V. N. Flerov, *Zh. Fiz. Khim.* **37**, 862 (1963).
83. R. L. Seth, *Electroplat. Metal Finish.* **25**, 5 (1972).
84. H. Brown, Proc. 8th Congress of Int. Union for Electrodeposition and Surface Finishing, Basel (1972), pp. 114–121, Forster Verlag A.G., Zurich (1972).
85. J. A. Harrison, R. P. J. Hill and J. Thompson, *J. Electroanal. Chem.* **44**, 445 (1973).
86. K. C. Narasimham and H. V. K. Udupa, *Metal. Finish.* **18**(204), 11; (205), 32 (1972).
86a. R. Adzic, R. Yeager and D. D. Cahan, *J. Electrochem. Soc.* **121**, 474 (1974).
87. D. L. Cox, *Plating* **51**, 976 (1964).
88. V. Q. Kinh, *Corrosion* **19**, 93 (1971); **20**, 226 (1972).
89. R. Sivakumar and T. L. Rama Char, *Electroplat. Metal Finish.* **24**, 14 (1971).
90. G. C. Wilson, *Trans. Inst. Metal Finish.* **50**, 109 (1972).
91. A. Brenner, *Plating* **52**, 1249 (1965).
92. H. J. Wiesner, W. P. Frey, R. R. Vandervoort and E. L. Raymond, *Plating* **57**, 358 (1970).
93. R. R. Vandervoort, E. L. Raymond, H. J. Wiesner and W. P. Frey, *Plating* **57**, 362 (1970).
94. N. V. Korovin, *Electroplat. Metal Finish.* **17**, 151 (1964).
95. H. Benninghoff, *Galvanotechnik* **62**, 5 (1971).
96. H. Benninghoff, *Galvanotechnik* **63**, 466 (1972).
97. A. Lafaye, *Galvano* **42**, 759 (1973).
98. M. Rubinstein, *Proc. Finish.* **24**, 39 (1971).
99. Hyogo Prefectural Government, French Patent 1,539,596 (1967).
100. Hyogo Prefectural Government, U.S. Patent 3,905,878 (1967).
101. Conversion Chemical Corp., U.S. Patent 3,785,939 (1970).
102. G. B. Rynne, *Plating* **58**, 867 (1971).
103. R. Fath and F. Hasko, *Zoroze Ochrana Mater.* **15**, 91 (1971).
104. M. and T. Chemicals Inc., U.S. Patent 3,749,649 (1971); U.S. Patent 3,926,749 (1975); British Patent 1,411,970 (1975).
105. Philips Electronic and Associated Ind. Ltd., British Patent 1,351,875 (1971).
106. R. O. Hull and Co Ltd., U.S. Patent 3,875,029 (1974).
107. A. Lafaye, *Galvano* **39**, 740 (1970).
108. A. Lafaye, *Galvano* **41**, 132 (1972).
109. A. Lafaye, *Galvano* **41**, 505 (1972).
110. J. B. Mohler, *Metal Finish.* **69**, 45 (1971).
110a. J. W. Thorp, *Products Finish.* **37**(11), 54 (1973).
111. A. K. Graham and H. L. Pinkerton, *Plating* **54**, 367 (1967).
112. G. C. Wilson, *Electroplat. Metal Finish.* **25**, 32 (1972).
113. Product Finishing, "Finishing Handbook and Directory". Sawell Publications, London (1975).
114. H. Benninghoff, *Galvanotechnik* **63**, 236 (1972).
115. F. Haskó, P. Bujtás, R. Fath and I. Karl, *Galvanotechnik* **66**, 478 (1975).

116. W. T. Hobson, Proc. 4th Symp. on Plating in the Electronics Industry, 161 (1973).
117. M. Paunovic and R. Oechslin, *Plating* **58**, 599 (1971).
118. H. Benninghoff, *Galvanotechnik* **63**, 347 (1972).
119. J. Vaid and T. L. Rama Char, *J. Sci. Ind. Res.* **16A**, 324 (1957).
120. T. L. Rama Char and J. Vaid, *Metal Finish.* **12**, 44 (1961).
121. T. L. Rama Char, *Elect. Metal. Finish.* **10**, 391, 408 (1957).
122. F. Campbell and J. A. von Fraunhofer, *Surface Technol.* **5**, 235 (1977).
123. A. K. Graham and H. L. Pinkerton, *Plating* **52**, 309 (1965).
124. North American Rockwell Corp., U.S. Patent 3,554,878 (1968).
125. B. F. Rothschild and D. Sanders, *Plating* **56**, 1363 (1969).
126. F. Haskó and R. Fath, *Metalloberflache* **21**, 113 (1967).
127. F. Haskó and P. Bujtás, *Trans. Inst. Metal. Finish.* **54**, 35 (1976).
128. E. Raub, *Plat. Surface Finish.* **63**, 30 (1967).
129. W. Machu and M. Lofty Ali el Sayed Badawy, *Werkst. Korros.* **16**, 554 (1965).
130. H. V. K. Udupa, K. C. Narasimham and P. S. Gomathi, *Plat. Surface Finish.* **62**, 1150 (1975).
131. A. A. Qureshi, R. A. Sherwani and Aziz-ur-Rehman, *Pakistan J. Sci. Ind. Res.* **13**, 463 (1970).
132. N. V. Korovin, *Electroplat. Metal Finish.* **17**, 188 (1964).
133. J. W. Dini and J. R. Helms, *J. Electrochem. Soc.* **117**, 269 (1970).
134. V. S. Galinker and A. I. Saprykin, *Ukr. Khim. Zh.* **31**, 578 (1965).
135. J. Akmens *et al.*, USSR Patent 217,853 (1967).
136. V. T. Pustovit, USSR Patent 153,820 (1961).
137. R. Sivakumar and T. L. Rama Char, *Electroplat. Metal Finish.* **24**(1), 14 (1971).
138. K. Mitsuhashi, Japanese Patent 4462 (1963).
139. N. A. Solov'ev, N. V. Merkulova, U. D. Arshchinov and A. M. Popov, *Avtomob. Prom.* **32**, 35 (1966).
140. Institute of General and Inorganic Chemistry, Academy of Sciences, Ukraine, SSR, USSR Patent 223,558 (1967).
141. P. F. Kalyuznhaya and V. B. Sobko, *Ukr. Khim. Zh.* **35**, 801 (1969).

7 Electrochemistry of Organolead Compounds

A. J. OWEN

Research and Development Department, The Associated Octel Company Ltd., Cheshire.

Introduction

Electrochemical methods have found widespread use for the manufacture of inorganic chemicals but have found little application in the manufacture of organic and, especially, organometallic chemicals. The commercial significance of organolead compounds, particularly tetramethyl lead (TML) and tetraethyl lead (TEL), as anti-knock additives for petrol has prompted a large research effort directed towards the development of commercially viable processes for their production. With the exception of the Nalco process, these research efforts have met with no success, but a substantial body of information on the electrochemistry of organoleads has been built up. The present work is intended as a brief but well-referenced guide to the literature.

The material is divided somewhat arbitrarily into three sections covering synthetic methods, electrochemical reactions and industrial processes. The first section, on synthetic methods, deals only with syntheses from metallic lead. Syntheses which involve the conversion of one organolead compound into another are dealt with in the second section.

Synthesis of Organolead Compounds

Cathodic syntheses

Organometallic compounds are formed at the cathode as a result of reduction of the electrolyte to hydrocarbon radicals which then attack the cathode metal forming organometallics.

$$R^+ + e \rightarrow \dot{R}$$

$$\dot{R} + M \rightarrow RM$$

Thus the cathode is consumed during electrolysis. Side reactions such as combination or further reduction of the radicals can also occur.

$$2\dot{R} \rightarrow R-R$$

$$H^+ + e \rightarrow \dot{H}$$

$$\dot{R} + \dot{H} \rightarrow RH$$

The extent of these side reactions depends on cathodic potential, type of electrolyte, type of solvent and other physical parameters.

Electrolysis of Carbonyl Compounds

The first synthesis of an organolead compound by an electrolyte method was reported by Tafel in 1906.[1] He found that upon electrolysis of a solution of methyl ethyl ketone in 30% sulphuric acid, a brownish-red oil was formed at the lead cathode. Treatment of this oil with chlorine or bromine was later found to give tri-sec-butyl lead chloride and di-sec-butyl lead bromide, respectively. Similarly, electrolysis of acetone in acidic solution gave a red organolead oil which was later shown to contain tetra-isopropyl lead and other very unstable compounds which are presumed to be di- or tri-isopropyl leads.[2-5]

A mechanism to account for the formation of organolead compounds has been proposed.[3,5] Thus, taking acetone as an example, the initial step is the generation of a radical:

$$(CH_3)_2CO + H^+ \xrightarrow{+e} (CH_3)_2\dot{C}OH$$

This radical may undergo self-combination or be further reduced to an anion:

$$(CH_3)_2\dot{C}OH \xrightarrow{+e} (CH_3)_2\bar{C}OH$$

This anion then reacts with lead cations thought to be present to generate a weakly bounded organometallic which can undergo further reduction to give isopropanol or di-isopropyl lead which itself may be further reduced to give propane and a lead cation.

$$(CH_3)_2COH + Pb^{2+} \rightarrow$$

$$\xrightarrow[2e]{2H^+} 2(CH_3)_2CHOH$$

$$\rightarrow [(CH_3)_2COH]_2Pb \xrightarrow[4H^+]{4e} [(CH_3)_2CH]_2Pb \xrightarrow{2H^+} Pb^{2+}$$

$$+ 2CH_3CH_2CH_3$$

As a result of the sort of side reactions indicated in the above mechanism the yields of organoleads from this method are generally very poor.

Examples of electrolyses of carbonyl compounds that give organo-lead compounds are collected in Table I.

Table I. Synthesis of organolead compounds by electrolysis of carbonyl compounds.

Compound	Solvent	Temperature ($^\circ$C)	Current density (A cm^{-2})	Product	Reference No.
Acetone	20% H_2SO_4	20	0·05	PbPr$_4$	2–5, 11, 12
Methyl ethyl ketone	20% H_2SO_4	20	0·05	Pb (*sec*-Bu)$_4$	1, 6
Diethyl ketone	20% H_2SO_4	20	0·05	Pb(3-pentyl)$_4$	6
Methyl isoamyl ketone	20% H_2SO_4			Organolead	7
Citral and other unsaturated ketones	20% H_2SO_4			Organolead	8, 9
Propionaldehyde				Organolead	10

Electrolysis of Halogenated Compounds

Calingaert reported the first electrosynthesis of an organolead from an alkyl halide in 1923.[13] Ethyl iodide dissolved in caustic aqueous alcohol gave TEL on reduction at a lead cathode. Mead[14] later proposed a similar synthesis but using aqueous caustic containing casein as the solvent. Yields from these processes were always low and it was hypothesized[15] that hydrogen formed at the cathode reduced the reaction mass forming alkyl derivatives of lower oxidation states of lead which underwent various secondary reactions.

Silversmith and Sloan[15] proposed the use of non-hydroxylic solvents such as acetonitrile and dimethylformamide (DMF) to avoid the formation of hydrogen at the cathode. If these poorly conducting solvents are used then a support electrolyte such as a trialkylammonium or a trialkylsulphonium is required. The electrolysis of methyl and ethyl bromides and iodides in a large number of solvent-support electrolyte systems were examined and yields of 12–100% alkyl lead based on lead consumed were obtained. Smeltz[18] found that the addition of small amounts of hydroxylic solvents (water, alcohols) to the acetonitrile systems results in lower power consumption, less gassing, higher yields and higher rates of production of tetra-alkyl leads.

It is clear that the nature of the solvent-support electrolyte system is very important to high yields of alkyl lead production. Galli[19–21] has experimented with propylene carbonate as a solvent for the electrolysis of ethyl bromide. In this solvent, tetraethylammonium perchlorate gave 100% yield of TEL, tetra-alkyl ammonium halides gave slightly lower yields but inorganic support electrolytes such as potassium chloride and lithium perchlorate gave no TEL formation. In contrast, the use of inorganic support electrolytes in acetonitrile was found to give high yields of TEL[15] and good yields were reported with DMF as solvent.[23]

Yang et al.[22,23] electrolysed ethyl bromide in solvent systems of water and acetone, tetrahydrofuran (THF) or acetonitrile with tetrabutyl phosphorium or tetrabutyl ammonium bromide as support electrolyte. They found that the formation of TEL was strongly dependent on the ratio of water to support electrolyte, the yield increasing rapidly towards 100% as the ratio approached 80 mol/mol then becoming substantially independent of the ratio. They also found that tetrabutyl ammonium bromide was much superior in promoting TEL yields than tetraethyl ammonium bromide.

Galli's work illustrates the influence of cathode potential on the formation of alkyl lead compounds.[20] At cathode potentials up to 500 mV, the electrolysis of ethyl bromide in propylene carbonate with tetraethyl ammonium bromide as support electrolyte did not give TEL, all the current being used in side processes due to impurities and dissolved oxygen. At potentials higher than 500 mV, one-electron reduction of ethyl bromide took place and TEL was formed, the yield increasing with the potential up to 100% at 1400 mV. At greater potentials the formation of TEL decreased as two-electron reduction of the ethyl bromide, decomposition of radicals to hydrocarbons C_2–C_4 and other reactions began to take place.

Ulery[24] has studied the polarographic reduction of simple alkyl halides at a stationary lead electrode and found a significant difference in the behaviour of bromides and iodides. The bromide results suggest a "backside attack" by the electrode somewhat analogous to an SN2 reaction with breaking of the C—Br bond with simultaneous formation of the new C—M bond.

Alternatively, bond breaking without new bond formation could occur giving a free alkyl radical which subsequently attacks the electrode metal.

The results for iodides suggested that the initial attack occurred at different orientations from that for bromides. Two mechanisms were proposed:

Molecular models show that (a) should be valid for MeI and EtI, and (b) should start to prevail with increasing chain length.

Britz and Knittel[145] have suggested that the onium support electrolytes used in many of the electrolyte systems were effective as a result of being absorbed on the electrodes and, presumably, aiding processes like those described above. They found that on addition of a small amount of a Kryptate complex (1 mM) to act as absorbate, the yield of TEL from electrolysis of ethyl bromide in water/acetonitrile solution increased from 8 to 90% based on lead loss.

Tetramethyl lead and a small number of other alkyl leads have been prepared by processes similar to those discussed above for TEL. The

Table II. Synthesis of organolead compounds by electrolysis of organic halides.

Halide	Support electrolyte	Solvent	Anode	Temperature (°C)	Current density $(A\ cm^{-2})$	Product	Yield (%)	Reference No.
MeBr	Et_4NBr	Acetonitrile	Pt	45	0·06	TML	81[a] 73[b]	15, 18
MeBr	Et_4NBr	Acetonitrile/ 0·7% H_2O	Pt	45	0·1	TML	95[a] 92[b]	18
				45	0·06	TML	93[a]	
MeBr	Et_4NBr	Acetonitrile/H_2O	Pt	45	0·025–0·15	TML	84[b]	16
MeBr	Et_4NBr	Acetonitrile/ MeOH	Pt	41		TML	83[a] 89[b]	18
MeBr	Bu_4NBr	Acetone/H_2O	Carbon	25	0·067	TML	98[b]	22, 23
MeBr	NaBr	DMF	Pt	—	—	TML	80[a] 85[b]	148, 149
MeBr	Bu_4NBr	—	Pt	—	0·012	TML	73[b]	18
MeBr	Bu_4NBr	5% H_2O	Pt	—	0·015–0·075	TML	79[b]	18
MeI	Me_4NBr	Acetonitrile	Pt	25–60	0·02–0·03	TML	61[a] 64[b]	15
MeI	Me_3SI	Acetonitrile	Pt	25–60	0·02–0·03	TML	86[a] 85[b]	15
MeI	NaI	DMF	Pt	—	—	TML	90[b] 90[b]	148, 149
EtCl	Et_4NBr	Acetonitrile	Pt	25–60	0·02–0·03	TEL	93[a]	15
EtBr	Et_4NBr	Acetonitrile	Pt	25–60	0·02–0·03	TEL	80[a] 70[b]	15
EtBr	LiBr	Acetonitrile	Pt	25–60	0·02–0·03	TEL	100[a]	15
EtBr	NaI	Acetonitrile	Pt	25–60	0·02–0·03	TEL	76[a]	15
EtBr	KI	Acetonitrile	Pt	25–60	0·02–0·03	TEL	84[a]	15
EtBr	$CaBr_2$	Acetonitrile	Pt	25–60	0·02–0·03	TEL	100[a]	15
EtBr	KSCN	Acetonitrile	Pt	25–60	0·02–0·03	TEL	100[a]	15

EtBr	LiClO$_4$	Acetonitrile	Pt	25–60	0·02–0·03	TEL	100[a]	14
EtBr	Et$_3$SBr	Acetonitrile	Pt	25–50	0·02–0·03	TEL	81[a]	15
EtBr	Ph$_3$EtPI	Acetonitrile	Pt	25–60	0·02–0·03	TEL	86[a]	15
EtBr	Ph$_3$MeBr	Acetonitrile	Pt	25–60	0·02–0·03	TEL	72[a]	15
EtBr	Et$_4$NBr	DMF	Pt	25–60	0·02–0·03	TEL	46[a]	15
EtBr	Et$_4$NBr	DMSO	Pt	25–60	0·02–0·03	TEL	16[a]	15
EtBr	Et$_4$NBr	1,2-Dimethoxyethane	Pt	25–60	0·02–0·03	TEL	47[a]	15
EtBr	Et$_4$NBr	Methylene chloride	Pt	25–60	0·02–0·03	TEL	71[a]	15
EtBr	Et$_4$NBr	Ethyl thiocyanate	Pt	25–60	0·02–0·03	TEL	84[a]	15
EtBr	Et$_4$NBr	N,N-Dimethyl cyanamide	Pt	25–60	0·02–0·03	TEL	44[a]	15
EtBr	Et$_4$NBr	Propylene carbonate	Pt	25–60	0·02–0·03	TEL	40[a]	15
EtBr	Et$_4$NBr	Propylene carbonate	Pt	—	0·003–0·005	TEL	up to 100[b]	19, 20
EtBr	Et$_4$NI	Propylene carbonate	Pt	—	0·006–0·008	TEL	71[b]	21
EtBr	Et$_4$NClO$_4$	Propylene carbonate	Pt	—	0·013	TEL	100[b]	21
EtBr	Bu$_4$NBr	Propylene carbonate	Pt	—	0·005	TEL	87[b]	21
EtBr	Bu$_4$NBF$_4$	Propylene carbonate	Pt	—	0·005	TEL	80[b]	21
EtBr	Me$_2$EtSBr	Propylene carbonate	Pt	—	0·014	TEL	77[b]	21
EtBr	LiClO$_4$	Propylene carbonate	Pt	—	0·0001–0·0005	TEL	0	21
EtBr	NaClO$_4$	DMF	Pt	—	0·005	TEL	78[a] 80[b]	148, 149

continued

Table II. (continued)

Halide	Support electrolyte	Solvent	Anode	Temperature (°C)	Current density (A cm^{-2})	Product	Yield (%)	Reference No.
EtBr	NaBr	DMF	Pt	—	0·005	TEL	75[a] 92[b]	148, 149
EtBr	LiBr	Acetone	Carbon	—	0·10	TEL	41[b]	23
EtBr	Bu$_4$PBr	H$_2$O/Acetone	Carbon	—	0·15	TEL	100·1[b]	23
EtBr	Bu$_4$PBr	H$_2$O/THF	Carbon	—	—	TEL	87·7[b]	23
EtBr	Bu$_4$NBr	H$_2$O/THF	Carbon	—	—	TEL	84·8[b]	23
EtBr	Bu$_4$PBr	H$_2$O/MeOH	Carbon	—	0·15	TEL	17[b]	23
EtBr	Bu$_4$NBr	H$_2$O/EtCN	Carbon	—	—	TEL	96[b] 89[b]	23
EtBr	Me$_2$(C$_{18}$H$_{37}$)NCl	Acetonitrile	Pt	25–60	—	TEL	26[a] 23[b]	15
EtI	EtOH/H$_2$O	Caustic aq. alcohol	Carbon	55	—	Pb loss	—	26
EtI		NaOH/H$_2$O	Carbon	—	—	TEL	—	13
EtI	Casein	DMF	Carbon	—	—	TEL	—	14
EtI	NaClO$_4$	DMF	Pt	—	0·012	TEL	60[a] 68[b]	148, 149
EtI	NaI	DMF	Pt	—	0·05	TEL	70[a] 89[b]	148, 149
EtI	Et$_4$NBr	Acetonitrile	Pt	—	0·02–0·03	TEL	72[a]	15
MeCl	Et$_4$NBr	Acetonitrile	Pt	25–60	0·02–0·03	Mixed methyl ethyl lead	93[a] 75[b]	15

MeBr+EtBr	Bu₄NBr	H₂O/Acetone	Carbon	—	0·114	Mixed methyl ethyl lead	—	23
MeI+EtI	NaClO₄	DMF	Pt			Mixed methyl ethyl lead	82[b]	148, 149
ICH₂CH₂CN		H₂SO₄/H₂O	Cu		0·035	Pb(CH₂CH₂CN)₄	13[a]	25
n-PrBr	Bu₄NBr	H₂O/Acetone	Carbon		0·114	PbPr₄ +Pr₃PbPbPr₃	88[a] 78[b]	23
iso-PrBr	Et₄NBr	Acetonitrile/H₂O				Organo-lead	30[a]	24
CH₂=CH CH₂Cl	LiCl	DMF	Pt	30	0·012	Pb(CH₂CH=CH₂) (supposed)	65[a]	15

[a] Yield based on lead consumed.
[b] Current yield.

results of the preparation of alkyl leads by electrolysis of organic halides are summarized in Table II.   ·

Electrolysis of Onium and Miscellaneous Compounds

Transfer of electrons to onium ions at room temperature and above results in carbonheteroatom bond scission forming an alkyl radical which reacts with the electrode metal to give an organometallic:

$$R_nA^+ + e \rightarrow [R_n\dot{A}] \rightarrow R_{n-1}A + \dot{R} \xrightarrow{M} RM$$

where A = S, P or N.

Tetraethyl lead has been synthesized by reduction of triethyl sulphonium bromide in acetonitrile[15] and by reduction of triethyl sulphonium iodide in DMF with tetraethyl ammonium bromide as support electrolyte.[27] Tetramethyl lead has been similarly prepared by reduction of trimethyl sulphonium iodide in DMF with tetraethyl ammonium bromide as support electrolyte.[27] Tetrabenzyl lead has been prepared by reduction of $PhCH_2S(CH_3)_2OTos$ in DMF.[27]

Tetraethyl lead has been synthesized by reduction of diethyl sulphate and ethyl acetate in a process similar to that used for reduction of alkyl halides.[15]

Information on the synthesis of organolead compounds from onium and other compounds is summarized in Table III.

Anodic synthesis

In principle, the anodic synthesis is similar to the cathodic process. Radicals, formed at the anode by oxidation of an organometallic electrolyte, attack the metal of the anode to form the organometallic:

$$R^- \rightarrow \dot{R} + e$$

$$\dot{R} + M \rightarrow RM$$

As in the cathodic process the radicals can take part in side reactions but conditions can be selected which largely eliminate them.

The fact that anodic processes start from organometallics is a serious disadvantage as they are normally difficult and costly to make but they have the advantage of giving high yields.

Table III. Synthesis of organolead compounds by electrolysis of onium and miscellaneous compounds.

Compound	Support electro- lyte	Solvent	Cathode	Tempera- ture (°C)	Current density (A cm^{-2})	Product	Yield (%)	Ref- erence No.
Me_3SI	Et_4NBr	DMF	—	—	—	TML	—	27
Et_4NBr	Et_4NBr	Acetonitrile	Pt	25–60	0·02–0·03	TEL	12[a]	15
Et_3SBr	Et_4NBr	Acetonitrile	Pt	25–60	0·02–0·03	TEL	33[a]	15
Et_3SI	Et_4NBr	DMF	—	—	—	TEL	50[a]	27
$PhCH_2S$ Me_2OTos	Et_4NBr	DMF	—	—	—	$Pb(C_6H_5$ $CH_2)_4$	9[a]	27
Et_2SO_4	Et_4NBr	Acetonitrile	Pt	25–60	0·02–0·03	TEL	81[a]	15
EtOAc	Et_4NBr	Acetonitrile	Pt	25–60	0·02–0·03	TEL	10[a]	15

[a] Based on lead consumed.

Electrolysis of Organomagnesium Compounds

The electrolysis of the most important group of organomagnesium compounds, Grignard reagents, forms the basis of an important industrial process and is discussed on pp. 188–191. The present section is concerned with electrolyses of sundry other organomagnesium compounds and a mechanistic discussion.

Giraitis[31] has produced TEL by the reduction of diethyl magnesium formed in the cell, being reconverted to diethyl magnesium by reaction with hydrogen and ethylene *in situ.*

Kobetz and Thomas[86] have produced mixed methyl vinyl- and phenyl vinyl-leads by electrolysing the complexes formed between vinyl magnesium compounds and alkylaluminium compounds. Thus, electrolysis of a mixture of divinyl magnesium and trimethyl aluminium in the dimethyl ether of diethylene glycol gives a product containing mixed methyl vinyl lead compounds.

The available data on the electrolyses of organomagnesium compounds at lead anodes to give organolead compounds are collected in Table IV.

Organometallic compounds have been synthesized by the electrolysis of Grignard reagents at sacrificial anodes since the early work of French and Drane.[120] However, despite several studies of the process,[87–91] both their structure and the mechanism of their electrolysis are still imperfectly understood.

The low electrical conductivity of Grignard reagents in ether solution has been attributed to the ionization processes:

$$RMgX \ \rightleftarrows \ R^- + MgX^+$$

Plausible anode and cathode reactions are then:

$$\text{Anode:} \ \ R^- \rightarrow e + \overset{M}{\dot{R}} \rightarrow RM$$

$$\text{Cathode:} \ \ 4\,MgX^+ + 4e \rightarrow 2\,Mg + 2\,MgX_2$$

Alternatively, it has been suggested[91] that the ions involved are $RMgCl_2^-$ and MgR^+. In this case the anode and cathode reactions would be:

$$\text{Anode:} \ \ RMgCl_2^- \rightarrow \dot{R}MgCl_2 + e \rightarrow \overset{\downarrow M}{\dot{R}} + MgCl_2$$

$$RM$$

$$\text{Cathode:} \ \ 2\,MgR^+ + 2e \rightarrow 2\,\dot{M}gR \rightarrow Mg + MgR_2$$

The dialkyl magnesium formed at the cathode would react with the magnesium formed at the anode to regenerate part of the Grignard. This mechanism is plausible but does not account for the detailed transport phenomena observed.[91]

Electrolysis of Organoaluminium Compounds

The extensive research into the use of aluminium compounds as precursors of lead alkyls was initiated in the 1950s by the discovery of new processes for their manufacture.[33,34]

Aluminium alkyls or their solutions in polar solvents exhibit very poor electrical conductivity and so cannot be electrolysed directly. However, certain complexes which are formed with alkali metal halides or alkylalkali metal compounds give highly conductive metals or solutions. In 1955 Ziegler reported the use of the sodium fluoride bis (triethyl aluminium) complex, $Na[Et_3AlFAlEt_3]$ in a synthesis of tetra-alkyl lead.[35,37] This complex melts at slightly above room temperature and at 63°C has an electrical conductivity of $0.02 \text{ ohm}^{-1} \text{ cm}^{-1}$. Electrolysis of this complex between a copper cathode and a lead anode proceeds according to the equations:

$$\text{Anode:} \quad [Et_3AlFAlEt_3]^- \longrightarrow \dot{E}t + Et_3Al + Et_2AlF + e$$

$$\text{Cathode:} \quad [Et_3AlFAlEt_2]^- \xrightarrow{+3e} Al + Et_3AlF^- + 3\,Et^-$$

The ethyl radicals so formed attack the lead anode with the consequent formation of TEL. The overall process can be represented by:

$$4\,Na[Et_3AlFAlEt_3] + 3\,Pb \longrightarrow 3\,PbEt_4 + 4\,NaFEt_3Al + 4\,Al$$

The TEL formed is insoluble in the molten electrolyte and separates out as a denser liquid phase. The $NaFEt_3Al$ formed has a melting point of 72–73°C and the cell is operated at a higher temperature, so it is formed as a melt which can be recycled in situ by reaction with added triethyl aluminium.

There are a number of problems associated with this process. The aluminium produced at the cathode is formed as a loose deposit which is difficult to separate from the electrolyte. The use of a high current density caused the aluminium to be formed as easily separable dendrites

Table IV. Synthesis of organolead compounds by electrolysis of organomagnesium compounds.

Compound	Solvent	Cathode	Temperature (°C)	Current density (A cm^{-2})	Product	Yield (%)	Reference No.
MeMgCl + MeCl	THF + Ether II	Steel	50	0·028	TML	87[a] / 180[b]	106
MeMgCl + MeCl	THF + Ether III	Steel	43	0·016	TML	98[a] / 167·5[b]	106
MeMgCl + MeCl	THF + Ether IV	Pb	30	—	TML	83[a] / 153[b]	133, 116
MeMgCl + MeCl	THF + Ether V	—	46	0·018	TML	100[a] / 174[b]	105
MeMgCl + MeCl	THF + Tetrahydrofurfuryl ether	Steel	42	0·030	TML	91[a] / 186[b]	106
MeMgCl + MeCl	THF + Benzene + Ether II	—	23	0·027	TML	81[a]	97
MeMgCl + MeCl	THF + Benzene + Ether IV	—	30	—	TML	81[a] / 164[b]	110, 116
MeMgCl + MeCl	THF + Benzene + Ether VI	Steel	23	0·027	TML	81[a] / 150[b]	98, 102
		Steel	41	0·029	TML	99[a] / 161[b]	110, 111, 116
MgEt$_2$	THF + Diethyl ether	Pt	30	0·07	TEL	—	31, 150
EtMgCl + EtCl	Ether I	Steel	50–65	—	TEL	81[a]	95
EtMgCl + EtCl	Ether IV	Steel	33–38	0·003	TEL	84[a]	93, 95
EtMgCl + EtCl	Ether IV	Steel	55–60	0·002	TEL	80·5[a]	93, 95
EtMgCl + EtCl	THF + Tetrahydrofurfuryl ether	Steel	48	0·02	TEL	98[a] / 173[b]	106
EtMgCl + EtCl	THF + Benzene + Ether II	Steel	25	—	TEL	—	110
EtMgCl + EtCl	THF + Benzene + Ether VI	Steel	40	—	TEL	—	98
EtMgCl + EtCl	THF + Toluene + Ether IV	—	33	0·071–0·037	TEL	82[a]	116
EtMgBr	Diethyl ether	Pb	—	—	TEL	—	92, 94, 150
MeMgCl + EtCl	THF + Benzene + Ether VI	Steel	38	—	Mixed methyl ethyl leads	—	100, 108
MeMgCl + t-BuCl	THF + Benzene + Ether VI	Steel	38	—	Mixed methyl t-butyl leads	—	100, 108

Reactants	Solvent	Cathode	Yield		Product		Ref.
MeMgCl + EtMgCl + t-BuCl	THF + Benzene + Ether IV	Steel	30	—	Mixed ethyl t-butyl leads	—	108
MeMgCl + cyclo-C_6H_{12}	Ether IV	Steel	38	—	Mixed methyl cyclo-hexyl leads	—	108
PhMgBr + PhBr	Ether VI	Steel	55	—	$PbPh_4$	—	107
MeMgCl + PhCl	Ether IV	Steel	38	—	Mixed methyl phenyl leads	—	107, 108
$CH_2=CHMgCl$	THF	Pb	—	—	$Pb(CH=CH_2)_4$	—	117, 119
$CH_2=CHMgCl$ + MeCl	THF	Pb	—	—	Mixed methyl vinyl leads	178[b]	117, 119
$CH_2=CHMgCl$ + EtCl	THF	Pb	—	—	Mixed ethyl vinyl leads	—	117, 119
$Mg(C_2H_3)_2$ + $AlMe_3$	Ether I	Steel	50	0·005	Mixed methyl vinyl leads	—	86
$Mg(C_2H_3)_2$ + BMe_3	Ether I	Steel	75	—	Mixed methyl vinyl leads	—	86
$Mg(C_2H_3)$ + $AlMe_2$ OMe	THF + Ether I	Steel	65	—	Mixed methyl vinyl leads	—	86
$Mg(C_2H_3)_2$ + $Al(C_2H_3)_2Me$ + MeCl	Ether I	Steel	65	—	Mixed methyl vinyl leads	—	86
$Mg(C_2H_3)_2$ + $AlEt_3$	Ether I	Steel	75	—	Mixed ethyl vinyl lead	—	86
$Mg(C_2H_3)_2$ + $AlPh_3$	Ether I	Steel	50	—	Mixed phenyl vinyl leads	—	86
$Mg(C_2H_3)_2$ + $AlOPh_3$ + MeCl	THF + Ether I	Steel	65	—	Mixed phenyl vinyl leads	—	86
$CH_3CH_2=CHMgCl$	4-Methoxytetrahydrofuran	Pb	50	—	$Pb(CH=CHCH_3)_4$	—	117, 119
$Mg(CH_3CH=CH_2)_2$ + $AlMe_3$ + MeCl	Ether I	Steel	50	—	Mixed	—	86

[a] Yield based on Grignard conversion.
[b] Current yield.

I = Dimethyl ether of diethylene glycol.
II = Diethyl ether of diethylene glycol.
III = Diethyl ether of tetraethylene glycol.
IV = Dibutyl ether of diethylene glycol.
V = Benzyl ethyl ether of diethylene glycol.
VI = Hexyl ethyl ether of diethylene glycol.

but within a short time they became long enough to short out the cell thereby stopping the electrolytic reaction. The aluminium also tends to react with the product alkyl lead so the anode and cathode have to be separated by a suitable membrane[38] or the cell operated under reduced pressure[39] to distill off the product before reaction can occur.

The use of the $NaF.2AlEt_3$ complex in the synthesis of TEL has also been reported by Roethli and Simpson[40] and Giraitis.[41] The latter produced tetraphenyl lead by electrolysing $NaF.SAl(C_6H_5)_3$ in benzene solution.

In an effort to overcome some of the problems outlined above, Ziegler in 1959 reported the electrolysis of molten complexes of the type $NaAlR_3Et$.[12,43] The overall process produced TEL, sodium and trialkyl aluminium:

$$4 \, NaAlR_3Et + Pb \rightarrow 4 \, Na + PbEt_4 + 4 \, R_3Al$$

Ziegler[44,45] used sodium tetraethyl aluminium with a little added sodium triethyl aluminium alkoxide as the electrolyte. Sodium tetraethyl aluminium melts at 124°C, therefore the sodium is formed in the cell as a molten deposit and drips off the cathode creating an ever-fresh cathode surface. The electrolyte is regenerated by reaction of recovered sodium with hydrogen, addition to the triethyl aluminium and finally addition of ethylene:

$$4 \, Na + 2 \, H_2 \rightarrow 4 \, NaH$$

$$4 \, NaH + 4 \, AlEt_3 \rightarrow 4 \, NaAlEt_3H$$

$$4 \, NaAlEt_3H + C_2H_4 \rightarrow 4 \, NaAlEt_4$$

The disadvantages of the process are that the sodium reacts with the products and the cell must be compartmented. Additionally, the high cell-operating temperature necessitates working under reduced pressure to avoid excessive decomposition of the product which is thermally unstable at such a temperature. Another problem is that TEL and triethyl aluminium have very close boiling points and are not easily separated. Ziegler's solution to this problem was to add a sodium trialkyl alkoxide to the electrolyte. This undergoes an exchange reaction with triethyl aluminium:[48]

$$NaAlR_3(OR) + Et_3Al \rightarrow NaAlR_3Et + Et_2AlOR$$

Thus, addition of sodium triethyl aluminium butoxide to the cell results in the formation of diethyl aluminium butoxide which is less volatile than TEL and separation is easily performed by distillation.[49] Using this system, 95% yields of TEL are achieved.[51,66] An alternative method uses the sodium trialkylalkoxide itself as the electrolyte, this process also giving high yields of TEL.[51,64,66] The electrolyte is regenerated by a process similar to that described above for regeneration from triethyl aluminium. A disadvantage is that the alkoxide has lower electrical conductivity.

A considerable advance was the use of a pool of mercury as the cathode.[52,53] The advantage here was that the sodium formed during the electrolysis immediately formed an amalgam with the mercury which did not react with the organolead product, thereby avoiding the need for a compartmented cell. The sodium was recovered from the amalgam by a secondary electrolysis employing the sodium amalgam as the anode and sodium as the cathode with sodium tetraethyl aluminium as the electrolyte. If sodium was merely converted to sodium hydroxide, then the net cell reaction claimed for TEL production would be:

$$Pb + 4\,Na + 4\,C_2H_4 + 4\,H_2O \rightarrow PbEt_4 + 4\,NaOH$$

Hydrogen being recycled for electrolyte regeneration does not appear.

The use of potassium cation electrolytes has been investigated[50,55] because although they are more expensive, they have better electrical conductivities and lower melting points than their sodium analogues. Sodium tetraethyl aluminium melts as $124°C$, but after reaction with potassium chloride an electrolyte which melts at $75°C$ is obtained.[50] The lower cell-operating temperature reduced the decomposition of the organolead product. A disadvantage of using potassium is the difficulty of separating the by-product from the amalgam obtained when a mercury cathode is used. Ziegler avoided this problem by developing a method of regenerating the electrolyte which employed the potassium amalgam itself in reaction with sodium tetraethyl aluminium.[54,56,68,72] The electrolysis and regeneration processes are:

Electrolysis: $4\,KAlEt_4 + Pb \rightarrow PbEt_4 + 4\,AlEt_3 + 4K(Hg)_x$

Regeneration: $\begin{cases} AlEt_3 + Na(Hg)_x + \frac{1}{2}H_2 + C_2H_4 \rightarrow NaAlEt_4 \\ NaAlEt_4 + K(Hg)_x \rightarrow KAlEt_4 + Na(Hg)_x \end{cases}$

Various mixed electrolyte systems have been used for the production of TEL. A 1:1 mixture of sodium tetramethyl aluminium and sodium tetraethyl aluminium forms a low melting eutectic which upon electrolysis gives TEL as the only product.[57,58] Either TEL is formed preferentially in the electrolysis process, or a mixed product rapidly undergoes an exchange reaction to give TEL only. Such reactions have been reported.[59] Information on other mixed electrolyte systems which have been used for the production of TEL is collected in Table V.

A recent development is the use of potassium diethyl aluminium dichloride as the electrolyte in a process for the preparation of TEL.[147] The main advantage of this electrolyte is that no by-product is deposited at the cathode if an alkyl halide is added to react with the aluminium formed. The probable anode and cathode reactions were given as:

$$\text{Anode:} \quad Pb + 4\,AlEt_2Cl_2^- \rightarrow PbEt_4 + 2\,AlEt_2Cl + 4e$$

$$\text{or} \rightarrow PbEt_4 + 4\,AlEtCl_2 + 4e$$

$$\text{Cathode:} \quad 4e + \tfrac{4}{3}AlEt_2Cl_2^- \rightarrow \tfrac{4}{3}Al + \tfrac{8}{3}Et^- + \tfrac{8}{3}Cl^-$$

$$\text{followed by} \quad \tfrac{4}{3}Al + 2\,EtCl \rightarrow \tfrac{2}{3}AlEt_2Cl + \tfrac{2}{3}AlEtCl_2$$

At the cell temperatures used, 70–90°C, the groups attached to the aluminium atom are very labile and the diethyl aluminium dichloride could disproportionate to give diethyl aluminium chloride and aluminium trichloride. Therefore the overall reaction could be given as:

$$Pb + KAlEt_2Cl_2 + 2\,EtCl \rightarrow PbEt_4 + AlCl_3 + KCl$$

TEL yields of 98% based on lead loss were obtained.

Tetramethyl lead is more difficult to synthesize than TEL because sodium tetramethyl aluminium has a high melting point (240°C) and a solvent has to be used. Ziegler[62,63] obtained a 90% yield of TML based on lead loss by electrolysing a solution of sodium tetramethyl aluminium in THF using a lead anode and a mercury cathode:

$$4\,NaAlMe_4 + Pb \xrightarrow[\text{THF}]{\text{electrolysis}} PbMe_4 + 4\,AlMe_3 + 4\,Na(Hg)_x$$

The by-product trimethyl aluminium formed an addition compound with the THF which was easily separated from the TML by distillation. The electrolyte was regenerated by reacting the trimethyl aluminium-THF adduct with sodium and methyl chloride in THF solution:

$$2\,Na + Me_3Al.THF + MeCl \rightarrow NaCl + NaAlMe_4 + THF$$

TML has also been prepared by electrolysis of sodium trimethyl aluminium alkoxides[63,64] and various mixed electrolytes.[58,63]

Lemkuhl *et al.*[65] have proposed a process for the combined production of TML and TEL in a dual electrolysis system. TEL is produced in one cell by electrolysis of sodium tetraethyl aluminium using a mercury cathode. The sodium amalgam formed at the cathode is reacted with methyl chloride and trimethyl aluminium to generate sodium tetramethyl aluminium which is electrolysed in diglyme solution in the second cell to form TML.

Although Ziegler[55] claims to have developed a commercially viable process for the manufacture of lead alkyls from aluminium alkyls, the process has never been used commercially. Presumably it does not show sufficient advantage over the other well established process to attract the necessary large investment.

Several other lead alkyls, including tetrapropyl, tetrabutyl and tetraphenyl lead, have been prepared by electrolysis of aluminium alkyls and data on these is collected in Table V.

Electrolysis of Organoboron Compounds

Tetralkyl lead compounds have been formed by electrolysis of organoboron compounds. Compounds of organoboron and alkyl lead have certain advantages over those of organoaluminium in that they are easily handled and have larger differences in boiling points so separations are more easily effected. Ziegler has described the use of alkyl boron compounds for separating alkyl leads in admixture with trialkyl aluminium.[79] A disadvantage is that their alkali metal complexes have high melting points and so they must be used in a solvent or in admixture with organoaluminium complexes.

Ziegler[74] in 1962 reported the synthesis of TEL by electrolysis of sodium tetraethyl boron in aqueous solution using a mercury cathode and a rotating lead anode. The cell was operated at 20°C with a current density of $0 \cdot 035 - 0 \cdot 1$ A cm^{-2}, and TEL was obtained in 92% yields based on the current used. In later work[77] current densities of 10 A cm^{-2} were used with a slightly lower current yield of 90%. Pinkerton[76] electrolysed a solution of sodium tetraethyl boron in the dimethyl ether or diethylene glycol between a copper cathode and a lead anode at a current density of $0 \cdot 01$ A cm^{-2} and obtained tetramethyl lead in 100% yield based on current consumption.

Table V. Synthesis of organolead compounds by electrolysis of organoaluminium compounds.

Electrolyte	Cathode	Temperature (°C)	Current Density (A cm^{-2})	Product	Yield (%)	Reference No.
NaAlMe$_4$	Hg	—	0·05	TML	—	70
NaAlMe$_4$ in THF	Hg	90	0·14	TML	90[b]	62, 63
NaAlMe$_4$ in Diglyme	Hg	100	0·30	TML	96[b]	63, 65
NaAlMe$_3$OEt	Cu	—	—	TML	—	64
NaAlMe$_3$OBu	Hg	—	—	TML	94[b]	63
NaAlMe$_4$ + NaAlMe$_3$OC$_8$H$_{17}$	Hg	90	—	TML	—	63
NaAlMe$_4$ + NaBMe$_4$	Hg	100	0·15	TML	94[b]	63
NaAlMe$_4$ + RbAlMe$_4$(1:5)	Cu, Steel	100	0·015	TML	—	57,58
AlEt$_3$ in Diethyl ether	Pt	25	0·007	TEL	—	67, 69
NaAlEt$_2$Me$_2$	Steel	—	—	TEL	—	58
NaAlEt$_4$	Pt	—	—	TEL	—	150
NaAlEt$_3$OEt	Cu	105	0·005	TEL	—	58, 64
NaAlEt$_3$OBu	Cu	90	0·004	TEL	93[b]	47, 51, 64, 66
NaF.AlEt$_3$	Pt	80	0·25	TEL	—	67
NaF.2AlEt$_3$	Cu	28–45	0·022	TEL	87[b]	35, 37, 41
KAlEt$_4$	Hg	100	0·465	TEL	100[a]	68, 72
KAlEt$_3$OBu	Cu	90	0·002	TEL	—	47, 64, 66
LiAlEt$_3$OC$_6$H$_5$	Cu	25	0·001	TEL	—	64
NaAlEt$_4$ + NaAlMe$_4$ (1:1)	Cu, Steel	100	0·25	TEL	77[b]	57, 58
NaAlEt$_4$ + NaAlMe$_4$ (3:1)	Cu	100	0·25	TEL	89[b]	57
NaAlEt$_4$ + NaAlEt$_3$OBu (1:1)	Cu	100	0·053	TEL	95[b]	51, 66
NaAlEt$_4$ + NaAlEt$_3$OC$_6$H$_5$	Cu	—	—	TEL	—	51
NaAlEt$_4$ + NaAlEt$_3$O(iso-C$_6$H$_{11}$)	Cu	—	—	TEL	—	51
NaAlEt$_4$ + NaAlEt$_3$OC$_{10}$H$_{21}$	Cu, Hg	—	—	TEL	—	51
NaAlEt$_4$ + NaF.AlEt$_3$	Hg	70	0·10	TEL	98[b]	55, 70
NaAlEt$_4$ + NaF.2AlEt$_3$	Hg	112–117	0·25–0·5	TEL	100[a]	52, 55
KAlEt$_2$Cl$_2$ + EtCl	Pt	70–90	—	TEL	98[a]	147

Reactants	Electrode	Temperature	Current	Product	Yield (%)	Reference
KAlEt4 + KF.AlEt3	Hg	85	0·185	TEL	98·5[b]	60, 70
NaAlEt4 + KAlEt4 (1:1)	Cu, Hg	70, 100	—	TEL	99[b]	66
MAlEt4 + MAlMe4 (M=25% K, 75% Na)	Cu, Steel	100	0·5	TEL	82[b]	57, 58
KAlEt4 + NaAlEt3OBu	Cu	70	0·04	TEL	93[a]	66
CsAlEt4 + NaAlMe4 (1:2)	Cu	80	0·25	TEL	—	57
NaAlEt4 + AlEt2OBu + NaAlEt3OBu	—	100	—	TEL	71	71
NaAlEt4 + NaFAlEt3 + NaF2AlEt3	Iron/Cu	70	0·08	TEL	96[b]	46, 61
KAlEt4 + KF.AlEt3 + KF.2AlEt3				TEL	—	60
KAlEt4 + NaAlEt4 + NaAlEt3OEt	Cu	130–140	0·1–0·3	TEL	—	66
KAlEt4 + NaAlMe4 + LiAlEt4 (1:1:1)	Cu	100	0·25	TEL	—	57
NaAlMe4 + NaAlMe4 + CaAlEt4	Steel	100	—	TEL	—	58
NaF.2Al(iso-C3H7) in Dioxane	Cu	50	0·022	Pb(iso-C3H7)4	—	41
NaAl(C3H7)4 + NaAlEt3OC6H5	Cu	100	0·04	Pb(C3H7)4	80[a]	47
NaAl(C3H7)4 + NaAl(C3H7)3OC6H11 (1:1)	Cu	—	—	Pb(C3H7)4	90[b]	51, 66
KAl(iso-C3H7)4 + KAlMe4 (8:1)	Cu, Steel	180	0·25	Pb(iso-C3H7)4	—	57, 58
NaAl(C3H7)4 + KAl(C3H7)4 (1:1)	Cu	—	—	Pb(C3H7)4	95[a]	66
NaAlEt4 + NaAl(C3H7)4	Steel	—	—	Mixed ethyl propyl lead	—	75
KAlBu3OBu	Cu	90	0·002	PbBu4	—	64
NaAlBu4 + NaAlEt4	Cu	—	—	PbBu4	—	47
NaAlBu4 + NaAlBu3OC10H21	Cu	70	0·04	PbBu4	80[b]	66, 51
NaAlBu4 + KAlBu4 (1:1)	Cu	—	—	PbBu4	—	66
NaAl(iso-Bu)4 + KAl(iso-Bu)4	Steel	80	—	Pb(iso-Bu)4	—	58
NaF 2Al(C6H5)3 in benzene	Cu	220	—	Pb(C6H5)4	—	41
NaAl(C6H5)4 + NaAlMe4	Cu	—	—	Pb(C6H5)4	—	57
NaAl(cyclo-C6H11)4 + KAl(cyclo-C6H11)4	Steel	100	0·25	Pb (cyclo-C6H11)4	—	58
NaAl(C6H5CH2)4 + KAl(C6H5CH2)4 + 10% aromatic	Steel	—	—	Pb(C6H5CH2)4	—	58
NaAl(β-CH3CH2C6H4) + KAl(β-CH3CH2C6H4)	Steel	—	—	Pb(CH3CH2C6H5)4	—	58
NaAl(C10H21)4 + LiAlMe4 (10:3)	Cu	40	—	Pb(C10H21)4	—	57
NaAlEt4 + NaBPh4	Steel	110	—	Mixed ethyl phenyl lead	—	58

[a] Yield based on lead consumed.
[b] Current yield.

Kobetz and Pinkerton[75,76] have reported the production of TEL from a large number of mixtures containing organoboron complexes and organoaluminium complexes both with and without solvents. For example, a mixture containing 10% NaBEt$_4$ and 45% each of NaAlMe$_4$ and NaAlEt$_4$ by weight gave upon electrolysis almost pure TEL with a current efficiency of 90%. Only triethyl boron is formed as a by-product of the cell therefore either NaBEt$_4$ is selectively electrolysed or, more probably, an exchange reaction occurs between triethyl aluminium and NaBEt$_4$ to form BEt$_3$:

$$NaBEt_4 + AlEt_3 \rightarrow Et_3B + NaAlEt_4$$

Triethyl boron has a much lower boiling point than TEL so it was easily removed from the cell by operating at reduced pressure. The electrolyte was regenerated by reaction of the triethyl boron with recovered sodium and hydrogen, followed by addition of ethylene:

$$Et_3B + Na + \tfrac{1}{2}H_2 \rightarrow NaBEt_3H$$

$$NaBEt_3H + C_2H_4 \rightarrow NaBEt_4$$

Tetramethyl lead has been synthesized by electrolysis of methyl boron complexes in either water[74,77] or THF solutions.[74,78] Yields of up to 94% were obtained.

Several other organolead compounds have been prepared by the electrolysis of organoboron compounds; information on these and the various systems used for the synthesis of TEL and TML are collated in Table VI.

Electrolysis of Miscellaneous Compounds

Hein has synthesized TEL by electrolysis of a mixture of diethyl zinc and ethyl sodium at a lead anode.[80,81] Diethyl zinc alone is a very poor electrical conductor but with ethyl sodium it forms the complex sodium triethyl zinc which is an excellent conductor. Zinc metal was deposited at the cathode. Giraitis[69] has reported a synthesis of TEL using diethyl zinc in diethyl ether solution, the electrolyte being continuously regenerated *in situ* by reaction of the zinc deposited at the cathode with hydrogen and ethylene. Tetracyclohexyl lead was prepared in a similar process.

A novel method for the electrolyte synthesis of organolead compounds has recently been reported by Mengoli and Daolio.[83-85] Alkyl bromides are electrolysed in DMF or dimethyl sulphoxide solution in an undivided cell between a zinc cathode and a lead anode. A small amount (0·06–0·25 M) of an alkyl iodide is added to the electrolyte and this induces an autocatalytic reduction of the alkyl bromide by the zinc cathode producing alkyl zinc intermediates which are subsequently oxidized at the lead anode to give high yields of organolead compounds. When the iodide is excluded the yield is poor. Tetramethyl lead, tetrapropyl lead and tetrabutyl lead were produced with yields of 63, 69·4 and 75·4% respectively based on lead consumed. A cadmium cathode has also been used successfully.

Electrolytic Reactions of Organolead Compounds

At the cathode

A small number of cathodic reactions of organolead compounds have been studied. Tetraethyl lead has been reported to undergo cathodic reduction at −0·65 to −0·7 V but no reaction products were reported.[131] Similarly, hexaethyl di-lead was reported to undergo reduction at −1·8 to −2·0 V in ethanol solution.[131]

Electrolysis of triethyl lead hydroxide at a lead cathode in alkaline solution has been found to give hexaethyl di-lead.[28-30,132,133] A recent patent reports that electrolysis of triorganolead salts in aqueous solution at a lead or carbon cathode gives tetraorganolead compounds.[134] This process has an application in the treatment of organolead containing effluent and is discussed further in a later section.

Several lead acetate compounds have been studied. Diethyl lead acetate in aqueous perchlorate solution gave tetraethyl lead on reduction at a mercury cathode.[137] The experimental results showed that the aquodiethyl lead (IV) cation underwent a two-electron reduction forming a diethyl lead(II) diradical, the organolead analogue of carbene, as a transitory intermediate species. The diradical then reacted either by disproportionation or transmetallation:

$$Et_2Pb^{2+} + 2e \rightarrow Et_2Pb:$$
$$2\,Et_2Pb: \rightarrow Et_4Pb + Pb^0$$
$$Et_2Pb: + Hg \rightarrow Et_2Hg + Pb^0$$

Table VI. Synthesis of organolead compounds by electrolysis of organoboron compounds.

Compound	Solvent	Cathode	Temperature °C	Current density (A cm^{-2})	Product	Yield %	Reference No.
NaBMe$_4$	H$_2$O	Hg	20	0.035–0.1	TML	78^b	74, 77
NaBMe$_4$	Ether I	Cu	—	—	TML	High	76
NaBMe$_4$	THF	Hg	90	—	TML	94^b	74, 78
NaBMe$_3$OMe	THF	Hg	—	—	TML	—	78
KBMe$_4$	THF	Hg	20	0.035	TML	78^b	74, 77
NaBEt$_4$	H$_2$O	Cu	20	0.03–0.1	TEL	91^b	74, 77
NaBEt$_4$	Ether I	Cu	20	—	TEL	100^b	76
NaBEt$_4$	Ether I + H$_2$O	Cu	—	—	TEL	100^b	76
NaBEtF$_3$	THF	Cu	50	—	TEL	High	76
NaBEtBu$_3$	THF	Cu	60	—	TEL	High	76
NaBEtPh$_3$	Ether II	Cu	100	—	TEL	High	76
NaBEt(C$_{18}$H$_{37}$)$_3$	Ether II	Cu	110	—	TEL	High	76
LiBEt$_4$	Amyl ether	Cu	—	—	TEL	High	76
MgBEt$_4$	Ether I	Cu	—	—	TEL	High	76
NaNH$_2$BEt$_3$	Pyridine	Cu	70	0.01	TEL	High	76
Et$_4$N.BEt$_4$	Triethylamine	Cu	—	—	TEL	100^b	76
BEt$_3$ + KCN		Cu	65	0.005	TEL	—	76
BEt$_3$ + 1-hexynylsodium (3:1)	Toluene	Cu	20	0.001	TEL	High	76
NaBEt$_4$ + NaAlMe$_2$Et$_2$(1:4)		Cu	110	—	TEL	High	75
NaBEt$_4$ + NaAlEt$_3$PBu	THF	Hg	90	—	TEL	93^b	77
NaBEt$_4$ + 1% KI	Ether III	Cu	30	—	TEL	High	76
NaBEt$_4$ + NaF.2AlEt$_3$(1:2)		Cu	—	—	TEL	High	75
NaBEt$_4$ + BEt$_3$ (1:1)		Cu	—	—	TEL	High	76
NaBEt$_4$ + BBu$_3$ (2:1)	THF	Cu	—	—	TEL	High	76
NaBEt$_4$ + KBEt$_4$	Ether II	Cu	100	—	TEL	High	76

KBEt$_4$ + KAlEt$_3$OEt(1:2)		Cu	110	—	TEL	High	75
NaBEt$_4$ + NaAlEt$_4$ + NaAlMe$_4$		Cu	100	0.3	TEL	90[b]	75
NaBEt$_4$ + NaAlEt$_4$ + NaAlEt$_3$OEt (2:7:1)		Cu	—	—	TEL	—	75
NaBPr$_4$ + NaBPr$_3$OMe	THF	Hg	90	—	PbPr$_4$	—	77
NaB(C$_6$H$_5$)$_4$	Dioxane	Cu	70	—	Pb(C$_6$H$_5$)$_4$	High	76
NaB(C$_6$H$_5$)$_4$ + NaAl(C$_6$H$_5$)$_4$ + KAl(C$_6$H$_5$)$_4$ (1:1:1)	—	Cu	160–180	—	Pb(C$_6$H$_5$)$_4$	—	75
KB(cyclo-C$_6$H$_{11}$)$_4$	Benzene	Cu	75	—	Pb(cyclo-C$_6$H$_{11}$)$_4$	High	76
KB(cyclo-C$_6$H$_{11}$)$_4$ + B(cyclo-C$_6$H$_{11}$)$_3$ (2:1)	Benzene	Cu	—	—	Pb(cyclo-C$_6$H$_{11}$)$_4$	High	76
BEt$_3$ + Sodium methyl-cyclopentadienyl	Ether I	Cu	20	—	Dimethyl cyclopentadienyl lead	High	76
NaB(C$_6$H$_5$CH$_2$)$_4$ + NaAl(C$_6$H$_5$CH$_2$)$_4$ + KAl(C$_6$H$_5$CH$_2$)$_4$(1:4:4)	—	Cu	—	0.004	Pb(C$_6$H$_5$CH$_2$)$_4$	—	75
NaB(C$_8$H$_{15}$)$_4$	Ether I	Cu	—	—	Pb(C$_8$H$_{15}$)$_4$	High	76
NaB(C$_8$H$_{17}$)$_4$	Ether I	Cu	20	—	Pb(C$_8$H$_{17}$)$_4$	High	78

Ether I = Dimethyl ether of diethylene glycol.
Ether II = Diethyl ether of diethylene glycol.
Ether III = Hexyl ethyl ether of diethylene glycol.
[a] Yield based on lead consumed.
[b] Current yield.

The disproportionation reaction predominated but it was found that transmetallation was favoured by increasing concentrations of depolarizer. Triphenyl lead acetate in dimethoxyethane solution with a current carrier was found to give either diphenyl mercury or the triphenyl lead anion, depending on the potential at the mercury cathode.[138]

$$Ph_3PbOAc \xrightarrow[-1\cdot4\ V]{e} Ph_3\dot{P}b \xrightarrow{Hg} Ph_2Hg$$

$$\xrightarrow[-2\cdot2\ V]{2e} Ph_3Pb:^-$$

Diphenyl lead acetate gave only diphenyl mercury,[138] being reduced in two one-electron steps which probably correspond to:

$$Ph_2Pb(OAc)_2 \xrightarrow[-1\cdot1\ V]{e} Ph_2\dot{P}bOAc + OAc^- \xrightarrow[-1\cdot6\ V]{e} Ph_2Pb:+OAc^-$$

Interestingly, in contrast to the diethyl lead (II) diradical, trans-metallation predominates over disproportionation for the diphenyl lead (II) diradical. The triphenyl lead anion has been produced by reduction of a variety of other bimetallic compounds.[139]

The available data on the electroreduction of organolead compounds is summarized in Table VII.

At the anode

Very little work has been done on the anodic electrochemistry of organolead compounds.

The reactions of lead acetate at the anode have been reported. The products of the reaction depend on the conditions used. In glacial acetic solution with potassium acetate, the product was lead tetracetate in 98% yield based on the current used.[135] In THF solution with a depolarizer and at a tin oxide anode, the product was metallic lead in 82% yield based on lead acetate consumed.[136] In dilute solution of hydroquinone or concentrated solution of monoglycol ether, both in THF, an amorphous Pb^{4+} deposit was formed on the anode.[151]

Industrial Processes

The Nalco process

The only commercially significant electrochemical process for the production of organolead compounds is that developed by the Nalco

Table VII. Electroreduction of organolead compounds.

Compound	Support electrolyte	Solvent	Cathode	Cathode potential (V)	Current density (A cm^{-2})	Product	Yield (%)	Reference No.
Me$_3$PbCl		H$_2$O	Pb or C	—	0·03–0·78	TML	51–84[b]	134
TEL				−0·65 to −0·7				131
Et$_3$PbPbEt$_3$		EtOH	Pb	−1·8 to −2·0				131
Et$_3$PbOH		H$_2$O/EtOH	Pb		0·01	Et$_3$Pb PbEt$_3$	—	132
Et$_3$PbCl		Caustic			—	Et$_3$Pb PbEt$_3$	80–90[a]	133
Et$_3$PbCl		H$_2$O	Pb or C		0·03–0·78	TEL	—	134
Et$_2$Pb(OAC)$_2$	HClO$_4$	H$_2$O	Hg	−0·07		TEL + HgEt$_2$	60–92[a] 8–40[a]	137
Ph$_2$PbOAC$_2$	Bu$_4$NClO$_4$	Dimethoxyethane	Hg	−1·6		HgPh$_2$		138
Ph$_3$PbOAC	Bu$_4$NClO$_4$	Dimethoxyethane	Hg	−1·4		HgPh$_2$		138
Ph$_3$PbOAC	Bu$_4$NClO$_4$	Dimethoxyethane	Hg	−2·2		Ph$_3$Pb$^-$	138	
Ph$_3$PbPbPh$_3$	Bu$_4$NClO$_4$	Dimethoxyethane	Hg	−2·0		Ph$_3$Pb$^-$		138, 139
CpFe(CO)$_2$PbPh$_3$	Bu$_4$NClO$_4$	Dimethoxyethane	Hg	−2·1		Ph$_3$Pb$^-$ + CpFe(CO)$_2^-$		139
(CO)$_5$MnPbPh$_3$	Bu$_4$NClO$_4$	Dimethoxyethane	Hg	−2·1		Ph$_3$Pb$^-$ + (CO) Mn^{-5}		139
(CO)$_5$RePbPh$_3$	Bu$_4$NClO$_4$	Dimethoxyethane	Hg	−2·4		Unknown		139
(CO)$_5$MnPbEt$_3$	Bu$_4$NClO$_4$	Dimethoxyethane	Hg	−1·8	—	Et$_3$Pb$^-$ + (CO)$_5$Mn$^-$		139
CpMo(CO)$_3$PbPh$_3$	Bu$_4$NClO$_4$	Dimethoxyethane	Hg	−2·2	—	Ph$_3$Pb$^-$ CpMo(CO)$_3^-$		139
Lead tetra-n-PrClO$_4$ phenyl-porphyrin		THF/DMSO	Hg or Pt	−1·30		Lead tetra phenyl-porphyrin anion		140

[a] Yield based on lead consumed.
[b] Current yield.

Chemical Company in collaboration with Standard Oil of Indiana for the production of TML and TEL using Grignard compounds as the alkylating agent. The process is described in a series of patents by Braithwaite and co-workers assigned to Nalco[92-108] and by Linsk and co-workers assigned to Standard Oil.[109-116] A small number of patents describing the production of vinyl lead compounds by a similar process have been taken out by Ethyl Corporation.[117-119]

The overall reaction on electrolysis of a Grignard reagent at a lead anode is:

$$Pb + 4\,RMgCl \rightarrow PbR_4 + 2\,Mg + 2\,MgCl_2$$

The poor electrical conductivity of Grignard reagents (in ether solution $\sim 10^{-4}\,ohm^{-1}\,cm^{-1}$) and the problems caused by the deposition of magnesium at the cathode eventually short-circuiting the cell were the major difficulties to be overcome before the process could become commercially viable. The addition of excess alkyl halide to the electrolyte was found to increase conductivity and remove the magnesium by reacting with it to regenerate the Grignard:

$$Mg + RCl \rightarrow RMgCl$$

The overall process then became:

$$2\,RMgCl + 2\,RCl + Pb \rightarrow PbR_4 + 2\,MgCl_2$$

The choice of the most suitable solvent for the Grignard reagent has been the subject of many of the Braithwaite patents. The solvent must, as far as possible, combine good solubility of the Grignard with reasonable electrical conductivity and be readily separable from the products. The preferred solvents are high boiling ethers, such as the benzyl ethyl ether of triethylene glycol, mixed with aromatic hydrocarbons, such as benzene or toluene, and tetrahydrofuran.[101]

The electrolysis cells are of complicated design.[126-129] Briefly, the cells are vertical cylinders using lead shot as the sacrificial anode and the steel walls of the cell as the cathode. The two electrodes are separated by a screen made of polyethylene, polypropylene, Teflon or a ceramic material. The lead consumed during the electrolysis is replaced from a hopper without necessitating the opening of the cells. The cells are operated at a temperature of about 35°C, the excess heat which results from the high resistance of the cell being removed by a refrigeration unit which uses the alkyl halide to make the Grignard as the coolant. An improved version of the cell has recently been patented.[130]

Yields based on the consumption of lead are claimed to be as high as 100%.[94] The factor that gives this process a crucial advantage over other electrochemical processes is the exceptionally high current yield, up to 173% for TEL.[106] Clearly either there is a parallel non-electrochemical process occurring in the cell[121] or diethyl lead is formed electrochemically and then reacts with the Grignard chemically to give TEL.[116] Similarly high yields are obtained for TML production.[106]

After electrolysis the cell solution is passed to a stripper where the alkyl halide is separated off and recycled. The remaining mixture of alkyl lead, solvent and magnesium chloride is separated by a combination of distillation and solvent extraction. The solvent is purified and recycled and the magnesium chloride is electrolysed as a fused melt to regenerate the magnesium.

The process is particularly flexible and the same plant can be used to produce TEL (feeding ethyl chloride), TML (feeding methyl chloride) and any desired mixture of methyl ethyl lead (feeding the required amount of ethyl chloride and methyl chloride).

Few precise details of the Nalco plant are known.[122-125] Construction of the first plant at Freeport, Texas, began in 1961 and it was put on stream by the end of 1962. The problems associated with working with large quantities of ethers were dramatically illustrated in 1963 when an explosion and fire destroyed the plant. The plant was rebuilt in 1964 with redesigned solvent and purification systems and was increased to $29 \cdot 5$ m kg year^{-1} capacity in 1966. Since then it has apparently been operating with no further major problems producing mainly TML which is chemically combined with TEL to produce methyl ethyl leads.

Effluent treatment

Recently the use of electrochemical methods for the treatment of organolead-containing effluents has been investigated. The effluents from industries manufacturing organolead anti-knock compounds contain soluble trialkyl and dialkyl lead compounds which present a serious disposal and control problem.

In the chemical process, tetra-alkyl leads are produced by the reaction of sodium–lead alloy with alkyl chloride in an autoclave. The autoclave product is a complex mixture from which the alkyl lead is separated by steam distillation. Considerable quantities of water used in this and other processes are fed to settling lagoons from which the overflow generally contains 5–200 p.p.m. lead. Electrolytic treatment of this effluent usually results in the production of water-soluble dimers and trimers of the organolead compounds, but under certain conditions water-insoluble products can be formed. Milam[141] found that alkyl lead halides could be converted to insoluble organoleads by electrolysis between platinum electrodes using alternating current. Typically, a feed containing 5–20 p.p.m. of soluble lead yielded after electrolysis an effluent containing 1–5 p.p.m. of soluble lead, the insoluble leads compounds being removed by filtration, sedimentation or settling. Milam and Estep[142] reported that electrolysis using direct current at a porous carbon or graphite cathode was effective in reducing soluble lead concentrations. Typically lead concentrations of 14 p.p.m. could be reduced to less than 0·35 p.p.m. The porous cathode eventually became blocked by particulate lead but could be regenerated by circulation of an acid. Carlin[143] found that electrolysis of soluble organolead at a metal cathode having a hydrogen overvoltage in excess of 1·6 V resulted in the formation of metallic lead. Various metal cathodes were suitable but as lead was being produced, lead shot was preferred. A reduction in soluble lead levels similar to those reported for the previous processes were obtained.

Aqueous effluents containing very high levels of triorganolead salts, up to 20–30% by weight, are produced during the process whereby TML and TEL are chemically mixed to give methyl ethyl lead compounds. Such high levels of triorganoleads can be treated by electrolytically converting them to tetra-alkyl compounds.[134] For example, electrolysis of trimethyl lead chloride between a lead or carbon cathode and a carbon anode with voltages of 1–30 V and current densities of 0·03–0·77 A cm^{-2} gave TML in 99% yield at current efficiencies up to 84%. It was found that the best results were obtained when the solution was maintained in the pH range 6–8 during the electrolysis. The insoluble tetra-alkyls can then be removed by settling or filtration.

In all these processes it is probable that the primary process is the formation of an organolead radical at the cathode:

$$R_3Pb^- \rightarrow R_3Pb + e$$

The behaviour of this radical then depends on the particular conditions used. It may dimerize, disproportionate or decompose to give hydrocarbons, tetra-alkyl leads or metallic lead.

Conclusion

The driving force behind the research effort expanded on the electrochemistry of organolead compounds has been their commercial significance. The research reached its peak in the late fifties and early sixties when the demand for lead alkyls as anti-knock compounds was increasing. That demand has now stabilized and it is very unlikely that a new electrochemical process could compete with the already existing efficient, paid-out plants, and this is reflected in the current low level of research.

There remains much work to be done on the electrochemical reactions of organolead compounds and research in this area may well be stimulated by the increasing emphasis on effective effluent treatment.

Acknowledgment

The author would like to thank Dr J. R. Grove and colleagues in the Research and Development Department of Associated Octel for their help in the preparation of this work and Associated Octel for permission to prepare and publish it.

References

1. J. Tafel, *Ber.* **39**, 3626 (1906).
2. J. Tafel, *Ber.* **44**, 323 (1911).
3. T. Sekine, A. Yamura and K. Sugino, *J. Electrochem. Soc.* **112**, 439 (1965).
4. A. P. Tomilov and B. L. Klyuev, *Electrokhimiya* **2**, 1405 (1966).
5. A. P. Tomilov and B. L. Klyuev, *Electrokhimiya* **3**, 1168 (1967).
6. G. Renger, *Ber.* **44**, 337 (1911); **45**, 3321 (1912).
7. J. Tafel, *Ber.* **42**, 3146 (1909).
8. H. D. Law, *J. Chem. Soc.* **101**, 1016, 1544 (1912); *Proc. Roy. Soc.* **28**, 98, 162 (1912).
9. A. I. Lebedeva, *Zh. Obshch. Khim.* **18**, 1161 (1948).
10. W. Schepps, *Ber.* **46**, 2564 (1913).
11. F. Fichter and I. Stein, *Helv. Chim. Acta* **14**, 1205 (1931).
12. A. Goldach, *Helv. Chim. Acta* **14**, 1436 (1931).
13. G. H. F. Calingaert (to General Motors), U.S. Patent 153,297 (1923).
14. B. Mead (to General Motors), U.S. Patent 1,567,159 (1926).
15. E. F. Silversmith and W. J. Sloan (to DuPont), British Patent 949,925 (1964).

16. DuPont, Dutch Patent Appl. 6,508,049 (1965).
17. E. F. Silversmith and W. J. Sloan (to DuPont), German Patent 1,246,734 (1967).
18. K. C. Smeltz (to DuPont), U.S. Patent 3,392,093 (1968).
19. R. Galli, *Chem. Ind. (Milan)* **50**, 977 (1968).
20. R. Galli, *J. Electroanal. Chem.* **22**, 75 (1969).
21. R. Galli and F. Olivani, *J. Electroanal. Chem. Interfacial Electrochem.* **25**, 331 (1970).
22. K. Yang, J. D. Reedy, M. A. Johnson and W. H. Harwood (to Continental Oil Co.), German Patent 1,955,201 (1970).
23. K. Yang, J. O. Reedy, M. A. Johnson and W. H. Harwood (to Continental Oil Co), British Patent 1,285,209 (1972).
24. H. E. Ulery, *J. Electrochem. Soc.* **116**, 1201 (1969).
25. A. P. Tomilov, Y. D. Smirnov and S. L. Vershavskii, *Zh. Obshch. Khim.* **35**, 391 (1965).
26. R. E. Plump and L. B. Hammett, *Trans. Electrochem. Soc.* **73**, 523 (1938).
27. W. J. Setterini and R. A. Wessling, "Electroduction of Sulfonium Compounds", Paper delivered at the 8th Annual E. C. Britton Symposium on Industrial Chemistry, Midland, Michigan, 30 April, 1970.
28. G. A. Razuvaev, N. S. Vyazkin and N. N. Vyshinskii, *Zh. Obshch. Khim.* **30**, 967 (1960).
29. U. Belluco, G. Tagliavini and R. Barbieri, *Ric. Sci.* **30**, 1675 (1970).
30. I. A. Korshunov and N. I. Malyugina, *Zh. Obshch. Khim.* **31**, 1062 (1961).
31. A. P. Giraitis (to Ethyl Corp.), German Patent 1,046,617 (1958).
32. C. Randaccio, Italian Patent 548,183 (1956).
33. K. Ziegler, H. G. Gellert, K. Kosel, W. Lemkuhl and W. Pfohl, *Angew. Chem.* **67**, 424 (1955).
34. H. E. Redman (to Ethyl Corp.), U.S. Patent 2,787,626 (1957).
35. K. Ziegler and H. Lemkuhl, *Angew. Chem.* **67**, 424 (1955).
36. K. Ziegler and H. G. Gellert, *Z. Anorg. Allg. Chem.* **283**, 414 (1956).
37. K. Ziegler, British Patent 814,609 (1959).
38. K. Ziegler, Belgian Patent 543,128 (1955).
39. K. Ziegler, Canadian Patent 582,016 (1959).
40. B. E. Roethli and I. B. Simpson (to Esso Res. and Eng. Co.), British Patent 797,093 (1958).
41. A. P. Giraitis (to Ethyl Corp.), U.S. Patent 2,944,948 (1960).
42. K. Ziegler, *Brennstoff—Chem.* **40**, 209 (1959).
43. K. Ziegler, *Angew. Chem.* **71**, 628 (1959).
44. K. Ziegler, Belgian Patent 575,595 (1959).
45. K. Ziegler, H. Lemkuhl and W. Grimme, German Patent 114,330 (1959).
46. K. Ziegler, British Patent 848,364 (1960).
47. K. Ziegler and H. Lemkuhl, German Patent 1,127,900 (1962).
48. K. Ziegler, Belgian Patent 575,641 (1959).
49. K. Ziegler, British Patent 864,394 (1961).

50. K. Ziegler and H. Lemkuhl, German Patent 1,153,754 (1963).
51. K. Ziegler, British Patent 864,393 (1961).
52. K. Ziegler, *Angew. Chem.* **72**, 565 (1960).
53. K. Ziegler and H. Lemkuhl, Belgian Patent 590,753 (1960).
54. K. Ziegler and H. Lemkuhl, German Patent 1,153,371 (1963).
55. K. Ziegler and H. Lemkuhl, *Chem. Ing. Tech.* **35**, 325 (1963).
56. W. Grimme, H. Lemkuhl, K. Ziegler, K. Kosel, H-D. Kobs and R. Schaeffer, *Bull. Soc. Chim. Fr.* (7), 1456 (1963).
57. P. Kobetz and R. C. Pinkerton (to Ethyl Corp.), U.S. Patent 3,028,322 (1962).
58. T. W. McKay (to Ethyl Corp.), U.S. Patent 3,088,885 (1963).
59. E. M. Marlett, P. Kobetz and R. C. Pinkerton, 142nd Meeting of Amer. Chem. Soc., Atlantic City, N.J. (1962). Abstract of Papers, p. 25N.
60. K. Ziegler and H. Lemkuhl, German Patent 1,666,196 (1964).
61. K. Ziegler, U.S. Patent 3,069,334 (1963).
62. K. Ziegler, Belgian Patent 617,628 (1962).
63. K. Ziegler and H. Lemkuhl, U.S. Patent 3,254,008 (1966).
64. A. P. Giraitis (to Ethyl Corp.), U.S. Patent 3,177,130 (1965).
65. H. Lemkuhl, R. Schaeffer and K. Ziegler, *Chem. Ing. Tech.* **36**, 612 (1964).
66. K. Ziegler and H. Lemkuhl, U.S. Patent 3,254,009 (1966).
67. Ethyl Corp., British Patent 842,090 (1960).
68. K. Ziegler and H. Lemkuhl, German Patent 1,181,220 (1964).
69. A. P. Giraitis (to Ethyl Corp.), German Patent 1,046,617 (1958).
70. K. Ziegler, U.S. Patent 3,164,538 (1965).
71. D. K. Wunderlich and L. N. Fussell (to Sinclair Research), U.S. Patent 3,159,557 (1964).
72. K. Ziegler, U.S. Patent 3,372,097 (1968).
73. K. Ziegler, German Patent 1,144,490 (1963).
74. K. Ziegler, *Amer. Chem.* **652**, 1 (1962).
75. P. Kobetz and R. C. Pinkerton (to Ethyl Corp.), U.S. Patent 3,028,323 (1962).
76. R. C. Pinkerton (to Ethyl Corp.), U.S. Patent 3,028,325 (1962).
77. K. Ziegler and H. Lemkuhl, German Patent 1,212,085 (1966).
78. K. Ziegler, German Patent 1,220,855 (1967).
79. K. Ziegler, German Patent 1,149,005 (1963).
80. J. B. Honeycutt and J. M. Riddle, *J. Amer. Chem. Soc.* **83**, 369 (1961).
81. F. Hein, *Z. Elektrochem.* **28**, 469 (1922).
82. F. Hein, W. K. Petzschner and F. A. Segitz, *Z. Anorg. Allgem., Chem.* **141**, 161 (1924).
83. G. Mengoli and S. Daolio, *Electrochim. Acta* **21**, 889 (1976).
84. G. Mengoli, *J. Electrochem. Soc.* **124**, 364 (1977).
85. G. Mengoli and S. Daolio, *J. Organometal. Soc.* **131**, 409 (1977).
86. P. Kobetz and W. H. Thomas (to Ethyl Corp.), U.S. Patent 3,344,048 (1967).
87. W. V. Evans and E. Field, *J. Amer. Chem. Soc.* **58**, 720, 2284 (1936).

88. W. V. Evans and D. Braithwaite, *J. Amer. Chem. Soc.* **61**, 898 (1939).
89. W. V. Evans and D. Braithwaite, *J. Amer. Chem. Soc.* **62**, 534 (1940).
90. R. Pearson and W. V. Evans, *Trans. Amer. Electrochem. Soc.* 297 (1942).
91. R. E. Dessy and G. S. Handler, *J. Amer. Chem. Soc.* **80**, 5824 (1958).
92. D. G. Braithwaite (to Nalco Chem. Co.), British Patent 839,172 (1960).
93. D. G. Braithwaite (to Nalco Chem. Co.), Belgian Patent 590,453 (1959).
94. D. G. Braithwaite (to Nalco Chem. Co.), U.S. Patent 3,007,857 (1961).
95. D. G. Braithwaite (to Nalco Chem. Co.), U.S. Patent 3,007,858 (1961).
96. D. G. Braithwaite and W. Hanzel (to Nalco Chem. Co.), Belgian Patent 611,212 (1962).
97. D. G. Braithwaite (to Nalco Chem. Co.), Belgian Patent 6,13,892 (1962).
98. D. G. Braithwaite (to Nalco Chem. Co.), German Patent 1,197,086 (1965).
99. D. G. Braithwaite (to Nalco Chem. Co.), German Patent 1,202,790 (1965).
100. D. G. Braithwaite (to Nalco Chem. Co.), German Patent 1,226,100 (1966).
101. D. G. Braithwaite (to Nalco Chem. Co.), German Patent 1,231,242 (1966).
102. D. G. Braithwaite (to Nalco Chem. Co.), U.S. Patent 3,234,112 (1966).
103. D. G. Braithwaite (to Nalco Chem. Co.), U.S. Patent 3,256,161 (1966).
104. D. G. Braithwaite (to Nalco Chem. Co.), U.S. Patent 3,312,605 (1967).
105. D. G. Braithwaite (to Nalco Chem. Co.), U.S. Patent 3,380,899 (1968).
106. D. G. Braithwaite (to Nalco Chem. Co.), U.S. Patent 3,380,900 (1968).
107. D. G. Braithwaite (to Nalco Chem. Co.), U.S. Patent 3,391,066 (1968).
108. D. G. Braithwaite (to Nalco Chem. Co.), U.S. Patent 3,391,067 (1968).
109. Standard Oil Co., Belgian Patent 6,12,795 (1962).
110. J. M. Coopersmith, J. Linsk, E. Field, R. W. Carl and E. A. Mayerle (to Standard Oil Co.), German Patent 1,157,616 (1963).
111. J. Linsk (to Standard Oil Co.), U.S. Patent 3,116,308 (1963).
112. J. Linsk (to Standard Oil Co.), U.S. Patent 3,118,825 (1964).
113. J. Linsk and E. A. Meyerle (to Standard Oil Co.), U.S. Patent 3,155,602 (1964).
114. J. Linsk, R. W. Carl and E. Field (to Standard Oil Co.), U.S. Patent 3,164,537 (1965).
115. F. G. Pearce, L. T. Wright, H. A. Birkness and J. Linsk (to Standard Oil Co.), U.S. Patent 3,180,810 (1965).
116. J. Linsk (to Standard Oil Co.), U.S. Patent 3,298,939 (1967).
117. Ethyl Corp., Dutch Patent Appl. 6,507,727 (1964).
118. G. C. Robinson (to Ethyl Corp.), U.S. Patent 3,431,185 (1969).
119. Ethyl Corp., U.S. Patent 3,522,156 (1970).
120. H. French and M. Drane, *J. Amer. Chem. Soc.* **52**, 4904 (1930).
121. G. Calingaert and H. Shapiro (to Ethyl Corp.), U.S. Patent 2,535,193 (1950).
122. Anon., *Chem. Eng. News* **42**, 52 (1964).
123. L. L. Bott, *Hydrocarbon Proc. Petrol. Refin.* **44**, 115 (1965).

124. E. Guccione, *Chem. Eng.* **72**, 102, 249 (1965).
125. Anon., *Oil Gas J.* Feb., p. 82 (1968).
126. D. G. Braithwaite, J. S. D'Amico, P. L. Gross and W. Hanzel (to Nalco Chem. Co.), U.S. Patent 3,141,841 (1964).
127. Nalco Chem. Co., Belgian Patent 671,840 (1966).
128. Nalco Chem. Co., British Patent 1,071,322 (1967).
129. D. G. Braithwaite (to Nalco Chem. Co.), U.S. Patent 3,287, 249 (1966).
130. G. E. Blackmar (to Nalco Chem. Co.), U.S. Patent 3,573,178 (1971).
131. L. N. Vertyulina and I. A. Korshunov, *Khim. Nauk. Prom.* **4**, 136 (1959).
132. T. Midgeley, C. A. Hochwalt and G. Calingaert, *J. Amer. Chem. Soc.* **45**, 1821 (1923).
133. F. Hein and A. Klein, *Ber.* **71**, 2381 (1938).
134. E. A. Mayerle and J. R. Minderhout (to Nalco Chem. Co.), U.S. Patent 3,696,009 (1972).
135. M. Y. Fioshin and V. A. Guskov, *Dokl. Akad. Nauk., SSSR* **112**, 303 (1957).
136. C. W. Lewis and P. C. Edge, *Ind. Eng. Chem. Prod. Res. Dev.* **8**, 399 (1969).
137. M. D. Morris, *J. Electroanal. Chem. Interfacial Electrochem.* **20**, 263 (1969).
138. R. E. Dessy, W. Kitching and T. Chivers, *J. Amer. Chem. Soc.* **88**, 453 (1966).
139. R. E. Dessy, P. M. Weissman and R. L. Pohl, *J. Amer. Chem. Soc.* **88**, 5117 (1966).
140. R. H. Felton and H. Linschitz, *J. Amer. Chem. Soc.* **88**, 1113 (1966).
141. J. E. Milam (to P.P.G. Industries Inc.), U.S. Patent 3,799,851 (1974).
142. J. E. Milam and E. E. Estep (to P.P.G. Industries Inc.), U.S. Patent 3,799,852 (1974).
143. W. W. Carlin (to P.P.G. Industries Inc.), U.S. Patent 3,799,853 (1974).
144. A. P. Tomilov, Yu. D. Smirnov and S. L. Varshavskii, *Zh. Obshch. Khim.* **39**, 2174 (1969).
145. D. Britz and D. Knittel, *Elektrochim. Acta* **20**, 891 (1975).
146. K. Yang, J. D. Reedy and W. H. Harwood (to Continental Oil Co.), U.S. Patent 3,622,476 (1971).
147. W. H. Harwood (to Continental Oil Co.), U.S. Patent 3,655,536 (1972).
148. M. Fleischmann, D. Pletcher and C. J. Vance (to Associated Octel Co. Ltd.), British Patent 1,290,211 (1970).
149. M. Fleischmann, D. Pletcher and C. J. Vance, *J. Electroanal. Chem. Interfacial Electrochem.* **29**, 325 (1971).
150. M. Fleischmann and D. Pletcher, *J. Organometal. Chem.* **40**, 1 (1972).
151. C. W. Lewis and P. C. Edge, *Ind. Eng. Chem. Prod. Res. Dev.* **8**, 399 (1969).
152. N. G. Bachchitsaraitsyan, *Novosti. Electrochim. Org. Soedin., Tezisy Dokl. Vses. Soveshch. Electrochim. Org. Soedin* 8th (1973).
153. M. S. Sataev, *ibid.* (1973).

8 Hydrogen Evolution and Oxygen Reduction on Lead

M. HAYES* and A. T. KUHN‡

* Ecological Engineering Ltd., Macclesfield, Cheshire
‡ Department of Dental Materials, Eastman Institute of Dental Surgery, London

Hydrogen Evolution on Lead

Introduction

Studies of hydrogen evolution on lead have been reported since the end of the last century. However, it is only comparatively recently that the need for high purity has been recognized. Even so, there is still a wide discrepancy between the results of different workers, which may be caused by differences in purification and/or test procedures. Reproducibility from one experiment to the next is also difficult to attain. Problems arise because of the very low activity of lead for hydrogen evolution—the rate is easily affected by trace impurities which can either poison or promote the reaction. It is useful to consider the experimental problems encountered in the study of hydrogen evolution on lead. These can be split into specific areas, i.e. those associated with: (i) the electrolyte, (ii) the working electrode, (iii) the counter electrode, and (iv) the gases and gas lines.

(i) Electrolyte. Sulphuric and perchloric acids are the most commonly used although Smith[7] has reported that use of the former causes or gives rise to too many problems with reproducibility. Sulphamic acid[1,2,3] and potassium hydroxide solutions[4] have also been reported. Purification of the working strength electrolyte is by pre-electrolysis plus, in some cases,[5,22] circulation through activated charcoal. It is important to avoid dissolution of the pre-electrolysis electrodes, which is possible even with platinum electrodes[6] and care should be taken to use low anodic potentials. This is achieved by either using low current densities or simply allowing the electrode to stand on open circuit with hydrogen bubbling over it.[7,8] Obviously, the activated charcoal must be thoroughly purified before use; the method has been described by Hampson[9] and Smith.[5]

(*ii*) *Working electrode*. Lead has one of the lowest activities for hydrogen evolution and so the presence of other metals is likely to cause an increase in activity.

The chemical and electrochemical cleaning methods normally employed work by selective dissolution and it is quite possible in some cases for the impurity metals to remain but for the lead to dissolve. Metzler and Schwartz[10] showed that when lead containing 18 p.p.m. bismuth was etched in dilute nitric acid, the bismuth dissolved and reprecipitated, forming large crystals of bismuth on the surface. It therefore seems most probable that the majority of the early work really used a lead alloy of unknown composition. Lead has been available for several years as 99·9999% pure, i.e. the total impurities are 1 p.p.m.

It has been suggested[5] that an oxide forms quickly on a freshly prepared lead surface and is stable to cathodization, so care is needed to exclude air from the surface as the electrode is transferred to the test cell.

The potential range available for study is somewhat limited: at the positive end the Pb/Pb^{2+} potential must not be reached if lead dissolution is to be avoided (and, in H_2SO_4, sulphate formation), while at the cathodic end an ill-defined process known as hydride formation and/or disintegration of the cathode occurs. There is some disagreement as to when this phenomenon is observed, as will be discussed later. The former fact means that the electrode must be cathodically polarized from the instant it enters the solution and must remain so for the duration of the experiment.

(*iii*) *Counter electrode*. Since the counter electrode will be an anode, it is important that it is both inert and pure since any dissolution products will plate out onto the cathode. A high surface area electrode in a hydrogen-saturated solution will favour the preferred reaction, the oxidation of hydrogen, and so ensure that the potential is kept low.[22] Platinum and carbon are the usual materials.

(*iv*) *Gas*. The electrolyte must be free from even traces of oxygen[7] and, therefore, so must the hydrogen gas used for electrolyte saturation. Gas purification trains have been widely reported;[7,11] the gas lines should be all glass or other non-permeable material and, according to Smith,[7] taps, even water-sealed, should be avoided in the line after the purification train. The observation of Bockris and Srinivasan[12] that hydrogen bubbling affected their potential measurements has been ascribed by Smith[26] to contamination of the gas by atmospheric oxygen diffusing through their gas lines. Smith[7] has found that air diffuses through PTFE tubing.

The reported work may be conveniently divided under several headings: (*i*) Tafel slope measurements, (*ii*) isotopic separation, (*iii*) cathodic disintegration, (*iv*) industrial applications, and (*v*) miscellaneous. These are discussed in the following sections.

Tafel slope measurements

Polycrystalline Electrodes

Numerous workers have reported Tafel slopes in a variety of electrolytes.[4,5,7,12–22] In the main, the results fall into two categories:
  (*i*) One Tafel region with a "high" slope.
  (*ii*) Two Tafel regions, one slope, at lower currents, being "high", and the other, at higher currents, being "normal" (where a "normal" Tafel slope is 120 mV decade^{-1}).
Few workers have studied the reproducibility of their results and, although all results were obtained galvanostatically, other experimental details, such as the time spent at each current density, are rarely quoted.

Smith and Ives[5] have stated that high slopes at low current densities are caused by an oxide film on the lead surface, although Bockris favours organic adsorption as an explanation.[12] Prolonged cathodization was usually effective in its removal. Despite this, single slopes of 120 mV have been reported. Ruetschi *et al.*,[18] apparently without deoxygenating the solution, obtained a slope of 120 mV, albeit over only one decade of current (and with only three experimental points!). It has been widely considered that at high overpotentials, traces of oxygen do not affect the behaviour of the electrode[12,76] and so little attention has been paid to removal of the last traces. This assumption has been criticized by Smith.[7]

Generally, there seem to be few reports in which reproducibility has been investigated.

There have been several reports[3,5,7,17,22] of hysteresis between the results obtained with decreasing current and those with increasing current. The term "normal" was used by Weedon[22] to describe the hysteresis observed by him and others[5,7,16] in which the potential is more negative (i.e. a greater overpotential) on the decreasing current run than on the preceding (increasing current) run. He also described a "reverse" hysteresis in which the opposite was true, i.e. the electrode was more active on the subsequent decreasing sweep. A similar effect was also reported by Bicelli and Romagnani[3] but only if the maximum current exceeded 10 mA cm^{-2}. Weedon[22] also observed that electrodes showing reverse hysteresis behaved differently from "normal" electrodes on first being immersed in the electrolyte. The normal electrodes quickly (0·5 min) took up a potential around -1 V whereas the reverse-type electrodes took up to 30 min. It seems likely that these reverse-type electrodes were in some way contaminated. (Bicelli and Romagnani[3] reported a high Tafel slope at low current densities.)

Normal hysteresis is associated with a slow process. Steady-state scans[16,23] show that the transition from one region to another occurs at certain current densities. However, if a rapid scan is made in either region, the region is extended. Kolotyrkin[16] attributed the effect to the slow adsorption/desorption of anions. It should be noted that this implies that anion adsorption increases electrode activity. It seems[16] that each ion has a critical value of potential, E_{crit}, associated with it and that the polarization measurements at potentials below E_{crit} were dependent on the anion. At potentials more negative than this the results were independent of anion and time. The order for the halides is $E_I < E_{Br} < E_{Cl}$. According to Frumkin[23] the hysteresis is caused by the potential passing through the p.z.c. of lead ($-0·65$ V). However, this is not a complete explanation, since on the decreasing current scans, the potential becomes time-dependent while still much more negative than $-0·65$ V. Change in the p.z.c. with anion adsorption is not the answer either because in the high current density region the electrode is expected to be almost free of adsorbed anions. A possible explanation is the lowering of the p.z.c. with hydrogen absorption as found by Gileadi et al.[73] for platinum electrodes.

The adsorption of SO_4^{2-} on lead has been studied by Hampson et al.[24] who showed that polarization at highly negative potentials reduced the amount of SO_4^{2-} adsorption and that its subsequent readsorption at lower potentials was slow. They reported the p.z.c. to be -0.59 ± 0.2 V. Kolotyrkin[16] also reported the anomalous behaviour of the potential on stepping the current from one value to another provided the currents were in the lower Tafel region. His electrolyte was sulphuric acid; later Smith,[7] Weedon[22] and Rao[77] also obtained similar effects in perchloric acid. When the current was increased, the potential quickly became more negative and then slowly became more positive, eventually approaching a value at an appropriate point on the lower Tafel line. A similar, complementary, effect was obtained when the current was decreased.[7,22]

Again, slow anion adsorption/desorption was proposed by the Russian workers.[23] However, in this case, adsorption is claimed to lower electrode activity. Smith and co-workers[5,7,25,77] also noticed that, on extended cathodization, in perchloric acid (but not in sulphuric acid)[7] the potential became steadily less negative. They found that the capacitance and potential of the cathode increased linearly with the logarithm of the charge passed (i.e. the electrode was becoming more active) and that the Tafel slope decreased with cathodization. Rao[77] also observed an increase in electrode capacitance with increase in charge over a wide range of constant overpotentials. At constant overpotential, he found that capacitance and (log) current varied linearly with log Q (charge passed) and these two plots had a similar slope, e.g. parallel. Smith and co-workers[5,26] suggested that hydrogen was being *absorbed* by the lead and that the rate of charge transfer was linearly dependent on the bulk hydrogen content of the electrode. Overshoot hysteresis was explained[5,77] by the assumption that transfer of hydrogen atoms across the interface is a slow process. An increase in current will increase the surface coverage (θ) of hydrogen atoms: this in turn will cause an inward migration of hydrogen atoms to counteract the increased concentration gradient and so re-establish a new steady-state value of θ. They suggested that θ varies widely according to the current density; at low current densities it is small and transfer of hydrogen atoms to the metal is negligible. Thus there is a threshold current density below which the effects of this transfer are not observed. At high current densities, they proposed, θ approaches unity and produces a condition favourable to "avalanche penetration" and disintegration of the cathode.

The formation of hydrides by lead has been recognized only in the last few years. Hitherto it had been discounted on the basis that hydrogen atoms are only weakly adsorbed on lead. Absorption of hydrogen from the gas phase by evaporated thin films of lead has been reported by Wells and Roberts.[27] At 0°C a composition of $PbH_{0.19}$ was reached. Smith and co-workers[28] have demonstrated the permeation of electrolytic hydrogen through lead foils. They found a diffusion coefficient of 10^{-7}–10^{-6} $cm^2 s^{-1}$ at 25°C. The bulk composition was calculated to be $PbH_{0.00002}$, although they later thought[29] that these could be too low by several orders of magnitude for reasons discussed in the section on isotopic separation.

Single Crystals

Hydrogen evolution on single crystals of lead having the (100), (110) and (111) orientations has been reported by Ruetschi and Cahan[30] in sulphuric acid and by Bicelli and co-workers[1–3,31] in perchloric acid and sulphamic acid. It seems that Ruetschi and Cahan[30] did not deoxygenate their solutions. They obtained a Tafel region with a slope of 120 mV decade^{-1} but only over less than a decade of current. There was very little difference between the three crystallographic orientations. The Bicelli work[1–3,31] produced two linear regions, the steeper one being at lower current densities with $b \simeq 200$ mV and the higher current density line having $b \simeq 120$ mV. The only exception was in sulphamic acid, where the (110) face gave only one slope. They found that the overpotentials on the three orientations differed by several centivolts in the first (high slope) Tafel region, the order being (110) > (111) > (100); in the second region the (111) and (100) crossed over. Their experimental technique is open to criticism in at least two respects. In all their work the electrode potentials were allowed to become more positive than −400 mV and, in some cases, −200 mV, thereby permitting dissolution of the electrode and introduction of lead ions into the electrolyte. Also, after being chemically polished in acetic acid/hydrogen peroxide, the electrodes were dried (in a hydrogen atmosphere) before being transferred to the cell. Smith[7] has suggested that drying of the electrode leads to a contaminated surface when the electrodes are immersed in the electrolyte.

Alloys

Ruetschi and Cahan[30] found that antimony concentrations of up to 6% caused an increase in exchange current density and in Tafel slope.

These results were obtained in 3·85 M H_2SO_4 over less than one decade of current and apparently without deoxygenation of the solution. Aguf and Dasoyan[32] have reported a study of alloys of silver and antimony with lead. Their electrolyte was 5·6 M H_2SO_4. Good reproducibility between experiments was claimed. Pure lead electrodes gave two slopes, both over approximately one decade of current, with the slope in the low current density region being 120 mV and that in the high current density region being 160 mV. Addition of 1–3% silver lowered the overpotential progressively; but from 3 to 10% silver, the overpotential was unchanged and only one Tafel region, with slope 120 mV, was obtained. The antimonial alloys have lower overpotentials than the silver alloys, but addition of up to 5% silver lowered the overpotential still further.

Croatto and Via[33] examined Pb–Cd and Pb–Sb alloys in 6 N KOH at 20°C. However, the electrode potential was always more positive than the Pb/Pb^{2+} potential and so the results are of dubious value. Croatto[34] has also reported a study of a variety of lead binary alloys, the alloying element (Ag, Ni, Cu, Zn, Sb, Cd, Bi, Sn) having a concentration of either 0·08% or 0·16%. They were tested in 7·5 M sulphuric acid at 120°C. The overpotentials were somewhat larger in this report, being between 400 and 800 mV. There was little difference in overpotential between any of the electrodes in the current density range 0·05–10·0 mA cm^{-2}.

Other early studies have been reported in a review by Brooman et al.[35] A recent patent[36] describes how cathodes for sea water electrolysis consisting of a palladium coating on a base metal such as titanium may be given a longer life by incorporating up to 35 wt % lead in the palladium coating. Loss of coating is caused by blisters forming at the interface between the coating and the base. These blisters are thought to be caused by hydrogen that has diffused through the coating. Alloying with lead is believed to reduce the amount of hydrogen diffusing into the electrode.

Isotopic separation

Hydrogen/deuterium separation has been reported by Horiuti and Okamoto[37] and Lewis and Ruetschi.[38] The former workers found a separation factor, S_D, of 3·0 in 0·5 M H_2SO_4 although the purity of the system, and therefore the validity of the results, has been questioned by Bockris and Srinivasan.[12] Lewis and Ruetschi[38] tried to measure S_D in 6 M KOH for several metals as a function of potential. However, they used constant current polarization and observed changes in potential during the experiments which lasted 3–12 h. There were also differences in potential at the same current density for different electrodes of the same metal. They also reported that the electrodes became rough and corrosion was observed.

Most of the electrodes showed two maxima in the S_D/overpotential relationship. For lead at 30°C, they were: (1) $S_D = 9$ at -0.8 V (RHE) and (2) $S_D = 7$ at -0.61 V. At 50°C the maxima were: $S_D = 7$ at -0.82 V and -0.61 V. The minimum at both temperatures was at approximately -0.67 V.

There have also been only two reports of hydrogen/tritium separation. Bockris and Srinivasan[12] measured the ratio of hydrogen/tritium evolved at lead cathodes in 0.5 M H_2SO_4 and 0.5 M NaOH. They obtained an H/T separation factor, S_T, of 6.7 ± 0.8 in acid and 7.2 ± 0.8 in alkali.* Muju and Smith[29] reported their investigations of the permeation rate of hydrogen and tritium through lead foils. One side of the foil was cathodized at 53 mA cm^{-2} in 1 M perchloric acid. The other (diffusion) side of the foil was held at a constant potential of $+140$ mV vs a Pd/H reference electrode so that the hydrogen atoms were oxidized as soon as they reached this surface of the electrode. The electrolyte was 0.2 M KOH, and the current on the diffusion side should have been indicative of the total number of atoms reaching the surface. The amount of tritium was measured by a scintillation technique. They showed that permeation is controlled by diffusion through the lattice for both isotopes. The value of S_T was found to be 0.3 ± 0.15.† It was suggested[77] that this unusual figure is another example of the problems associated with working with lead—the results are very sensitive to impurities. If oxygen had not been completely removed, it was argued, then it is possible for both types of atom to interact with oxygen and enter the solution without undergoing the usual electrochemical oxidation step:

$$PbH_{ads} + OH^-_{aq} = Pb + H_2O + e$$

This would result in a lowering of the permeation current but not the radiochemical count, the net effect being to "lose" H atoms and so

* Rao[77] found a value of 7.2 ± 0.7 which was independent of current density in the range 5.85 to 30.5 mA cm^{-2}.
† Under similar conditions, Rao[77] found a value of 0.9 ± 0.5.

produce a low value for S_T. Bockris and Srinivasan[12] found that for the best results, the electrolyte should not be purged with hydrogen during the measurements. Smith[7,26] has experienced similar problems and attributes them to contamination of the hydrogen by air diffusing into the gas lines.

Cathodic disintegration*

The disintegration of cathodes was a phenomenon that attracted much interest around the end of the last century.[39,40] Reed[39] in 1895 reported a long list of substances which in aqueous solution were "active" for the disintegration of lead cathodes. He also found arsenic behaved in a similar manner to lead. Haber[40] also managed to bring about the disintegration of Sn, Bi, Tl, As, Sb and Hg cathodes. Solutions containing alkali metal cations were most active—the effect being achieved only with difficulty in pure acids, and then only for lead and bismuth. He suggested that the cause was alloy formation between the cathode metal and the cations in solution. These alloys were unstable and reacted with water, releasing the alkali metal back into solution and leaving either a roughened cathode surface in the case of dilute alloys, or particles of lead in suspension in the case of rich alloys. In pure acids an alloy was formed with hydrogen. Alloys prepared by thermal methods were observed to behave in this way when immersed in electrolyte. Platinum was found to become black and spongy after cathodization at high current densities in acid but not in alkali.

Van Muylder and Pourbaix[41] found that in unstirred solutions, lead disintegrated at a potential of $-2 \cdot 1$ V (NHE), independent of pH. The current density varied with pH, being constant at 10 mA cm^{-2} in the pH range of $3 \cdot 3 - 12 \cdot 8$, but higher at pH values outside this range. Salzberg[42] found that pH did not affect current density down from pH 12 to pH $3 \cdot 7$. When the electrolyte was stirred, Van Muylder and Pourbaix[41] observed a different behaviour. The disintegration potential/pH relationship, E (V) $= -1 \cdot 54 - 0 \cdot 050$ pH, was found. The current densities at disintegration showed large fluctuations but were about 10–20 mA cm^{-2} above pH 5. It is hard to understand why there should be such a difference in behaviour between the stirred and unstirred solutions. The two experiments used buffered solutions containing sodium ions (usually $0 \cdot 1$ M) and different anions (although these were not expected to have any effect). The stirred solutions were of different

---

* An important contribution to the whole question of cathodic disintegration has come in a recent paper by Rostami and Smith.[83]

composition from the static solutions and, apart from the stirring being by magnetic stirrer, no other details are known. The static solution results do, at least in part, confirm the findings of other workers that disintegration in acid solutions requires much higher current densities than in neutral and alkaline solutions. No one else has reported potential measurements in acid solutions, most effort being concentrated on strong alkaline solutions, so no comparison is possible. The results with the static solution were interpreted[41] as showing that a certain (high) pH is required to produce electrode disintegration. However, stirring an acid solution would be expected to help keep the pH low and so delay the onset of disintegration. Thus a more negative potential and a higher current density would be expected. This is not what is observed; the potential is more positive at low pH although the current density is higher. In the absence of further experiments, these results must remain an interesting but unexplained anomaly.

More recently, the Kabanov school have studied the electrochemical implantation of alkali metals in cathode materials.[46-59] They have shown that alkali metal atoms are incorporated in the electrode lattice to an extent dependent on solution concentration, potential and time. Astakov and Bogatyrev[60] showed that electrode disintegration takes place by dissolution of the surface and not along grain boundaries. In $10 M$ NaOH[60] lead forms $NaPb_3$ at around $-1 \cdot 3$ V (RHE) and NaPb at around $-2 \cdot 1$ V. At potentials more positive than $-2 \cdot 1$ V, the $NaPb_3$ is decomposed by the water but only the Na passes into solution. Electrode disintegration starts at $-2 \cdot 1$ V. The amount of metal incorporated is determined from the length of arrests obtained during an anodic charging transient. It was found[57] that cathodized electrodes absorb ions faster than fresh electrodes and this is explained as being due to formation of lattice vacancies during cathodization. However, a maximum effect is reached, the optimum length of polarization depending on solution concentration and electrode material.

The Russian workers were unable to obtain disintegration in acid solutions—neither could Angerstein[45]—and accused workers who did of using impure solutions. Salzberg, however, showed that disintegration in pure acids was possible[44] but very high current densities were required, and that the current densities of the unsuccessful workers were too low. He found that the rate of disintegration was proportional to current density (above a certain minimum value) and water activity, but inversely proportional to temperature. Increased acidity (below pH 3·7) displaced the rate curve towards lower values but did not change the slope. From this information he deduced that a hydride, originally suggested as being PbH_2, but later[43] changed to Pb_2H_4, was produced above a certain critical current density. A fraction of this left the surface and decomposed in solution giving gaseous hydrogen and lead in suspension; some decomposed on the surface. The acid decomposed the hydride before it had a chance to leave the surface and so, until the acid was depleted below a critical level, no hydride was evolved; after the depletion the rate of hydride evolution (and therefore, decomposition) proceeded at a rate independent of the initial acid concentration.

Liporetz and Lohonyai[74] have presented evidence for the electrochemical implantation of aluminium and potassium into lead. The rate of incorporation of aluminium was much lower than that of potassium.

Smith and Ives[5] and Rao[77] observed disintegration of a lead cathode in ultrapure perchloric acid at current densities as low as 10 mA cm^{-2}. The electrode had been cathodized for several days and it was suggested that the hydrogen content of the electrode had finally reached a critical level and made the electrode unstable.

Industrial applications

Lead–Acid Batteries

The development of maintenance-free lead–acid batteries requires the rate of gassing during charge and on open-circuit to be minimized to prevent excessive water loss. With this in mind, Caldwell et al.[61] measured the rate of hydrogen evolution on alloys of lead with calcium, calcium and tin, and antimony. The microstructure is influenced by the casting conditions, in particular the rate of solidification, and their main interest was the effect of microstructure on gassing rate. The electrolyte, 4·2 M H_2SO_4 at 21°C, was not deoxygenated but was pre-electrolysed for 72 h. By changing the casting conditions but keeping the composition constant, they were able to demonstrate a correlation between gassing rate and casting conditions. For alloys without well-developed dendritic structures (Pb–Ca, Pb–Ca–Sn), the electrochemical properties are directly related to the cooling rate. Lead–antimony alloys solidify with structures having well-defined secondary and higher-order dendrite arms. Dendrite arm spacing is of great importance: it affects the mechanical properties of the alloys and their corrosion resistance. It is a unique function of the local solidification time which is inversely proportional to local cooling rate. Change in cooling rate was found to have a marked effect on the gassing rate at Pb–2% Sb alloys, low gassing rates being associated with higher cooling rates.

Impurities in the electrolyte may affect the negative battery plate in two ways. The first is an electrochemical reaction between the impurity cation and the lead, resulting in reduction of the cation (either to the metal or to a lower valence state ion (e.g. Fe^{3+} to Fe^{2+}) and oxidation of the lead to lead sulphate. If the metal is deposited, a cell is set up between it and the lead, the result being hydrogen evolution on the metal and oxidation to sulphate of the lead. This reaction depends on the hydrogen overvoltage of the impurity metal and does not occur with all impurities. If a lower-valence cation is formed, it will diffuse to the positive plate, be re-oxidized, diffuse to the negative plate, be reduced and so on, each reaction producing sulphation of the electrodes. This chapter is only concerned with the reaction involving hydrogen evolution. These sulphation processes, termed self-discharge, occur when the battery is on open circuit. When the plated impurity has a low hydrogen overvoltage, problems are experienced in recharging because the potential of the negative plate is largely controlled by the activity of the impurity for hydrogen evolution—the potential does not become sufficiently negative for complete reduction of the sulphate.

Vinal[62] has reported the effect of several impurities in the electrolyte on the self-discharge of negative plates. The discharge was followed by the weight increase as lead was converted to lead sulphate. As would be expected from its low hydrogen overvoltage, platinum had a dramatic effect, with 100 p.p.m. producing violent gassing and loss of lead from the surface.[63] Copper and silver ions separately produced a small continuous weight gain; when present in large quantities they both plated out in a spongy mass that dropped off. Bismuth (III) was reduced to the metal and produced only the predicted gain in weight for this process.[64] Arsenic and antimony (III) were reported to cause continuous gains in weight.[64]

Dawson et al.[65] showed that addition of Sb(III) to the electrolyte increased the rate of sulphation, whereas Sb(V) had no effect. Ferric ions are reduced to ferrous at the negative plate, and so cause sulphation of the electrode,[62] but Lea and Crenell[66] found that open circuit gas evolution was not affected by the presence of either ferric or ferrous ions.

Vinal et al.[67] observed that the overpotential of the negative plate during charging was reduced by the presence of nickel ions, whereas iron and cobalt had no effect. Chan and Kuhn in these laboratories[75] have studied the effect of nickel on the hydrogen overvoltage on lead.

It was observed by Kugel in 1892[68] that combinations of impurities such as copper and tungsten result in a greater self-discharge than either component separately. It seems that the presence of the tungsten lowers the overvoltage for hydrogen evolution on copper. Vinal and Schramm[64] investigated the combination of copper with several metals. The procedure was to add 100 p.p.m. of the other metal and after a certain time add 500 p.p.m. of copper sulphate solution. The change in weight was monitored. The metals studied were W, Hg, Mo, Zn, As and Sb, and in all cases a rapid gain in weight occurred in the first few hours after the copper addition. This combination effect is of particular importance in the case of antimony because it is always present in solution and on the negative plate, having dissolved from the antimonial positive plate.[65]

Pierson et al.[69] measured the rate of gassing in batteries over a 4-h charging period after the addition of known amounts of impurity cation. They tested 24 elements at the 5000 p.p.m. (or saturation) level. Those ions that increased the gassing rate were then retested at successively lower concentrations until no effect was observed. Some elements lowered the gassing rate, P, Ca and Sn all having substantial effects. The largest effect at 5000 p.p.m. (a gassing enhancement factor of 24) was experienced with cobalt. The recommended maximum allowable concentration of cobalt from this work is 1 p.p.m. Other 1-p.p.m. metals are Sb, As and Ni. Manganese is 3 p.p.m., tellurium is 0·1 p.p.m., and iron is 160 p.p.m. Most metals show a limiting effect, e.g. Ni at 5000 p.p.m. had a gassing enhancement factor of only 4·7 but its effect was noticeable down to very low concentrations.

Electro-organic Processes

Electro-organic reductions are usually carried out on lead cathodes because of the high hydrogen overvoltage. Any metal ions in the electrolyte will plate on to the cathode and catalyse the evolution of hydrogen. One U.K. chemical company suffered drastic reductions in efficiency on changing from pilot scale to full scale production because the latter process used a feedstock contaminated with iron: removal of the iron raised the efficiency (see Chan and Kuhn[75] for further consideration of the problem).

Miscellaneous

Ostrovskaya et al.[70] have reported that during hydrogen evolution from alkaline NH_4Cl solutions the NH_4^+ discharge stage takes place irreversibly at lead electrodes but reversibly at mercury electrodes.

Conclusions

Fundamental studies on lead are very difficult: high purity is essential and exposure of the electrode to oxygen, either in solution or atmospheric, must be avoided. One way to achieve this would be the use of a "dry box"[71,72] in which to prepare and test the electrode. A number of workers have looked at lead while studying a series of high-overvoltage metals and do not seem to have appreciated the problems involved, especially the need for rigorous oxygen elimination. Hydride formation especially in alkali[4] and time effects do not seem to have been adequately considered in a number of cases.

Reduction of Oxygen on Lead

Oxygen is reduced on lead, as it is on many other metallic electrodes. The reaction appears to be the four-electron one:

$$O_2 + 4 H^+ + 4e^- = 2 H_2O,$$

with no evidence of the intermediate hydrogen peroxide which is formed on many other cathodes. Thus Harrison,[81] using a rotating disc electrode, found that the plot of current vs (rotation speed)$^{1/2}$ had a gradient twice that for a Pt electrode under otherwise identical conditions, thereby showing the four-electron reaction to be the predominant one. Other work with r.d.e. has been reported by Atkin and Bonnaterre.[78] The discussion section following Harrison's paper[81] is also of interest in that it refers to the early work of Delahay on the reaction as well as raising other aspects of the subject. A recent paper by Armstrong and Bladen[79] confirms what all other authors show, that the reaction is diffusion-controlled over most of the potential range. Thus a current density of 8 mA cm^{-2} is found over the potential range $-0\cdot35$ to $-0\cdot2$ V (vs RHE) at a rotating disc electrode with $53\cdot3$ Hz rotation frequency. At more anodic potentials, where the Pb dissolution reaction becomes significant, the reaction appears to be inhibited and at $-0\cdot15$ V there is no net current. Armstrong attributes this effect to anion adsorption and impedance studies of the system are reported in the same paper.

The practical importance of this reaction lies, of course, in the operating mode of the sealed (maintenance-free) lead–acid battery. Oxygen formed at the positive plate during overcharge is reduced back to water at the negative, care being taken that the latter electrode has a greater surface area than the former. A study of oxygen reduction of lead in a battery-type situation is reported by Hills and Chu,[80] with emphasis here on oxygen partial pressures and diffusion through the meniscus formed at partially immersed electrodes (see also Awad[82] for data in $HClO_4$ and H_2CrO_4).

References

1. I. M. Tordesillas and L. P. Bicelli, Z. Elektrochem. **63**, 1049 (1959).
2. L. P. Bicelli and B. Rivolta, WADC Tech. Note 59–393, PB 161, 801 (1960).
3. L. P. Bicelli and C. Romagnani, NRC-TT-1525 N72-19186 (1965).
4. T. S. Lee. J. Electrochem. Soc. **118**, 1278 (1971).
5. D. G. Ives and F. R. Smith, Trans. Far. Soc. **63**, 217 (1967).
6. E. W. Brooman, M. Hayes and A. T. Kuhn in "Electrochemistry—the Past 30 and the Next 30 Years" (H. Bloom and F. Gutmann, eds), p. 139. Plenum (1977).
7. F. R. Smith, Ph.D. Thesis, Birkbeck College, London (1964).
8. P. Malachesky, R. Jasinskii and B. Burrows, J. Electrochem. Soc. **114**, 1104 (1967).
9. N. A. Hampson and D. Larkin, J. Electrochem. Soc. **18**, 401 (1968).
10. H. Metzler and W. Schwarz, Electrochim. Acta **15**, 97 (1970).
11. S. Schuldiner and R. M. Roe, J. Electrochem. Soc. **110**, 332 (1963).
12. J. O. M. Bockris and S. Srinivasan, Electrochim. Acta. **9**, 31 (1964).
13. B. Kabanov and S. Jofa, Acta Physicochim. URSS **10**, 617 (1939).
14. A. Hickling and F. W. Salt, Trans. Far. Soc. **38**, 474 (1942).
15. J. O. M. Bockris, Far. Soc. Disc. **1**, 95 (1947).
16. Y. M. Kolotyrkin, Trans. Far. Soc. **55**, 455 (1959).

17. M. I. Gillibrand and G. R. Lomax, *Trans. Far. Soc.* **55**, 643 (1959).
18. P. Ruetschi, J. B. Ockerman and R. Amlie, *J. Electrochem. Soc.* **107**, 325 (1960).
19. U. V. Palm and V. E. Past, *Sov. Electrochem.* **1**, 527 (1965).
20. N. D. Tomashov, N. M. Strukov and L. P. Vershinina, *Sov. Electrochem.* **5**, 22 (1969).
21. H. Kita and T. Kurisu, *Res. Catal. Hokkaido Univ.* **18**, 167 (1970).
22. C. J. Weedon, Ph.D. Thesis, Leicester University (1972).
23. A. N. Frumkin, *Adv. Electrochem. Electrochem. Eng.* **3**, 296 (1963).
24. J. P. Carr, N. A. Hampson, S. N. Holley and R. Taylor, *J. Electroanal. Chem.* **32**, 345 (1971).
25. G. M. Rao and F. R. Smith, *Chem. Comm.* **5**, 266 (1972).
26. F. R. Smith, *Disc. Far. Soc.* **56**, 113 (1973).
27. B. R. Wells and M. W. Roberts, *Proc. Chem. Soc.* 173 (1964).
28. I. Cadersky, B. C. Muju and F. R. Smith, *Can. J. Chem.* **48**, 1789 (1970).
29. B. L. Muju and F. R. Smith, *Can. J. Chem.* **49**, 2406 (1971).
30. P. Ruetschi and B. D. Cahan, *J. Electrochem. Soc.* **104**, 406 (1957).
31. R. Piontelli and L. P. Bicelli, *Bull. Nat. Inst. Sci. Ind.* **29**, 80–87 (1965).
32. I. A. Aguf and M. A. Dasoyan, *J. Appl. Chem. USSR* **32**, 2071 (1959).
33. U. Croatto and M. da Via, *Gazz. Chim. Ital.* **73**, 117 (1943).
34. U. Croatto, *Gazz. Chim. Ital.* **73**, 113 (1943).
35. E. W. Brooman and A. T. Kuhn, *J. Electroanal. Chem.* **49**, 325 (1974).
36. U.S. Patent 4,000,048 (1976).
37. J. Horiuti and G. Okamoto, *Sci. Pap. Inst. Chem. Res. Tokyo* **28**, 231 (1936).
38. G. P. Lewis and P. Ruetschi, *J. Phys. Chem.* **66**, 1487 (1962).
39. C. J. Reed, *J. Franklin Inst.* **139**, 283 (1895).
40. F. Haber, *Trans. Amer. Electrochem. Soc.* **4**, 189–196 (1902).
41. Van Muylder and M. Pourbaix, CITCE 6th Meeting, p. 306. Butterworth, London (1955).
42. H. W. Salzberg, *J. Electrochem. Soc.* **100**, 146 (1953).
43. H. W. Salzberg, *J. Electrochem. Soc.* **100**, 588 (1953).
44. L. W. Gastwirt and H. W. Salzberg, *J. Electrochem. Soc.* **104**, 701 (1957).
45. H. Angerstein, *Bull. Acad. Polon. Sci.* **3**, 447 (1955).
46. I. G. Kiseleva and N. N. Tomashova, B. N. Kabanov, *Russ. J. Phys. Chem.* **38**, 648 (1964).
47. D. P. Aleksandrova, I. G. Kiseleva and B. N. Kabanov, *Russ. J. Phys. Chem.* **38**, 811 (1964).
48. B. N. Kabanov, I. I. Astakhov and I. G. Kiseleva, *Russ. Chem. Rev.* **34**, 775 (1965).
49. B. N. Kabanov, I. G. Kiseleva, I. I. Astakhov and N. N. Tomashova, *Sov. Electrochem.* **11**, 911 (1965).
50. I. G. Kiseleva, *Sov. Electrochem.* **3**, 241 (1967).
51. B. N. Kabanov, *Electrochim. Acta* **13**, 19 (1968).
52. A. Alumoa, U. Palm and V. Past, *Chem. Abstr.* **73**, 104917 (1970).
53. N. N. Tomashova, I. G. Kiseleva, I. I. Astakhov and B. N. Kabanov, *Sov. Electrochem.* **4**, 419 (1969).

54. G. L. Teplitskaya and I. I. Astakhov, *Sov. Electrochem.* **6**, 373 (1970).
55. A. I. Chernomonskii, T. G. Kiseleva and B. N. Kabanov, *Sov. Electrochem.* **6**, 424 (1970).
56. A. I. Chernomonskii and B. N. Kabanov, *Sov. Electrochem.* **6**, 1194 (1970).
57. A. I. Chernomonskii, I. G. Kiseleva and B. N. Kabanov, *Sov. Electrochem.* **6**, 271 (1970).
58. N. N. Tomàshova, I. G. Kiseleva and B. N. Kabanov, *Sov. Electrochem.* **7**, 423 (1971).
59. B. N. Kabanov and A. I. Chernomonskii, *Sov. Electrochem.* **8**, 138 (1972).
60. I. I. Astakhov and V. N. Bogatyrev, *Sov. Electrochem.* **11**, 308 (1975).
61. T. W. Caldwell and U. S. Sokolov *in* "Power Sources 5" (D. H. Collins, ed.), p. 73. Academic Press, London and New York (1975).
62. G. W. Vinal, "Storage Batteries", p. 131, Chapman and Hall (1955).
63. G. W. Vinal and F. W. Altrup, *Trans. Amer. Inst. Elec. Eng.* **43**, 709 (1924).
64. G. W. Vinal and G. W. Schramm, *Trans. Amer. Inst. Elec. Eng.* **44**, 288 (1925).
65. J. L. Dawson, M. I. Gillibrand and J. Wilkinson *in* "Power Sources 3" (D. H. Collins, ed.), p. 1. Oriel Press, Newcastle (1971).
66. F. M. Lea and J. T. Crennel, *Trans. Far. Soc.* **23**, 269 (1927).
67. G. W. Vinal, D. N. Craig and C. L. Snyder, *J. Res. Nat. Bur. Stand.* **25**, 417 (1940).
68. A. M. Kugel, *Elektrotech. Z.* **13**, 8 and 19 (1892).
69. J. R. Pierson, C. E. Weinlein and C. E. Wright *in* "Power Sources 5" (D. H. Collins, ed.), Academic Press, London and New York (1975).
70. S. E. Ostrovskaya, E. G. Tsventarnyi and Ya V. Durdin, *Sov. Electrochem.* **9**, 1275 (1973).
71. C. Huang, W. E. O'Grady and E. Yeager, *NASA Rep.* N76 25326 (1975).
72. P. N. Ross, *J. Electroanal. Chem.* **76**, 139 (1977).
73. E. Gileadi, S. D. Argade and J. O. M. Bockris, *J. Phys. Chem.* **70**, 2044 (1966).
74. I. Liporetz and N. Lohonyai, *Periodica Polytechnica*; *Chem. Eng.* **13**, 359 (1969).
75. C. Y. Chan and A. T. Kuhn, *Oberflaeche-Surface* (in press).
76. J. O. M. Bockris and R. Parsons, *Trans. Far. Soc.* **45**, 916 (1949).
77. G. M. Rao, Ph.D. thesis. St. John's Memorial University, Newfoundland (1973).
78. J. Atkin, R. Bonnaterre and J.-F. Laurent *in* "Power Sources 6" (D. H. Collins, ed.), pp. 91–102. Academic Press, London and New York (1977).
79. R. D. Armstrong and K. L. Bladen, *J. Appl. Electrochem.* **7**, 345 (1977).
80. S. Hills and D. L. Chu, *J. Electrochem. Soc.* **116**, 1155 (1969).
81. A. Fleming and J. Harrison, *in* "Power Sources 5" (D. M. Collins, ed.), pp. 1–14. Academic Press, London and New York (1975).
82. S. A. Awad, *J. Electroanal. Chem.* **34**, 431 (1972).
83. A. A. Rostami and F. R. Smith, *in* "Power Sources 7" (J. Thompson, ed.), Chapter 4. Academic Press, London and New York (1979).

9   Anodic Reactions on Lead Dioxide

T. H. RANDLE* and A. T. KUHN‡

* *Department of Chemistry, Swinburne College of Technology,
Hawthorne, Australia*
‡ *Department of Dental Materials, Eastman Institute of Dental
Surgery, London*

Introduction

We shall consider here what is known of the rate and mechanism of
Faradaic reactions, both anodic and cathodic, on lead dioxide anodes.
Deliberately, we have chosen to omit virtually all reference to electro-
organic reactions on lead dioxide. Though these reactions are of some
importance, both synthetically and industrially, too little work has been
done to enable much mechanistic information to be cited. It has come as
a surprise to us to learn that of the many inorganic reactions which take
place on PbO_2, only a handful have been the subject of any study except
for the simple "current density *vs* current efficiency" type called for in
industrial development work. Clearly this represents a rich and untilled
field of study for future workers. In considering the mechanisms of the
reactions we are here to describe, one question should, we feel, remain
open, and that is: when is the mechanism one of direct electron transfer
to an inert, electronically conducting electrode (as is the case with the Pt
anode) and when is it one we might term "immobilized indirect
oxidation"? (In the latter, PbO_2 is electrochemically formed, and then
reacts chemically with the substrate species, itself being reduced in the
process, whereupon the cycle can begin again.) Clearly these two
mechanistic possibilities mark a watershed in the conceptual approach
to the problem. Furthermore, in the case of immobilized indirect
oxidation, one might hope to look for an upper limiting reaction rate,
being that due to the rate of formation of PbO_2. Of course it may well be
that the subsequent chemical reaction is in all cases so slow that the
former limit is never reached, but one should bear this in mind. Another

question we may ask, and receive only a partial reply to, concerns the extent to which one anodic reaction can displace another. Thus, oxygen evolution can proceed simultaneously with most of the reactions described below. To what extent do the two processes interact? These are some of the questions which we shall raise, even if only to give the smallest fragment of an answer.

Ozone Formation

O_3 is formed by anodic oxidation of water at PbO_2 anodes. To the best of our belief, no study exists of this process beyond the recording of certain current efficiencies. On PbO_2/Pb anodes, the high potentials required may result in severe corrosion taking place during the process, which thus limits the commercial interest. Thus Stender[1] reports 3% O_3 in O_2 when Na_2SO_4 at 50–60°C is electrolysed at 80 mA cm^{-2}, 2·2% at 73 mA cm^{-2} and 0·33% at 48 mA cm^{-2}. Wabner[2] finds similar figures for PbO_2/Ti anodes at 25°C, namely 5·3% at 200 mA cm^{-2} and 3·1% at 100 mA cm^{-2}. Other work has been done by Semchenko[3] who reached up to 12% O_3 at 1·2 A dm^{-2} in orthophosphoric acid, stating that corrosion was "completely absent" under some conditions.

Oxygen Evolution Reaction (o.e.r.)

This is the most important single anodic reaction on lead dioxide anodes, being the "counter-reaction" in all metal electrowinning processes by design, as well as taking place concurrently with almost all other anodic reactions when these are driven too hard, or the reactant species is depleted. The same is true in lead–acid batteries on charge, for when the sulphate conversion to PbO_2 is near completion, oxygen evolution commences until at full charge, virtually all current goes to oxygen evolution.

Fundamentals of the reaction

Numerous early investigators of this reaction failed to obtain "rational" Tafel slopes, or found the current–voltage plots to vary from run to run.

It was Wynne-Jones and co-workers who first appeared to have realized[4] that the best procedure was to start at the most anodic limit chosen for the experiment and then to work downwards, that is, obtaining the current–voltage data in the cathodic sense. In this way they obtained the "rational" 0·12 V Tafel slopes which would be expected on the basis of simple theory. The same workers also showed that the magnitude of the current at a given potential was related to the length of time for which the electrode had been previously held at the anodic limit. Later work showed that, even after holding times of 12 h, a (small) effect could still be seen. Other workers found it was not necessary to "precondition" the electrode at the most anodic value of the experiment, provided that sufficient time was given at whatever condition of potential or current density was adopted. The results of Wynne-Jones remain undisputed although, as will be seen, the comparatively low value of the maximum current density shown by these workers fails to reveal the most interesting phenomena uncovered by later workers. Figure 1 shows data by Houghton and Kuhn[5] which, though resembling those of Wynne-Jones, goes to higher current densities.

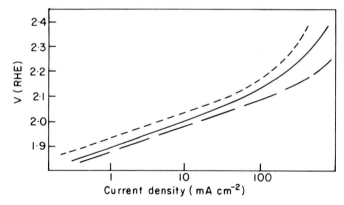

Fig. 1. Oxygen evolution curves on lead, 25°C, 1 M H_2SO_4. Anode pretreated anodically for 24 h. (a) Upper curve at 2 mA cm^{-2}, (b) middle curve at 10 mA cm^{-2}, (c) lower curve at 50 mA cm^{-2}.

The explanation of this type of behaviour is agreed to be the result of continuing growth of the lead dioxide film on the surface of the lead. This results in a "rougher" surface or possibly a more porous one with corresponding higher current density on a superficial area basis.

Confirmation of this hypothesis has been reached in terms of capacitance measurements.[26] More recently, Hyvarinen has published[9] a more
detailed study of the effects of pre-anodization procedures on current–
voltage data, while older publications by Ruetschi et al.,[25] Peters et al.[6]
and Feitknecht[27] again describe oxygen overvoltages both on electrodeposited PbO_2 on Pt and also on "battery plates". The results of
these authors broadly agree with Wynne-Jones et al.,[4] but data are also
included on the rates of polarization (following current application) and
potential decay (on open-circuiting). Porous anodes are only some 18
times more active than their smooth counterparts, notwithstanding
their vastly greater surface area as measured by B.E.T. This is undoubtedly due to the filling up of the majority of pores with gaseous
oxygen, thereby excluding electrolyte and further electrolysis. Data for
oxygen evolution on Pb single crystals has been given.[17] The 111, 100
and 110 behave almost identically. At the time of Wynne-Jones et al.,
the lead dioxide was treated as a single species. Only in subsequent
years was the existence of the α- and β-form of the dioxide recognized.
It fell to Ruetschi and Cahan[15] to show that the oxygen overvoltages on
these two species (Fig. 2) differed, the α-form being more active than

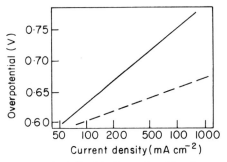

Fig. 2. Oxygen evolution on lead dioxide formed on Pt substrate. Solid line, β-PbO_2.
Dashed line, α-PbO_2.

the β. Ruetschi et al. also showed results for the rate of open-circuit
decay and potential build-up for both α and β modifications of the
dioxide when evolving oxygen. Unlike the Wynne-Jones' results, these
findings have been called into question on several occasions. Makrides
has claimed[7] that if a thin, non-porous film of the dioxide is laid down on
Pt, no difference between the α- and β-forms is seen. However, the
further details of this work (referred to by the author as "to be

published") have not appeared, 10 years after the earlier publication. Work from the Finnish school of Vuorio and Kivalo[8] (using sulphuric acid and nickel sulphate electrolyte) and again by Hyvarinen[9] similarly failed to detect any differences, and Puzey and Taylor[18] also reported a 120 mV slope on what was basically a β-PbO$_2$ anode. This disagreement calls for some comments. Hampson has suggested that unless the anode is at all times held anodic, there may be a reduction, followed by a reformation (in acid media) of the more stable β-form. Ruetschi[18] in the published discussion on the Puzey and Taylor paper, makes the very valid point that while the outer layer of dioxide formed on a massive lead electrode in acid will be of the β-variety, it is well known that the underlying deposits will be of the α-form. While only the outer layer controls the electrode kinetic behaviour (i.e. the slope) any X-ray analyses will record an overwhelming amount of the α-form. This comment in the literature back in 1964 appears to have been forgotten by later workers, and would appear to be as valid a criticism of Hyvarinen's work[9] as it was of Puzey and Taylor's, both these authors having published X-ray data for their anodes based on the use of massive lead substrates.

The effect of pH on rate of the reaction

Several authors have considered the effect of pH changes on the rate of the reaction. Convers has published a few current–voltage plots without drawing any conclusions.[10] Work has been done by Hyvarinen[9] (Fig. 3) who obtained the following relationships:

$$\left(\frac{dE}{d\,pH}\right)_{1\,mA} = -62 \text{ mV}; \qquad \left(\frac{d\log j}{d\,pH}\right)_{1\cdot65\,V} = 0\cdot62,$$

though he himself casts doubts as to whether these represent fully equilibrated conditions. Rather more straightforward work by Kokhanov et al.[11] considers the oxygen evolution process on a range of anode materials. These authors reach the important conclusion (Fig. 4) that the shape of the pH plots is very similar for all materials considered by them and they conclude that the oxygen evolution process changes over from a water-discharge to a hydroxyl ion discharge process at a given pH. The measurements of these three schools are in broad agreement though the Soviet workers present some interesting ideas regarding the inclusion of hydrogen atoms into the oxide film. This will be considered later. Puzey and Taylor[18] show Tafel slopes for oxygen evolution in a series of sulphuric acids up to 6 M. Several workers consider oxygen evolution on lead in strongly alkaline media. These include Hampson,[24]

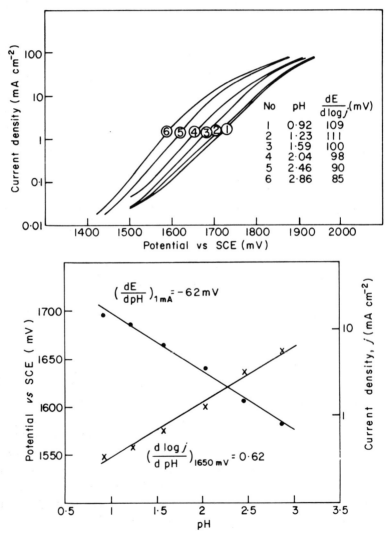

Fig. 3. Effect of pH on oxygen evolution kinetics. (a) j-V data. (b) E-pH, j-pH plots.

Wynne-Jones *et al.*[21] and, more recently, Fortunatov.[23] The latter author shows that about 50% of the total current goes to gas evolution, the remainder being corrosion current.

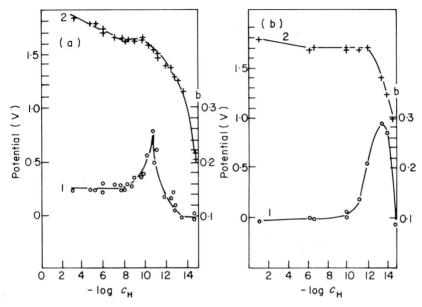

Fig. 4. Coefficient *b* in Tafel equation (1) and anode potential (2) as functions of solution pH at lead dioxide anode (a) and platinum anode (b): the left-hand ordinate represents the potential (V), while the right-hand ordinate represents the coefficient *b* in V; the abscissa represent $-\log c_H$.[11]

Effect of temperature

This has been considered by various authors over a range from −50 to 83°C. Wynne-Jones,[4] as well as Koch,[12] working in the range 0–60°C, show no effects other than the expected decrease in overvoltage and Hyvarinen[9] also agrees, although at 83°C some rather curious passivation effects are shown (Fig. 5). Puzey and Taylor also examine the effect of temperature over a more restricted temperature range[18] and publish current densities as a function of temperature. The most remarkable contribution here, however, is that of Izidinov *et al.*[13] (Fig. 6). In the first place, this is an interesting work in that current densities of up to 10 A cm^{-2} are shown. This is a current density now being used in the chloralkali industry and of the order of that considered for newer types of metal-winning electrolysis cell. It is perhaps surprising that the earlier workers referred to were content to restrict their measurements to $100 \, \text{mA cm}^{-2}$ or even less. The work of Izidinov, agreeing with Houghton and Kuhn, for example, then shows (Fig. 6) the onset of a

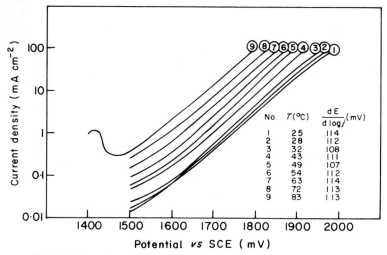

Fig. 5. Effect of temperature on oxygen evolution kinetics, 1 M H_2SO_4.

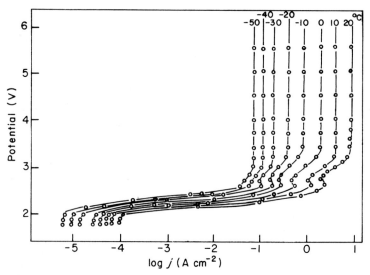

Fig. 6. Effect of temperature on oxygen evolution kinetics showing effects at low temperatures and high current densities (5 M H_2SO_4).

limiting current. Before going any further, one should recollect that at these high anodic potentials, processes other than oxygen evolution occur, and Izidinov discusses persulphate formation as well as ozone and hydrogen peroxide formation. However, the main conclusion in the present context is that, at high current densities, a change in the properties of the lead dioxide takes place, mainly a dramatic increase in the electrical resistance, leading to a "shut-off" of current. Izidinov did in fact measure the resistance of the system and showed it to increase considerably at the highest potentials. However, simple calculations would suggest that the values shown in his paper are insufficient to account for the current–voltage limit observed by him. The question therefore remains largely an open one, and before leaving the subject, it is worth recalling that the whole idea of changes in the stoichiometry of the PbO_2 with corresponding effects on its conductivity were earlier put forward by Feitknecht[14] (who found a 15 Ω film resistance, as compared with a 6 Ω value reported by Izidinov). However, Wynne-Jones[4] objected to Feitknecht's mechanistic involvement of this resistance increase, showing that it could not quantitatively explain the results, and the idea was withdrawn. Other interesting work by Feitknecht,[27] in which he applied slow anodic and cathodic pulses to lead dioxide electrodes, coupled with chemical analyses of the Pb : O ratio, must be interpreted with great care in the light of the known, highly porous nature of the dioxide and its ability to take up oxygen as the molecular gas by physical adsorption, rather than by chemisorptive forces and the implied modification of stoichiometry. A simple chemical analysis would fail to differentiate between these two, and the slow diffusion of oxygen into or out of the dioxide could account for all of Feitknecht's results. A simple experiment by Ruetschi,[15] in which a lead dioxide electrode was held at a high anodic potential, then stepped back to some lower value, when, after open-circuiting, a potential recovery took place. The authors state that this behaviour could be explained in terms of stoichiometry change, although they very properly point out that a number of other phenomena such as collapse of a sulphate/persulphate rich layer could explain it.

Stirring

The effect of stirring or other agitation appears to be insignificant in relation to the rate of the oxygen evolution reaction. This is a finding of

Puzey and Taylor[18] who used gas stirring and is supported by subsequent work of Clarke et al.[20] who used a rotating disc electrode. Such conclusions cannot, of course, hold at the highest current densities where mass transport limitation must occur. At current densities of up to 200 mA cm^{-2} there is no indication that this step is incipient.

Behaviour on open-circuiting

Because the lead dioxide frequently operates at much more anodic potentials than the theoretical reversible oxygen evolution potential, gas evolution will continue for some time after external current has ceased. This has been studied by several authors, including Ruetschi,[22] and Feitknecht and Gaumann,[14] whose work has already been referred to. It is suggested that theoretical interpretation of these results must be difficult, for several effects will be superimposed. These include the genuine coupling of a Faradaic reaction (oxygen evolution) with double-layer discharge, a similar coupling in which the reduction of the oxide balances gas formation. Also the nature of the substrate results in substantial volumes of gas being absorbed, to be slowly released as the equilibrium pressure of oxygen in the surrounding solution falls. Finally, Ruetschi again suggests that diffusion of sulphate anions has an effect on the process.

Oxygen evolution on alloys of lead

The data here are scant, and often far from rigorous. Ruetschi and Cahan[15] report that oxygen overvoltage is lowered as Sb content increases over the range 0–11%. However, they report a simultaneous increase in the corrosion rate and it is possible the two effects are connected and that a constant surface area is not being maintained. This work has been extended by Rogachev.[29] A number of authors have examined the Ag–Pb system. Bryntseva (see Chapter 13) finds that 1% Ag gives a 90 mV depolarization at 10 mA cm^{-2} dropping to 57 mV at 100 mA cm^{-2}. Houghton[5] reports a 20 mV depolarization (current density independent) with a 1% Ag–0·065% Ca ternary system. The fullest investigation is due to Hyvarinen[9,16] who reports a 100 mV depolarization for 0·5% Ag, and 200 mV for 2·5% Ag.

Effect of added metal salts

Addition of metal salts affects both oxygen overpotential and corrosion rates. To some extent these two phenomena may be linked and the section on p. 368 *et seq.* should be consulted. Nickel salts, added to a sulphuric acid system, have no effect according to Vuorio and Kivalo.[8] However, cobalt exerts a very considerable effect as shown by Koch, Hyvarinen and other workers. In kinetic terms, the results are best seen

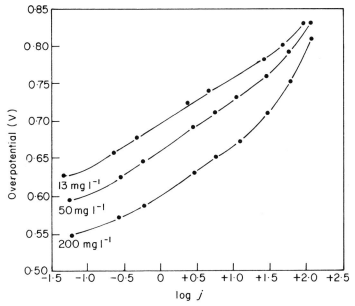

Fig. 7. Oxygen evolution in the presence of added Co^{2+} ions (mg l^{-1} Co^{2+}, 1 M H_2SO_4, 25°C).[12]

in Fig. 7. Completely different current–voltage characteristics are seen in the presence of Co, with a Tafel slope of 50–60 mV instead of the normal 120 mV, a greater depolarization under most conditions, and— perhaps surprisingly—a higher value of ΔH, the apparent enthalpy value. In morphological terms the reaction now takes place on a thinner, tighter layer of dioxide. The Finnish school have again made valuable contributions to the problem. Kivalo,[8] for example, found that addition of cobalt after formation of the dioxide layer gave even more effective depolarization than when added before dioxide formation. Agitation (which normally has no effect on oxygen evolution kinetics) now caused considerable effects. However, at lower pH values, this statement ceased to hold. Hyvarinen showed that there was approximately unit reaction order with respect to cobaltic ion concentration, although here some conflict appears to exist between this author and Koch.[12] The latter author shows the effect of cobalt to disappear at higher ($100 \, \text{mA cm}^{-2}$) current densities. Hyvarinen and Tikkanen[16] also demonstrated the absence of β-PbO_2 peaks in the presence of cobaltic species by X-ray analysis.

At this point one can point out the very close resemblance between the α–β dichotomy as shown by Ruetschi and Cahan on the one hand, and that of, say, Koch on the other for presence and absence of cobaltic salts. In both cases we have a high slope (120 mV) attributable to either the β-form or the absence of cobalt. The addition of this seems to give low slope and also the dominance of the α-PbO_2 which characterizes the low slope. The higher activation energy is also common to the low slope situation. The alternative approach is to postulate that the cobaltic ions enable an entirely different mechanism for oxygen evolution to occur. Cobaltous ions are oxidized to cobaltic ions. These react homogeneously with water, as is well known, to form oxygen, being themselves reduced to cobaltous. The cycle then recommences. The latter theory sits well with observations on the effect of stirring and of the first order dependence on Co concentration. The effect of both Co and Cu ions was studied by Ravindran and also by Gendron, both quoted on p. 368. Kokarev[26] showed that the surface area decreased during o.e.r. when Co was added.

Dunaev and Kiryakov[31] examined the effect of adding $\leqslant 0.06 \, \text{M} \, Ag^+$, Co^{2+}, Fe^{2+} and Ni^{2+} on Pb anodes and found their efficiency in lowering overvoltage to decrease in that order. Thallium increased the overvoltage, and was not beneficial in this sense, though it reduced corrosion.

Oxygen evolution on PbO₂ deposited on Ti and other substrates

In recent years, more and more use has been made of electrodes consisting of a stable basis material such as Ti or graphite, on which layers of PbO_2 are deposited. Few data are available for such anodes, but such work as has been published shows straightforward oxygen evolution occurring. The current density at a given potential is possibly somewhat lower than on PbO_2/Pb anodes, reflecting lower specific surface area,[28] but also the resistive loss at the substrate–PbO_2 interface. This is well illustrated in Fig. 8[2] which shows how oxygen overvoltage is low on graphite-coated anodes (where no oxide film forms) and higher on Ti and PbO_2 anodes, though the overvoltage can be reduced by special pretreatment (*see* p. 290). Wabner's work extends earlier findings by Udupa[32] who showed o.e.r. data on Pt, Pt/PbO_2 and graphite/PbO_2.

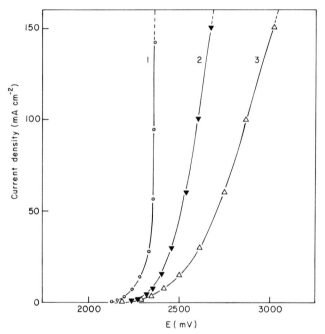

Fig. 8. Steady-state galvanostatic current–voltage curves for oxygen evolution in 1 M sulphuric acid, on PbO_2 surface with various substrates (jR corrected). 1 = graphite (Diabon R 200). 2 = Ti (Contimet 30) pretreated with Ti^{4+} perchlorate. 3 = as (2), but pretreated with Ti^{4+} oxalate.[2]

Simultaneous oxygen evolution with second Faradaic reaction

The situation frequently arises where oxygen evolution proceeds concurrently with another Faradaic reaction, either an organic or an inorganic oxidation. Not only in laboratory situations, but also in a wide range of industrial processes, this situation can occur. The co-existence of two reactions makes it impossible (unless certain assumptions are made) to estimate the rate of either reaction by current measurement alone. Certain workers have attempted to circumvent this by assuming that the two processes are additive, that is, that current flowing in the presence of the second reactant is simply the current flowing in its absence (due to o.e.r.) plus additional current due to oxidation of the second species. This assumption, as we shall see, is frequently false.

The problem should be considered in terms of the almost certainly activation-controlled nature of the o.e.r. (as it is up to several A cm^{-2}) and the possibility that the second oxidation may either be activation- or diffusion-controlled.

In the former case, there will be a substantial concentration of reactants at the electrode surface; in the latter case, the electrode surface will be essentially free of any steady-state concentration of second species. There are two publications which illustrate this situation. Zabotin[33] measured the rate of Cr(III) oxidation to Cr(VI) with simultaneous oxygen evolution. The partial current due to o.e.r. is clearly depressed in the presence of higher Cr concentrations, and as these increase from 3 to 30 g l^{-1}, the oxygen current declines to some 30% of its former value. Going from pure sulphuric acid to chromic acid, the effect would presumably be even greater. Nor is $i_{Cr} + i_{O_2}$ constant, being 1·1 A in strong chromic solution and only 0·8 A in the weaker Cr solutions (at 2·1 V). Zabotin's paper suffers in that the hydrodynamic regime was ill-defined, and it cannot be stated whether the Cr oxidation was mass-transport-controlled or activation-controlled over the range of potential studied. Another paper in which an attempt was made to remedy this, is that of Clarke et al.,[20] who studied anodic oxidation of benzene at a lead dioxide rotating disc electrode. In the activation-controlled region, the rate of benzene oxidation, j_{benz}, was nearly two orders of magnitude higher than j_{O_2}, presumably reflecting a higher j_0 value for the former process. The amount of oxygen evolved was thus negligible. At higher potentials and current densities, the benzene oxidation, notwithstanding the rotating disc electrode, moved into partial mass-transport control, and the fraction of oxygen increased accordingly, reaching 90% at 1·6 V where the measurements ceased. It is clear from this paper that benzene did suppress the o.e.r. in the activation-controlled region, for at 1·4 V, for example, there was 10 mA of o.e.r. in the absence of benzene and virtually none in its presence. Figure 5 of Clarke et al.[20] may perhaps have been drawn too confidently, in that one is tempted to adduce a total current efficiency for all products well in excess of 100% at the highest potentials. Presumably, as the limiting current for benzene oxidation was reached, the oxygen took up the remainder, apart from side reactions, and this then is the pattern of behaviour one might predict in similar situations.

Oxygen evolution in other solutions

In alkali, the reaction has been studied by Thirsk et al.[34] and more recently by Fortunatov.[36] In brine solutions, Mindowicz[35] has published data.

Mechanistic conclusions

We can begin by postulating (with Kokhanov) that water is the species from which oxygen is discharged. A key question which we appear to be unable to answer, is whether the PbO_2 takes part in the discharge mechanism. Thus it has been shown, for example, that oxygen reduction on platinum oxide involves the Pt—O bond and this was proved with radiolabelling experiments. Other studies point strongly to the fact that anodic oxidation of both organic and inorganic species by lead dioxide proceed by a chemical oxidation with PbO_2 as the oxidizing agent, followed by an electrochemical regeneration of the PbO_2. But is this the case in straightforward oxygen evolution, as has been suggested[26]? We appear not to know, although if Ruetschi and Cahan are correct and it is possible to evolve oxygen on, say, α-PbO_2, under acidic conditions where the β-form would take over if a continuous reduction–reoxidation cycle was involved, then it follows that the dioxide is not involved in the oxidation mechanism directly. Apart from this, it is simply a question of the nature of the intermediate species involved in the conversion from water to oxygen.

From the foregoing, it will be seen that the qualitative behaviour of oxygen evolution of PbO_2 is well-documented. Two areas can be singled out as being still unknown. The first relates to what one might describe as the "intimate" details of mechanism and intermediates. Apart from steady-state studies, Wabner[2] has reported a.c. and pulse measurements. Wabner studied both Pt and PbO_2 anodes. The former has been the subject of extensive researches and it must be stated that Wabner's results on Pt, which differ substantially from those obtained by well-known workers, cast doubt on his results for PbO_2. Using steady-state, Wabner's j_0 value equals 10^{-11} A cm^{-2}. Using non-steady state, $j_0 = 0.2$ A cm^{-2}. His argument that rate of gaseous O_2 diffusion limits the steady-state process would not explain how other gas-evolving reactions can have j_0 values of 10^{-3} A cm^{-2}. The second area of uncertainty relates to the absolute value of overpotential. Because of its porous and rough nature, it is virtually impossible to derive kinetic data on a true surface-area basis. This is so not only for lead, but also its alloys, and so we find it hard to reach unequivocal conclusions as to the electrocatalytic activity of the alloys compared with the parent lead. These questions must await further study, which will no doubt be forthcoming.

Oxidation of Chlorates to Perchlorates

$$ClO_3^- + H_2O \rightarrow ClO_4^- + 2\,H^+ + 2e \qquad E° = 1.19\ V^{37}$$

Conditions

Lead dioxide has been extensively investigated as an alternative anode material to platinum in electrolytic perchlorate manufacture. Most studies have been parametric, examining the influence of current density [both anodic (j_a)[38] and cathodic (j_c)[39]], pH,[38] temperature (T),[38] cathode material,[39] additives[38,39,41] and chlorate concentration[39,42–51] on cumulative current efficiency (C.E.). Table I summarizes the conditions defined in these studies. Anode stability has also been extensively investigated.[39,40,43,44,47–49] Chlorate oxidation occurs at current efficiencies of less than 100% because of the simultaneous evolution of oxygen.

$$2\,H_2O \rightleftharpoons O_2 + 4\,H^+ + 4e \qquad E° = 1.229\ V$$

Hence the process is operated at high current densities under constant current conditions, a high anodic potential (V_a) being required for a reasonable rate, e.g. $j_{ClO_3} = 10$ A dm^{-2} at $V_a = 2.2$ V* in 0.47 M ClO_3^-.[41]

* All potentials expressed with respect to S.H.E.

Table I. Conditions for the electrolytic production of perchlorate.

Electrolyte	Anode	Cathode	Anode current density (A dm⁻²)	Cathode current density (A dm⁻²)	Temperature (°C)	pH	Additive	Current efficiency (%)	Conversion (%)	Cell voltage (V)	Energy (d.c.) (kWh kg⁻¹)	Ref. No.
$NaClO_3$ (sat.)	PbO_2/Fe	Mild steel	20·0	—	15·0	—	None	72–79	—	5·05	—	50
$NaClO_3$ (4·7 M)	PbO_2/graphite PbO_2/Ni	—	12·0	—	5·0	—	None	58	100	—	—	51
$NaClO_3$ (sat.)	PbO_2-massive	Stainless steel (18 :8)	20·0	20·0	29·0–35·0	—	NaF (0·047 M)	99	82	4·5	3·18–3·30	52, 47
$NaClO_3$ (4·74 M)	PbO_2/Ni PbO_2/Pt-Ta	Stainless steel	15·5	7·25	30·0–45·0	6·5 (HCl)	NaF (0·012 M)	92	90	4·75	2·29	39, 42
$NaClO_3$ (5·65 M)	PbO_2/Ta	Stainless steel	28·0	—	33·0	—	None	53	91	6·3	—	48
$NaClO_3$ (5·7 M)	PbO_2/Ta	Stainless steel	30·0	—	25·0–35·0	—	$K_2S_2O_8$ (0·008 M)	75	84	5·0–6·5	—	43
$NaClO_3$ (5·75 M)	PbO_2/graphite	Stainless steel	23·2	—	28·0–30·0	6·6–6·8 (H_2SO_4)	NaF (0·047 M)	83	86	4·3	2·22	38
$NaClO_3$ (5·7 M)	PbO_2/graphite	Stainless steel	40·0	—	40·0	8·0–11·0	NaF (0·047 M)	83	86	4·7–5·1	2·37	
$NaClO_3$ (6·91 M)	PbO_2-massive	Stainless steel	25·0	35·0	45·0–52·0	—	NaF (0·047 M)	70	99	4·5–5·0	3·0	40
$NaClO_3$ (5·7 M)	PbO_2/graphite	Stainless steel	23·0	—	37·0–44·0	6·6–6·8 (HCl)	NaF (0·047 M)	67	94	4·2–4·6	—	46

Current efficiency decreases with decreasing chlorate concentration, falling markedly, e.g. 40%,[46] at concentrations below 1 M[43,46] unless the electrolyte contains an additive, e.g. NaF. Such a decrease in C.E. is not observed at platinum anodes.[43] De Nora *et al.*[41] have shown that F⁻ markedly inhibits oxygen evolution at PbO_2 anodes while only slightly retarding chlorate oxidation in the chlorate cell (*see also* p. 366). In contrast, sodium dichromate, a common additive with platinum anodes, decreases perchlorate current and increases oxygen current in the relevant potential range. Chlorate oxidation occurs at potentials 0·25 V less positive at PbO_2 than at Pt in the same electrolyte (1–6·4 M $NaClO_3$), and current increases more quickly with potential increase.[40] Hence, although C.E. is less with lead dioxide anodes,[41,51,43] the d.c. energy requirement is similar to that for platinum anodes (3·0–4·0 kW h kg⁻¹)[43,53,54] because of the lower cell voltage.

Most workers recommend slightly acid conditions for high C.E.; however, Narasimham *et al.*[38] claim that under otherwise similar conditions, current efficiencies are higher in alkaline solutions (pH 8–9). They observed surface pH to be lower than bulk pH, e.g. 1–2 at a bulk pH of 6·8 during electrolysis, with high surface acidity increasing decomposition of chloric acid and lowering current efficiency. Reported temperature effects are conflicting and may depend on pH. Narasimham *et al.*[38] report an optimum temperature of 30°C under acidic conditions (pH 6·6–6·8), with negligible temperature influence on C.E. under alkaline conditions (pH 8–11·5). Osuga *et al.*[40] observed a 20% increase in C.E. with increase in temperature (30–67°C), but have not specified the pH used.

Kinetics and mechanism

Few interpretative kinetic data exist for chlorate oxidation at lead dioxide electrodes. Most investigations of the reaction have been carried out in simple, undivided two-electrode cells under galvanostatic conditions.

From the observed increase in acidity of the anode surface during electrolysis, and the formation of perchlorate only at potentials positive to oxygen evolution, Narasimham *et al.*[38] postulated the reaction to proceed via adsorbed chlorate radicals, i.e.

$$ClO_3^- \rightarrow \dot{C}lO_3 + e$$
$$ClO_3 + H_2O \rightarrow ClO_4^- + 2\,H^+ + e$$

They also observed anode rotation to decrease C.E., particularly in alkaline solution, suggesting that anode rotation increased surface pH, thereby enhancing oxygen evolution.[55] Thermodynamically it is not possible to exclusively favour ClO_3^- oxidation or O_2 evolution by changing pH, since hydrogen ion concentration influences both equilibria identically. Kinetically, oxygen evolution at PbO_2 anodes has only been thoroughly investigated in H_2SO_4 (see p. 218) while the influence of pH on the kinetics of chlorate oxidation has not been investigated.

De Nora et al.[41] recorded polarization curves for both chlorate oxidation and the o.e.r. at PbO_2 in chlorate electrolyte (0·47–4·7 M). They observed that ClO_3^- inhibits oxygen evolution at lead dioxide, although to a lesser extent than at platinum anodes. In solutions of higher pH (11·4–12·0), Osuga et al.[40] also observed limiting oxygen currents at potentials between 1·24–1·84 V in the presence of 5 M $NaClO_3$, perchlorate formation occurring around 2·04 V. De Nora et al.[41] suggested that with increasing anode potential, ClO_3^- gradually replaces water at the anode surface. Their polarization data at PbO_2 anodes gave linear Tafel plots for chlorate oxidation, but not for the o.e.r. over the potential range 2·11–2·5 V, indicating that chlorate oxidation occurs independently of oxygen evolution, rather than by interaction with adsorbed oxygen species. Tafel slopes of 0·11–0·12 indicated the rate-determining step to be a simple charge transfer, namely:

$$ClO_3^- \rightarrow \dot{C}lO_3 + e$$

Hence De Nora et al.[41] implied that the lead dioxide acts as an inert anode, previously observed differences in C.E. between Pt and PbO_2 anodes possibly resulting from their differing abilities to adsorb ClO_3^-.

Oxidation of Chloride to Chlorate

$$Cl^- + 6\,OH^- \rightarrow ClO_3^- + 3\,H_2O + 6e \qquad E° = 0·63\ V^{57}$$

Conditions

Table II gives the conditions used in chlorate cells with lead dioxide anodes, and the cumulative current efficiencies obtained. PbO_2 anodes are reported to be capable of operating at higher current densities[58–61] and temperatures[58,60–62] than the more commonly used graphite anodes.

Table II. Conditions for the electrolytic production of chlorate.

Electrolyte	Anode	Cathode	Anode current density (A dm^{-2})	Cathode current density (A dm^{-2})	Temperature (°C)	pH	Additive	Current efficiency (%)	Conversion (%)	Cell voltage (V)	Energy (d.c.) (kW h kg^{-1})	Ref. No.
KCl (0·43 M)	PbO$_2$-massive	Fe	20·0	—	60·0–80·0	Slightly acid	Na$_2$Cr$_2$O$_7$ (0·001 M)	81	40	—	—	47, 79
KCl (3·50 M)	PbO$_2$/graphite	Stainless steel	18·0	—	65·0	6·6–7·0 (HCl)	Na$_2$Cr$_2$O$_7$ (0·01 M)	87	—	3·6	5·45	58
NaCl (3·67–4·27 M) NaClO$_3$ (1·70–2·25 M)	PbO$_2$/graphite	Stainless steel	3·36	—	37·0	6·6 (HCl)	Na$_2$Cr$_2$O$_7$ (0·008 M)	75–80	60	3·2	6·0–6·4	59
NaCl (3·76–4·27 M) NaClO$_3$ (1·77–1·88 M)	PbO$_2$/graphite	Mild steel	5·2–5·7	2·5–3·0	39·0	6·4–6·8 (HCl)	Na$_2$Cr$_2$O$_7$ (0·008 M)	68–73	60	3·2–3·3	6·8–7·2	62
NaCl (4·1–4·6 M)	PbO$_2$/graphite	Stainless steel	4·9–5·1	1·40	40·0–47·0	6·2–6·8 (HCl)	Na$_2$Cr$_2$O$_7$ (0·008 M)	72–78	60	3·4–3·7	7·0–7·6	61, 66
NaClO$_3$ (0·56–0·75 M)		Mild steel	10·5	1·13	40–45	6·2–6·8 (HCl)	Na$_2$Cr$_2$O$_7$ (0·008 M)	68–75	60	3·7–4·2	8·5–9·0	
NaCl (5·12 M)	PbO$_2$/graphite	—	20·0	—	40·0	Slightly acid	Na$_2$Cr$_2$O$_7$ (0·02 M)	93	—	—		80
NaCl (5·12 M)	PbO$_2$/?	—	20·0–30·0	—	40·0	6–7	Na$_2$Cr$_2$O$_7$ (0·008 M)	90–93	—	—	—	81
KCl (3·83 M) KClO$_3$ (0·27 M)	PbO$_2$/graphite	Stainless steel	5·0	—	55·0–60·0	6–7 (HCl)	Na$_2$Cr$_2$O$_7$ (0·008 M)	82–85	—	—	6·3–6·5	60

Similar current efficiencies are observed at both anode types.[63,64] Nam[64] measured a small increase in C.E. (10%) with increase in j_a (5–25 A dm^{-2}). Temperature effects appear conflicting. Nam[64] found C.E. to increase (59–74%) with increase in temperature (10–60°C), whereas Shlyapnikov[65] reported a decrease in C.E. with temperature (40–80°C). Higher operating temperatures decrease cell voltage.[58,62,65]

Current efficiency remains relatively high as the chloride concentration decreases,[59,61,69] e.g. 75–80% at 1·02 M chloride and 42–46% at 0·17 M chloride ($j = 5$ A dm^{-2}).[61] Graphite anodes disintegrate under such conditions.[61,69] Anode potentials are higher at PbO_2 than at graphite for the same current densities, e.g. by 0·2 V at 10 A dm^{-2}.[70] However, Indian workers[59] point out that the greater stability of PbO_2 electrodes does not allow the interelectrode gap to increase during cell operation as it does with graphite. The greater stability of PbO_2/graphite anodes (cf. graphite[59,61,66–68]) simplifies product purification.

Sodium dichromate is commonly used as a cell additive to maintain high efficiency in cells without diaphragms by preventing product reduction at the cathode. There has been no investigation of its effect on the anodic reaction at least with dioxide anodes (cf. chlorate oxidation), although Jaksic et al.[71] found that its presence decreases C.E. at Pt/Ti anodes by increasing oxygen evolution. Using PbO_2/Ti anodes, Shlyapnikov[63] found the optimum pH for chlorate formation at 6·2–6·8. At pH values of less than 6 he observed increased Cl_2 evolution, while at pH values greater than 7 there was increased oxygen evolution.

Kinetics and mechanism

It seems to be accepted that chlorate formation at PbO_2 anodes occurs by the same combination of electrochemical and chemical reactions assumed to occur at other inert anodes,[72–74] i.e.

Anode: $$2\,Cl^- \rightarrow Cl_2 + 2e \tag{1}$$

Cathode: $$2\,H_2O + 2e \rightarrow H_2 + 2\,OH^- \tag{2}$$

Solution: $$Cl_2 + OH^- \rightarrow HOCl + Cl^- \tag{3}$$

Reactions: $$HOCl + OH^- \rightarrow OCl^- + H_2O \tag{4}$$

$$2\,HOCl + ClO^- \rightarrow ClO_3^- + 2\,Cl^- + 2H^+ \tag{5}$$

Reaction (5) is favoured by high temperature and slight acidity.[78] Current efficiency is decreased by a series of anode reactions,[73] the main one being oxidation of OCl^-,[73,74]

$$6\,OCl^- + 3\,H_2O \;\rightarrow\; 2\,ClO_3^- + 4\,Cl^- + 6\,H^+ + \tfrac{3}{2}\,O_2 + 6e$$
$$E^\circ = 1\cdot 14 \text{ V}$$

Oxygen evolution by discharge of water is considered negligible in the electrolytes of chlorate cells.[73,74] In contrast to the above reaction scheme, Russian workers[75–77] consider the mechanisms of chlorate formation to be mainly electrochemical, especially at high temperatures. This conclusion is extended to lead dioxide anodes, although on what evidence is not entirely clear.

In his investigation of various anode materials for chlorate production (PbO_2, MnO_2, RuO_2, Pt, C), Shlyapnikov[63] observed the same type of variation of C.E. with change of pH, j_a and T for each anode type, and similar variations in current in the potential range 1·40–2·0 V. He concluded that electrode material has no influence on the mechanism of chlorate formation. A Tafel slope of 0·15 can be estimated from Shlyapnikov's data for chlorate formation at PbO_2 in the potential range 1·45–1·62 V ($j_a = 1\cdot 0$–10 A dm^{-2}). At a potential of 1·47 V, the reaction proceeds at a rate of 0·87 A dm^{-2}, assuming the 87% current efficiency given by Shlyapnikov. Izidinov and Veselovskii[82] have also investigated the mechanism of this reaction. They reported steady state and transient current–voltage data, also showing the effect of added Co^{2+} ions. However, in view of the various possible partial reactions, these data are of limited value. They do, however, show current efficiency at one pH value as a function of potential, with simultaneous presentation of homogeneous chlorate yields at the same time. The paper of Izidinov and Veselovskii[82] is difficult to understand (for example, why should chemical chlorate yield be potential dependent?) and dwells largely on changes occurring at the electrode surface, as inferred from other observations, and is thus somewhat speculative.

Oxidation of Iodate to Periodate

$$IO_3^- + H_2O \;\rightarrow\; IO_4^- + 2\,H^+ + 2e \qquad E^\circ = 1\cdot 6 \text{ V}^{[83]}$$

Conditions

Surprisingly, electrolytic oxidation of iodate occurs at a much higher current efficiency at PbO_2 than at platinum.[84-87] Some reaction at smooth platinum has been observed under alkaline conditions[88] (C.E. = 11% in 1 M KOH, 0·1 M KIO_3), with little reaction under acidic conditions (C.E. = 1% in 0·5 M H_2SO_4, 0·1 M KIO_3). No oxidation of iodate occurs at manganese dioxide or carbon anodes.[87,88]

Periodate is a selective oxidizer for cis-glycol configurations, electrolytically generated IO_4^- being used extensively to oxidize starch to dialdehyde starch.[89-97] Electrolytic regeneration of spent periodate greatly reduces total consumption of periodate.[90] Unfortunately, most investigators in this area have not focused specifically on the anodic reaction but rather on (other) plant design,[96] cell design[97] and yield of dialdehyde starch.[89-92,95] The anode used in most cases has been lead,[91-95,97] even though it is restricted to sulphate media, and is reported to give significant contamination in the final product.[90]

The various "conditions" studies are summarized in Table III. A considerable variety of conditions have been used, and reports of the effects of parameters such as pH, temperature and iodate concentration on C.E. are conflicting. Current efficiency depends on the degree of conversion of IO_3^-,[94,98] although an efficiency of 60–70% seems reasonable for complete conversion. Several workers[88,94,99] note that C.E. is markedly influenced by electrode roughness (surface area), e.g. PbO_2/Pb anodes require lower potentials at the same C.E. than PbO_2/graphite anodes.[94,99] Similar effects have been observed at platinum.

With increasing j_a, current efficiency reaches a maximum value,[94,99] then decreases as a result of increasing oxygen evolution.[93,100,101] Reported optimum current densities differ (different electrolysis conditions and electrode surfaces), but maximum C.E. would seem to occur around 2·0 V.[94,99] For lead anodes, j_a is often less than 5 A dm^{-2}.[94] The effect of iodate concentration is not clear. Dzhafarov and Efendieva[101] report a large increase in C.E. (9·9–58·6%) with increase in concentration (0·12–0·23 M) at $j_a = 10$ A dm^{-2}, with little further increase (10%) up to 0·47 M IO_3^-. In a pumped cell (41 h^{-1}, j_a = 3·9 A dm^{-2}), Branislav[99] observed a regular increase in C.E. (28·5–67·1%) with increase in concentration (0·28–1·13 M).

Table III. Conditions for the electrolytic production of periodate.

Electrolyte	Anode	Cathode	Anode current density ($A\,dm^{-2}$)	pH	Temperature (°C)	Additive	Current efficiency (%)	Conversion (%)	Cell voltage (V)	Energy (d.c.) ($kW\,h\,kg^{-1}$)	Ref. No.
KIO_3 (0·1 M)	PbO_2/Pt	Pt	2·5	(1) 0 (2) 14	18·0 18·0	KF (0·1 M)	6 38	—	—	—	88
$NaIO_3$ (1·57 M)	PbO_2/Pb	Graphite	11·4	Slightly acidic	50·0	—	45	97	—	—	100
HIO_3 (0·34–0·45 M)	PbO_2/Pb	Carbon steel	(1) 20–30 (2) 2–3		25·0–40·0	Na_2SO_4 (0·05 M)	75–90 80–90	—	—	—	93
HIO_3 (0·5 M)	PbO_2-massive	Graphite	15·0	Acidic	30·0–40·0	—	71	96	5·2–5·8	—	87
$NaIO_3$ (0·47 M)	PbO_2-massive	Graphite	5·0	—	22·0–38·0	—	69	95	4·8–6·3	—	
HIO_3 (0·43 M)	PbO_2/Pb	Al	5·0	0·5–4·0	35·0	—	81	—	—	—	94
KIO_3 (0·35 M)	$PbO_2/?$	—	20·0	—	60·0	$K_2Cr_2O_7$ (0·007 M)	70	96–98	—	—	101
KIO_3 (1 M)	PbO_2/graphite	Stainless steel	9·4	7·0	60·0	$K_2Cr_2O_7$ (0·007 M)	68·5	98	7·5	3·0	98

It is also difficult to specify an optimum pH. Acid or neutral solutions are commonly used. Variation in pH between 1·0 and 5·0 has little influence on C.E.[94,99] Nam and Kim[98] observed maximum C.E. at pH 7, while Hickling and Richards[88] recommended alkaline conditions (0·1 M KOH). Similarly some authors[88,101] claim increase in temperature (20–80°C) increases C.E., while others[94,99] state that T has a minor effect. In non-diaphragm cells, $2 \text{ g l}^{-1} \text{ Na}_2\text{Cr}_2\text{O}_7$ greatly increases current efficiency.[98]

Kinetics and mechanism

The lack of fundamental data on iodate oxidation is somewhat surprising in view of the unique position of lead dioxide as anode for this reaction. In a very early paper, Muller[84] suggested the oxidation was electrochemical, the high efficiency at lead dioxide resulting from its "catalytic activity". In contrast, Hickling and Richards,[88] in another early paper, postulated that IO_4^- was formed by chemical interaction between IO_3^- and higher oxides of lead.

Using a pumped cell, Branislav[99] found the conversion rate of IO_3^- to increase with increasing pump rate, suggesting the reaction may be subject to some mass-transport control. With a magnetically stirred cell, Nam and Kim[98] observed a limiting current region in the potential range 1·9–2·17 V, current increasing "exponentially" at higher potentials. However, the current plateau is not due to mass-transport limitations, since the magnitude of the limiting current decreases with increasing iodate concentration (0·5–1·0 M). This may indicate inhibition of oxygen evolution by adsorbed IO_3^-, or perhaps chemical interaction between IO_3^- and adsorbed oxygen species. No such limiting plateau currents are observed at Pt anodes.[98] Semchenko et al.[102] have recorded partial polarization curves for the oxidation of iodic acid and the o.e.r. in HIO_4 (1·5–3·0 M). Increasing the iodic acid concentration from 0·1 to 1·0 M caused increasing inhibition of oxygen evolution. An approximate Tafel slope of 0·12 can be obtained from the data of Semchenko et al. in the potential range 2·06–2·19 V.

Oxidation of Iodide (and Iodine) to Iodate

Included in this section is the oxidation of alkaline iodine solutions. In view of the quantitative nature of the reaction

$$3 \text{ I}_2 + 6 \text{ OH}^- \rightarrow 5 \text{ I}^- + IO_3^- + 3 \text{ H}_2\text{O},^{103}$$

the oxidation of alkaline iodine and iodide solutions can be given by:

$$I^- + 6\,OH^- \rightarrow IO_3^- + 3\,H_2O + 6e \qquad E^\circ = 0{\cdot}26\;V \text{[104]}$$

Conditions

Interest in IO_3^- formation stems mainly from its occurrence as an intermediate in the electrolytic production of periodate from iodide or iodine. Conditions formulated in the available parametric–current-efficiency studies are given in Table IV. Venkatachalapathy et al.[105] gave the optimum current density as 20 A dm^{-2}, smaller or larger current densities decreasing current efficiency. In contrast, Dzhafarov et al.[106] reported j_a to have little influence on current efficiency in the range 5–60 A dm^{-2}. Current efficiency may reach a constant value with increasing iodide concentration, Dzhafarov et al.[106] finding little change in C.E. above an iodide concentration of 0·3 M, while Venkatachalapathy et al.[105] observed a small decrease in C.E. at I_2 concentrations above 0·775 M. The reaction was carried out in alkaline solution; however, Nam and Kim[107] observed a decrease in C.E. at hydroxide concentrations above 0·1 M. Temperature had little influence on the reaction up to 80°C.[106,107] In cells without diaphragms, high C.E. was maintained by addition of dichromate. Under similar conditions the efficiency at lead dioxide anodes was comparable with that at platinum anodes.[106,108] Anode stability has been determined by Aiya et al.[108]

Kinetics and mechanism

No kinetic data are available on I^- oxidation at lead dioxide anodes in alkaline solutions. Some authors[106,109] imply the stoichiometric electrochemical–chemical reaction sequence below, alkaline conditions favouring reactions (2) and (3):

$$2\,I^- \rightarrow I_2 + 2e \tag{1}$$

$$I_2 + OH^- \rightarrow HOI + I^- \tag{2}$$

$$HOI + OH^- \rightarrow IO^- + H_2O \tag{3}$$

$$2\,HOI + IO^- \rightarrow IO_3^- + 2\,H^+ + 2\,I^- \tag{4}$$

Table IV. Conditions for the electrolytic production of iodate.

Electrolyte	Anode	Cathode	Anode current density (A dm⁻²)	pH	Temperature (°C)	Additive	Current efficiency (%)	Conversion (%)	Cell voltage (V)	Energy (d.c.) (kW h kg⁻¹)	Ref. No.
I_2 (0·79 M) NaOH (0·83 M)	PbO_2/Pb	Graphite	11·4	Basic	50·0	Na_2SO_4 (0·07 M)	95	—	6·0	—	112, 100
I_2 (0·31 M) NaOH (0·75 M)	PbO_2/Pb	Carbon steel	20–30	Basic	—	None (diaphragm)	75–90	—	—	—	93
I_2 (0·33 M) in bromine water (0·2 M)	PbO_2-massive	Cu	14·0	—	40·0–50·0	None (diaphragm)	94	96	3·6–4·2	—	108
I_2 (0·78 M) NaOH (2·0 M)	PbO_2-massive	Stainless steel	25·0	8·0–9·5	50·0–60·0	$Na_2Cr_2O_7$ (0·006 M)	98	97	—	—	108
I_2 (0·79 M) NaOH (2·0 M)	PbO_2/graphite	Stainless steel	20·0	10·0–11·0	45·0–50·0	None (diaphragm)	73	77	4·0	3·4	105
KI (1·26 M)	PbO_2/graphite	Stainless steel	20·0	9·0	70·0	$K_2Cr_2O_7$ (0·006 M)	95	—	3·6	—	109
KI (1·8 M) KOH (0·1 M)	PbO_2-massive	Stainless steel	20·0	13·0	60·0	$K_2Cr_2O_7$ (0·0003 M)	93	93	4·0	4·5	107

Nam and Kim[107] have recorded total current–potential data in $0\cdot1$ M KOH at various concentrations of KI ($0\cdot5$–$2\cdot0$ M), which show the electrode reaction to commence at potentials around $0\cdot5$–$0\cdot6$ V.

In the absence of alkali, Nam and Kim[107] found iodide current to reach a limiting value in the potential range $0\cdot7$–$0\cdot9$ V, the current being linearly proportional to iodide concentration. No limiting currents were observed at platinum anodes, or at lead dioxide anodes in the presence of $0\cdot1$ M KOH. Current limitation was said to result from chemical interaction between I^- and PbO_2, namely:

$$PbO_2 + 2\,I^- + 2\,H^+ \rightarrow PbO + I_2 + H_2O,$$

the anode not acting as an inert anode.

Semchenko et al.[110,111] have studied iodine oxidation at lead dioxide and platinum anodes in acidic solution ($1\cdot5$–$3\cdot0$ M HIO_4),* recording partial polarization curves for both iodine oxidation (4×10^{-3} M) and the accompanying oxygen reaction. For iodine oxidation, two linear Tafel regions were obtained at lead dioxide, the first ($2\cdot04$–$2\cdot16$ V) being assigned (on rather qualitative evidence) to HIO_3 formation, the second ($2\cdot16$–$2\cdot26$ V) to HIO_4 formation. The first Tafel region is described by:

$$V_a = 2\cdot127 + 0.11 \log j,$$

the slope suggesting a charge transfer as the rate-determining step. Iodine was found to inhibit oxygen evolution at lead dioxide, unlike platinum where both iodine oxidation and the o.e.r. proceeded independently. Semchenko et al. postulated that oxidation products of water participated in iodine oxidation, although presumably iodine adsorption could also hinder oxygen evolution.

Oxidation of Iodide (and Iodine) to Periodate

Anodes of lead dioxide have been used, in a one-stage combination of the previous two reactions, to prepare periodate from iodide

---

* $\frac{1}{2}\,I_2 + 3\,H_2O \rightarrow IO_3^- + 6\,H^+ + 5e$, $E° = 1\cdot195$ V.[104]

solutions.[89,93,108,112,113] This process is more practical than the two-stage process, originally formulated by Willard and Ralston,[86] which used platinum anodes to oxidize iodide to iodate, and lead dioxide anodes to oxidize iodate to periodate. Present studies have been restricted to simple current efficiency studies, and are adequately summarized by Nam and Kim.[113] In general, the appropriate solution of iodide or iodine can be oxidized to periodate at current densities of 3–15 A dm^{-2} with current efficiencies of 75–85% for 96–98% conversion. In non-diaphragm cells, dichromate is used as additive to prevent cathodic reduction of periodate. Reported energy consumption (d.c.) is 6–7 kW h kg^{-1} IO$_4^-$. See Aiya *et al.*[108] and Nam and Kim[113] for comment on anode stability.

Oxidation of Bromide to Bromate

$$Br^- + 6\,OH^- \rightarrow BrO_3^- + 3\,H_2O + 6e \qquad E^\circ = 0{\cdot}61\ V^{114}$$

Conditions

Relatively few workers have examined PbO$_2$ as a substitute anode material for graphite in the oxidation of bromide. Conditions and current efficiencies from available studies are given in Table V, and are similar to those for graphite.[115] Sundarajan *et al.*[116] have shown that C.E. varies little with j_a (10–40 A dm^{-2}), T (40–60°C) or pH, provided the electrolyte is initially alkaline. If the electrolyte is maintained acidic, C.E. drops considerably due to evolution of Br$_2$, possibly as a result of reaction between Br$^-$ and BrO$_3^-$,[114] or a decrease in the solution reactions (2) and (3) below. Dichromate additive is required to maintain high C.E. in cells without diaphragm.[116]

Kinetics and mechanism

Again there is a lack of kinetic data on this reaction at lead dioxide anodes. The combined electrochemical–chemical sequence of reactions

Table V. Conditions for the electrolytic production of bromate.

Electrolyte	Anode	Cathode	Anode current density (A dm^{-2})	pH	Temperature (°C)	Additive	Current efficiency (%)	Conversion (%)	Cell voltage (V)	Energy (d.c.) (kW h kg^{-1})	Ref. No.
KBr (1·80 M) KBrO$_3$ (0·13 M)	PbO$_2$-massive	Stainless steel	20·0-22·0	1·0	65·0-72·0	K$_2$Cr$_2$O$_7$ (0·006 M)	90-92	97-90	3·5-3·8	3·84	117
NaBr (2·3 M) NaBrO$_3$ (0·46 M)	PbO$_2$-massive	Stainless steel	20·0-22·0	9·0-9·7	75·0-85·0	None (no diaphragm)	70-73	99	3·8-4·3	—	119
KBr (1·9 M) KBrO$_3$ (0·14 M)	PbO$_2$/graphite	Stainless steel	22·4	8·0→	55·0-60·0	K$_2$Cr$_2$O$_7$ (0·006 M)	95·2		3·4-3·7	3·60	116
NaBr (3·7 M)	PbO$_2$/graphite	Stainless steel	22·4	8·0→	55·0-60·0	K$_2$Cr$_2$O$_7$ (0·006 M)	95·4		3·9-4·5	4·45	116
KBr (0·63 M) KOH (0·18 M)	PbO$_2$/Ni	—	20-50	Basic	25·0	K$_2$Cr$_2$O$_7$ (0·006 M)	95	—	—	—	120

postulated in early work[118] is assumed to also apply at lead dioxide,[117] i.e.

$$2\,Br^- \rightarrow Br_2 + 2e \tag{1}$$

$$Br_2 + OH^- \rightarrow HBrO + Br^- \tag{2}$$

$$Br_2 + OH^- \rightarrow BrO^- + H_2O + Br^- \tag{3}$$

$$2\,HBrO + BrO^- \rightarrow BrO_3^- + 2\,H^+ + 2\,Br^- \tag{4}$$

with reactions (2) and (3) being favoured by high pH and, it is thought, high temperature.[121]

Oxidation of Sulphate to Perdisulphate

$$2\,SO_4^{2-} \rightarrow S_2O_8^{2-} + 2e \qquad\qquad E° = 2·01\ V^{122}$$

Conditions

Early attempts[123] to produce perdisulphate by oxidation of $(NH_4)_2SO_4$–H_2SO_4 solution at anodes of PbO_2 ($j_a = 50$ A dm^{-2}) failed because of oxygen evolution. By making the electrolyte 0·5–3·0 M in HF, and working at lower current densities (10 A dm^{-2}), Japanese workers[47] obtained a current efficiency of 60% for perdisulphate formation. Later workers[124–126] have all used HF presumably as a means of retarding the o.e.r. General conditions for the oxidation of sulphate (or bisulphate) at lead dioxide anodes can be summarized as: $j_a = 10$–20 A dm^{-2}, $T \simeq$ 25°C, with the electrolyte consisting of $(NH_4)_2SO_4$ (4–4·5 M), H_2SO_4 (1–2 M) and HF (0·5–3·0 M). Current efficiencies are 60–70%.

Nam and Kim[127] observed the percentage ozone in the anode gases to decrease with increasing T (10–40°C) at PbO_2, the reverse occurring at platinum anodes. Hence higher temperature may increase current efficiency. These authors also measured total anode current as a function of potential in a saturated $(NH_4)_2SO_4$ solution, 2 M H_2SO_4, in the presence and absence of HF. With increasing HF concentration (1·2–2·4 M), the polarization curves shifted to more positive potentials approaching those recorded at platinum. In the presence of 2·4 M HF, perdisulphate formation proceeded at a rate of 6 A dm^{-2} (assuming efficiency = 60%) at a potential of 2·41 V.[127]

Kinetics and mechanism

As part of a wider study on the lead anode, Izidinov and Rakhmatul-lina[128] monitored the formation of $H_2S_2O_8$ in 5 M H_2SO_4 at $-20°C$ over the potential range 2·0–6·0 V. Although the major reaction was oxygen evolution (C.E. = 95–96%), they detected $H_2S_2O_8$ formation at low rates (10^{-2} A dm^{-2}) at 2·1 V, the rate reaching a maximum of approximately 0·25 A dm^{-2} at 2·3 V, with a C.E. of 2·3%. At potentials above 2·3 V, $H_2S_2O_8$ formation decreased, this change coinciding with increased rate of anode corrosion and ozone formation. Since the electrode surface was covered with adsorbed oxygen at 2·1 V,[129] these authors gave the mechanism as:

$$PbO_2(O)_x + HSO_4^- \rightarrow PbO_2(O)_xHSO_4 + e$$

$$PbO_2(O)_xHSO_4 + HSO_4^- \rightarrow PbO_2(O)_x + H_2S_2O_8 + e$$

At potentials above 2·3 V it was postulated that bond formation occurred between lead and the adsorbed bisulphate anion-radicals, giving surface compounds, e.g. $PbO(HSO_4)_2$, whose hydrolytic decomposition resulted in anode corrosion:

$$PbO(HSO_4)_2 + H_2O \rightarrow PbO_{2\,sol} + 2\,H_2SO_4$$

Estimation of the Tafel slope from the current potential data for $H_2S_2O_8$ formation gave a value around 0·11 in the potential range 2·1–2·3 V, indicating charge transfer as possibly the rate-determining step.

Oxidation of Cr(III) to Cr(VI)

This has recently been the subject of a comprehensive review by one of us[56] and will not therefore be discussed at any length. From such data as are available, there are indications that the reaction proceeds via an e.c. mechanism, that is, electrochemical PbO_2 formation followed by chemical oxidation of the Cr. Analysis of data in this chapter suggests strongly that the reaction is not diffusion-controlled and that the limiting currents observed are due to hindrance of one or other of the e.c. steps described above. Two further papers have been published by Lee and Sekine,[30] these relating to the operation of an undivided cell.

Readers should also refer to p. 405 where corrosion of lead anodes in chromic acid is discussed.

Miscellaneous

This section summarizes the findings on reactions studied much less extensively at lead dioxide anodes than the previous reactions.

Oxidation of sulphite

Strafelda and Krofta[130] have briefly studied sulphite oxidation on a platinum-based lead dioxide anode in 0·5 M H_2SO_4. They obtained a well-defined current potential wave in a magnetically stirred solution ($E_{1/2} = 1·64$ V). Limiting currents were linearly proportional to SO_3^{2-} concentration in the range $1·53 \times 10^{-3}$–$3·85 \times 10^{-2}$ M, suggesting an analytical application for the lead dioxide anode. In contrast, oxide formation rapidly inhibits SO_3^{2-} oxidation at Pt and Au.

Current efficiency studies showed the reaction to involve sulphate formation and not dithionate or perdisulphate, i.e.

$$SO_3^{2-} + H_2O \rightleftharpoons SO_4^{2-} + 2\,H^+ + 2e \qquad E° = 0·17\ V^{131}$$

No mechanistic analysis was made, but the large difference between $E_{1/2}$ and $E°$, and the steepness of the current potential curve ($E_{3/4} - E_{1/4} \approx 62·5$ mV) would strongly suggest sulphate formation via chemical reaction between PbO_2 and SO_3^{2-}, current flow resulting from electrolytic regeneration of PbO_2. At Pt and Au, sulphite oxidation commenced at much less positive potentials, e.g. 0·6 V. At sulphite concentrations above $3·85 \times 10^{-2}$ M, current decreased with increasing concentration and no longer varied linearly with concentration.

Oxidation of chloride to chlorine

Thangappan et al.[46] formed Cl_2 with 96–98% current efficiency by electrolysing a saturated NaCl solution in a diaphragm cell at 85–90°C with an anode of PbO_2/graphite.

Current efficiency was reported to be independent of j_a in the range $5 \cdot 1$–$15 \cdot 3$ A dm^{-2}. Mraz et al.[132] obtained Tafel plots over a range of temperature (15–75°C) for the oxidation of Cl$^-$ at PbO$_2$/Pt anodes in a 6 M NaCl solution saturated with Cl$_2$. Activation energy (E) for the reaction varied linearly with overpotential (η), with $E_{\eta=0}$ being determined by extrapolation. The Tafel parameters, a and b, and the activation energy, varied from one lead dioxide electrode to another; however, $E_{\eta=0}$ was independent of the electrode, the value of $18 \cdot 2$ kcal mol^{-1} being higher at PbO$_2$ than at graphite or γMnO$_2$. Tabulated b values were $0 \cdot 150$ and $0 \cdot 201$ with exchange current densities being given as $6 \cdot 74 \times 10^{-3}$ and $7 \cdot 95 \times 10^{-3}$ A dm^{-2} for the two lead dioxide anodes used. For a purely activation controlled process, $E = E_{\eta=0} - \alpha Z F_Z$, slopes of E vs η plots were negative but not equal to $\alpha Z F$, suggesting the electrode reaction was influenced by a control other than that of activation.

Oxidation of manganese (II)

Lead has been used extensively in the electrolysis of manganous sulphate to produce manganese dioxide. Fleischmann et al.[133] have shown the mechanism of MnO$_2$ formation at platinum anodes to be quite complex. Studies of MnO$_2$ formation at lead anodes have only examined factors affecting current efficiency. As the lead anode is probably functioning as a manganese dioxide electrode, these studies will not be considered here.

A few efficiency studies of Mn^{2+} oxidation at PbO$_2$ anodes have been made under conditions where MnO$_2$ does not form on the anode surface, i.e. under conditions which stabilize the Mn^{3+} species (high concentration of H$_2$SO$_4$ and low temperature).[134] Potdar and Udupa,[135] using a PbO$_2$/graphite anode, formed MnO$_2$ as a fine suspension in solution by electrolysing a solution of MnSO$_4$ (2 M) and H$_2$SO$_4$ (1·6 M) at 30°C, with a current density of 5 A dm^{-2}. Current efficiency obtained was 75%. Similarly, Solanki and Prabhanjanmurty,[136] using a lead anode, formed a mixture of soluble higher manganese products (not specifically identified) by electrolysing a solution of MnSO$_4$ (0·5 M) and H$_2$SO$_4$ (5·2 M) at 60°C, with a current density of 300 A dm^{-2}. Current efficiency for manganese oxidation was again 75%.

Oxidation of cerium (III) to cerium (IV)

Again little information is available for this reaction, other than that it has been carried out at lead dioxide anodes. Ramaswarmy et al.[137] carried out a brief parametric–current efficiency study of the reaction in H_2SO_4 at a variety of lead dioxide anodes. They observed C.E. to decrease with increasing concentration of H_2SO_4 ($0 \cdot 26 \rightarrow 2 \cdot 6$ M) and with increasing current density ($1 \rightarrow 3 \cdot 0$ A dm^{-2}) in a solution $0 \cdot 176$ M in Ce(III). Current efficiency was measured at 54% for 98% conversion in an undivided cell ($j = 1$ A dm^{-2}, $T = 30°C$). At the anode potentials required ($2 \cdot 0$–$2 \cdot 1$ V) there was considerable oxygen evolution. A cell diaphragm had little influence on C.E. provided cathode current density was high, e.g. 15 A dm^{-2}, the Ce(IV) existing in solution as an anionic sulphate complex. Anode rotation increased C.E., indicating the reaction to be subject to some mass-transport control. Ishino and Shiokawa,[138] in a similar study, also found the best electrolysis conditions to be low anodic current density (2 A dm^{-2}) and low sulphuric acid concentration ($0 \cdot 43$ M).

References

1. W. W. Stender, *Trans. Amer. Electrochem. Soc.* **68**, 510 (1935).
2. D. W. Wabner, Habil Thesis, T.U. Munich (1976).
3. D. P. Semchenko, *Sov. Electrochem.* **9**, 1641 (1973).
4. P. Jones, R. Lind and W. F. K. Wynne-Jones, *Trans. Far. Soc.* **50**, 972 (1954).
5. R. W. Houghton and A. T. Kuhn, unpublished data.
6. M. Barak, M. I. Gillibrand and K. Peters in Symposium on "Batteries", Christchurch, Hants., 21–23 Oct. (1958).
7. A. C. Makrides, *J. Electrochem. Soc.* **113**, 1158 (1966).
8. P. Kivalo and V. Vuorio, *Suom. Kemistil.* **34**, 179 (1961).
9. O. D. Hyvarinen, Thesis, Helsinki University of Technology (1972).
10. M. Convers, *Bull. Chim. Soc. France* 792 (1959).
11. G. N. Kokhanov and N. G. Baranova, *Sov. Electrochem.* **8**, 838 (1972).
12. D. F. A. Koch, *Aust. J. Chem.* **12**, 127 (1959); *Electrochim. Acta* **1**, 31 (1959).
13. S. O. Izidinov and E. Kh. Rakhmatullina, *Sov. Electrochem.* **4**, 647 (1968).
14. W. Feitknecht and A. Gaumann, *J. Chim. Phys.* **49**, C135 (1952).
15. P. Ruetschi and B. D. Cahan, *J. Electrochem. Soc.* **105**, 369 (1958).
16. G. Hyvarinen and M. H. Tikkanen, *Acta Polytechnica Scand.* No. 89 (1969) (*Chem. Abstr.* **73**, 30957).

17. P. Ruetschi and B. D. Cahan, *J. Electrochem. Soc.* **104**, 406 (1957).
18. J. E. Puzey and R. Taylor in "Batteries 2" (D. H. Collins, ed.). Pergamon Press, Oxford (1965).
19. P. Ruetschi, *J. Electrochem. Soc.* **107**, 325 (1960).
20. J. S. Clarke, R. Ehigiamusoe and A. T. Khun, *J. Electroanal. Chem. Interfacial Electrochem.* **70**, 333 (1976).
21. P. Jones, H. R. Thirsk and W. F. K. Wynne-Jones, *Trans. Far. Soc.* **52**, 1002 (1956).
22. P. Ruetschi, *Electrochim. Acta* **8**, 333 (1963).
23. S. S. Popova and A. V. Fortunatov, *Sov. Electrochem.* **4**, 444 (1968).
24. J. G. Farr and N. A. Hampson, *Electrochem. Technol.* **6**, 10 (1968).
25. P. Ruetschi, J. P. Ockermann and R. Amlie, *J. Electrochem. Soc.* **107**, 328 (1960).
26. G. A. Kokarev and N. G. Bakhchisaraitsyan, *TR Mosk. Khim. Tekhnol.* **54**, 161–168 (1967).
27. W. Feitknecht, *Z. Elektrochem.* **62**, 6/7, 799 (1958).
28. R. Huss and D. W. Wabner, *Metalloberfläche* **8**, 305 (1974).
29. T. Rogachev and S. Ruenski, *J. Appl. Electrochem.* **6**, 33 (1976).
30. J. Lee and T. Sekine, *Denki Kagaku* **44**, 357 and 821 (1976).
31. Y. D. Dunaev and G. Z. Kiryakov, *Chem. Abstr.* **61**, 6636g (1964).
32. M. S. V. Pathy and H. V. K. Udupa, *Electrochim. Acta* **10**, 1185 (1965).
33. P. I. Zabotin and N. F. Razina, *Trudy Inst. Khim. Nauk, Akad. Nauk Kaz. SSR* **9**, 49 (1962).
34. P. Jones, H. R. Thirsk and W. F. K. Wynne-Jones, *Trans. Far. Soc.* **52**, 1003 (1956).
35. J. Mindowicz and Biallozor, *Electrochim. Acta* **9**, 1129 (1964).
36. S. S. Popova and A. V. Fortunatov, *Vop. Elektrokhim.* 3–21 (1968).
37. W. M. Latimer, "Oxidation Potentials", 2nd edn, p. 56. Prentice-Hall, Englewood Cliffs, N.J. (1952).
38. K. C. Narasimham, S. Sundararajan and H. V. K. Udupa, *J. Electrochem. Soc.* **108**, 798 (1961).
39. J. C. Schumacher, D. R. Stern and P. R. Graham, *J. Electrochem. Soc.* **105**, 157 (1958).
40. T. Osuga, S. Fujii, K. Sugino and T. Sekine, *J. Electrochem. Soc.* **116**, 203 (1969).
41. De Nora, P. Gallone, C. Traini and G. Meneghini, *J. Electrochem. Soc.* **116**, 147 (1969).
42. D. R. Stern and J. C. Schumacher, U.S. Patent 2,840,519, June 24 (1958).
43. J. C. Grigger, N. C. Miller and F. D. Loomis, *J. Electrochem. Soc.* **105**, 100 (1958).
44. K. C. Narasimham, S. Sundararajan and H. V. K. Udupa, *Bull. Nat. Inst. Sci. Ind.* **29**, 279–288 (1961).
45. A. Kormatsu and Y. Yamanoto, Japanese Patent 74,32,197.
46. R. Thangappan, S. Nachippan and S. Sampath, *Ind. Eng. Chem. Prod. Res. Develop.* **9**, 563 (1970).
47. K. Sugino, *Bull Chem. Soc. Jap.* **23**, 115–120 (1950).

48. H. C. Miller and J. C. Grigger, U.S. Patent 2,872,405, Feb. 3 (1959).
49. T. Yokoyama, K. Nishibe and Y. Nakamura, *Denki Kagaku* **41**, 784 (1973).
50. G. Angel and H. Mellquist, *Z. Elektrochem.* **40**, 702 (1934).
51. Y. Kato and K. Koizumi, *J. Electrochem. Ass., Jap.* **2**, 309 (1934).
52. K. Sugino and M. Yamashita, *J. Electrochem. Ass., Jap.* **15**, 61 (1947).
53. A. T. Kuhn "Industrial Electrochemical Processes", p. 98. Elsevier (1971).
54. C. A. Hampel (ed.), "Encyclopedia of Electrochemistry", p. 856. Reinhold (1964).
55. K. C. Narasimham, S. Sundararajan and H. V. K. Udupa, *J. Electrochem. Soc.* **108**, 798 (1961).
56. R. S. Clarke, *J. Appl. Chem. Biotechnol.* **26**, 407 (1976).
57. F. A. Cotton and G. Wilkinson, "Advanced Inorganic Chemistry", 2nd edn, p. 568. Interscience, New York (1966).
58. N. Ramachandran, V. Dhuruvan, S. Sampath and H. V. K. Udupa, *Ind. Chem. Eng.* **8**, 6 (1966).
59. H. V. K. Udupa *et al.*, *Ind. J. Technol.* **4**, 305 (1966).
60. H. V. K. Udupa *et al.*, *J. Appl. Chem. Biotechnol.* **24**, 43 (1974).
61. H. V. K. Udupa *et al.*, *Ind. J. Technol.* **9**, 257 (1971).
62. H. V. K. Udupa *et al.*, *Chem. Agc. Ind.* **22**, 21 (1971).
63. V. A. Shlyapnikov, *Zh. Prikl. Khim.* **49**, 90 (1976).
64. C. W. Nam, *Daehan Hwahak Hwoejee* **13**, 165 (1969).
65. V. A. Shlyapnikov, *J. Appl. Chem. USSR* **47**, 2774 (1974).
66. M. Nagalingam *et al.*, *Chem. Age. Ind.* **16**, 491 (1965).
67. E. A. Dzhafarov, F. G. Bairamov and F. A. Rzaev, *Issled. Obl. Neorg. Fiz. Khim.* 185 (1971).
68. V. I. Skripchenko, K. G. Il'in, G. M. Zhitnyi, E. P. Drozdetskaya, K. L. Ushakova and T. V. Varsh, *Tr. Novocherkask Politekh. Inst.* 23 (1973).
69. V. I. Skripchenko, E. P. Drozchetskaya and K. G. Il'in, *Khim. Prom.* **47**, 910 (1971).
70. D. Stojkovic, M. M. Jaksic and B. Z. Nikolic, *Glas. Hem. Drus. Beograd.* **34**, 211 (1969).
71. M. M. Jaksic, A. R. Despic and B. Z. Nikolic, *Elektrokhimiya* **8**, 1573 (1972).
72. F. Foerster and E. Muller, *Z. Elektrochem.* **9**, 171 (1903).
73. L. Hammar and G. Wranglen, *Electrochim. Acta* **9**, 1 (1964).
74. D. Landolt and N. Ibl, *Electrochim. Acta* **15**, 1165 (1970).
75. V. A. Shlyapnikov and T. S. Fillipov, *Sov. Electrochem.* **5**, 866 (1969).
76. V. A. Shlyapnikov, *Zh. Prikl. Khim.* **10**, 2182 (1969).
77. V. A. Shlyapnikov, *Sov. Electrochem.* **7**, 1128 (1971).
78. W. M. Latimer and J. H. Hildebrand, "Reference Book of Inorganic Chemistry", 3rd edn, p. 177. Macmillan, New York (1958).
79. K. Sugino and M. Yamashita, *J. Electrochem. Soc. Jap.* **16**, 123 (1948).
80. E. A. Dzhafarov and F. G. Bairamov, *Mater. Konf. Molodykh. Uch. Inst. Neorg. Fiz. Khim. Akad. Nauk Azerb, SSR* 226 (1968).

81. E. A. Dzhafarov and F. G. Bairamov, *Issled. Obl. Neorg. Fiz. Khim.* 364 (1970).
82. S. O. Izidinov and V. I. Veselovskii, *Sov. Electrochem.* **6**, 1552 (1970).
83. W. M. Latimer, "Oxidation Potentials", 2nd edn, p. 66. Prentice-Hall, Englewood Cliffs, N.J. (1952).
84. E. Muller and A. Friedberger, *Ber.* **35**, 2652 (1902).
85. E. Muller, *Z. Elektrochem.* **10**, 49, 753 (1904).
86. H. H. Willard and R. R. Ralston, *Trans. Electrochem. Soc.* **62**, 239 (1932).
87. Y. Aiya, S. Fujii, K. Sugino and K. Shirai, *J. Electrochem. Soc.* **109**, 419 (1962).
88. A. Hickling and S. H. Richards, *J. Chem. Soc.* 256 (1940).
89. V. F. Pfeifer, V. E. Sohns, H. F. Conway, E. B. Lancaster, S. Dabic and E. L. Griffin, *Ind. Eng. Chem.* **52**, 201 (1960).
90. C. L. Mehltretter, J. C. Rankin and P. R. Watson, *Ind. Eng. Chem.* **49**, 350 (1957).
91. W. Dvonch and C. L. Mehltretter, U.S. Patent 2,648,629, Aug. 11 (1953).
92. C. L. Mehltretter, U.S. Patent 2,713,533, July 19 (1955).
93. C. L. Mantell, *Ind. Eng. Chem.* **1**, 144 (1962).
94. R. Ramaswamy, M. S. Venkatachalapathy and H. V. K. Udupa, *Ind. J. Technol.* **1**, 125 (1963).
95. W. Dvonch and C. L. Mehltretter, *J. Amer. Chem. Soc.* **74**, 5522 (1952).
96. H. F. Conway, E. B. Lancaster and V. E. Sohns, *Electrochem. Technol.* **2**, 43 (1964).
97. H. F. Conway and E. B. Lancaster, *Electrochem. Technol.* **2**, 46 (1964).
98. C. W. Nam and H. J. Kim, *Daehan Hwahak Hwoejee* **15**, 324 (1971).
99. N. Branislav, *Tehnika* **25**, 779 (1970).
100. C. L. Mehltnetter and C. S. Wise, *Ind. Eng. Chem.* **51**, 511 (1959).
101. E. A. Dzhafarov and Sh. M. Efendieva, *Azerb. Khim. Zh.* 104 (1967).
102. D. P. Semchenko, V. I. Lyubushkin and Sh. Sh. Khibirov, *Sov. Electrochem.* **7**, 932 (1971).
103. F. A. Cotton and G. Wilkinson, "Advanced Inorganic Chemistry", 3rd edn, p. 477. Interscience, New York (1972).
104. W. M. Latimer, "Oxidation Potentials", 2nd edn, p. 65. Prentice-Hall, Englewood Cliffs, N.J. (1952).
105. M. S. Venkatachalapathy, S. Krishnan, M. Ramachandran and H. V. K. Udupa, *Electrochem. Technol.* **5**, 399 (1967).
106. E. A. Dzhafarov, Sh. M. Efendieva, F. G. Bairamov and A. M. Musaev, *Azerb. Khim. Zh.* 125 (1966).
107. C. W. Nam and A. J. Kim, *Daehan Hwahak Hwoejee* **17**, 378 (1973).
108. Y. Aiya, S. Fujii, K. Sugino and K. Shirai, *J. Electrochem. Soc.* **109**, 419 (1962).
109. Ching Fa Teng and Yuan-Pu' Lee, *Hua Hsueh Tung Pao* 51 (1962).
110. Sh. Sh. Khidirov, D. P. Semchenko and V. I. Lyubushkin, *Tr. Novocherk. Politekhnich. Inst.* **197**, 24 (1969).

111. D. P. Semchenko, V. I. Lyubushkin and Sh. Sh. Khidirov, *Sov. Electrochem.* **7**, 932 (1971).
112. C. L. Mehltretter, U.S. Patent 2,830,941, April 15 (1958).
113. C. W. Nam and H. J. Kim, *Daehan Hwahak Hwoejee* **18**, 373 (1974).
114. W. M. Latimer, "Oxidation Potentials", 2nd edn, p. 61. Prentice-Hall, Englewood Cliffs, N.J. (1952).
115. C. A. Hampel (ed.), "Encyclopedia of Electrochemistry", p. 127. Reinhold, New York (1964).
116. S. Sundarajan, K. C. Narasimham and H. V. K. Udupa, *Chem. Proc. Eng.* **43**, 438 (1962).
117. T. Osuga and K. Sugino, *J. Electrochem. Soc.* **104**, 448 (1957).
118. W. C. Bray, *J. Amer. Chem. Soc.* **32**, 932 (1910); **33**, 1485 (1911).
119. C. A. Hampel (ed.), "Encyclopedia of Electrochemistry", p. 130. Reinhold, New York (1964).
120. E. A. Dzhafarov and Sh. M. Efendieva, *Azerb. Khim. Zh.* 166 (1967).
121. W. M. Latimer and J. H. Hildebrand, "Reference Book of Inorganic Chemistry", 3rd edn, p. 181. Macmillan (1958).
122. W. M. Latimer, "Oxidation Potentials", p. 78. Prentice-Hall, Englewood Cliffs, N.J. (1952).
123. F. W. Skirrow and E. R. Stein, *Trans. Amer. Electrochem. Soc.* **38**, 209 (1920).
124. E. A. Dzhafarov and F. G. Bairamov, *Issled. Obl. Neorg. Fiz. Khim.* 355 (1970).
125. E. A. Dzhafarov and F. G. Bairamov, *Issled. Obl. Beorg. Fiz. Khim.* 151 (1971).
126. N. P. Fedotev and Yu. M. Pozin, *Sbornik. Stud. Rabot, Leningrad* 59 (1956).
127. C. W. Nam and H. J. Kim, *Daehan Hwahak Hwoejee* **15**, 223 (1971).
128. S. O. Izidinov and E. Kh. Rakhmatullina, *Elektrokhimiya* **4**, 647 (1968).
129. R. T. Ruetschi, R. T. Angstadt and B. D. Cahan, *J. Electrochem. Soc.* **106**, 547 (1959).
130. F. Strafelda and J. Krofta, *Coll. Czech. Chem. Commun.* **36**, 1634 (1971).
131. W. M. Latimer, "Oxidation Potentials", 2nd edn, p. 95. Prentice-Hall, Englewood Cliffs, N.J. (1952).
132. R. Mraz, V. Srb and S. Tichy, *Electrochim. Acta* **18**, 551 (1973).
133. M. Fleischmann, H. R. Thirsk and I. M. Tordesillas, *Trans. Far. Soc.* **58**, 1865 (1962).
134. J. Y. Walsh, *Electrochem. Technol.* **5**, 504 (1967).
135. M. G. Potdar and H. V. K. Udupa, *Bull. Acad. Polon. Sci.* **16**, 39 (1968).
136. D. N. Solanki and M. Prabhanjanmurty, *J. Ind. Chem. Soc.* **19**, 473 (1942).
137. R. Ramaswamy, M. S. Venkatachalapathy and H. V. K. Udupa, *Bull. Chem. Soc. Jap.* **35**, 1751 (1962).
138. T. Ishino and J. Shiokawa, *Technol. Rep., Osaka Univ.* **10**, 261 (1960).

10   The Lead–Platinum Bielectrode

L. L. SHREIR

Department of Metallurgy and Materials, City of London Polytechnic

Cathodic Protection

Cathodic protection[1,2] is an electrolytic process in which the potential of the metal/solution interface of the structure to be protected is made sufficiently negative to arrest the anodic corrosion reaction

$$M \rightarrow M^{2+} + 2e \tag{1}$$

Under these conditions the whole of the metal surface will become cathodic and will sustain the cathodic reactions

$$H_2O + 2e \rightarrow H_2 + 2\,OH^- \tag{2}$$

and

$$H_2O + \tfrac{1}{2} O_2 + 2e \rightarrow 2\,OH^- \tag{3}$$

This transfer of electrons requires an anode, which must be in both electrolytic and electronic contact with the structure to be protected. Depending on the nature of the anode, two types of cathodic protection systems may be distinguished. In the "sacrificial anode" system, the structure is connected to a metal of a more negative corrosion potential, so that spontaneous electron transfer occurs; thus for protecting steel, anodes of aluminium, magnesium or zinc may be used, and steel anodes may be used for protecting copper and its alloys. On the other hand, the anodes used in the "impressed current" system may be any electronically conducting material, metal or non-metal, and electron transfer is achieved by a source of e.m.f.—usually a transformer-rectifier which converts 240 V a.c. to 10–50 V d.c. Anodes for this system range from massive scrap iron (disused railway lines, sunken barges and ships) to small electrodes of solid platinum, and include graphite, magnetite, Fe–Si–Mo and Fe–Si–Cr alloys and lead alloys of a variety of compositions. If the anode is inert and electron-conducting, the anode reaction will be O_2 evolution in fresh water:

$$H_2O \rightarrow \tfrac{1}{2}O_2 + 2\,H^+ + 2e \qquad (4)$$

and Cl_2 evolution in water containing Cl^- ions:

$$2\,Cl^- \rightarrow Cl_2 + 2e \qquad (5)$$

It is relevant at this point to consider the properties of an ideal anode for impressed current cathodic protection systems, which may be summarized as follows:

(i) it should be completely resistant to corrosion in a variety of natural aqueous environments ranging from sea water to soils of low water content (and consequently high resistivity);

(ii) it should operate at high current densities with little polarization, irrespective of the electrode reaction (oxygen evolution and/or chlorine evolution);

(iii) it should have a prolonged life (5 to 30 years according to application), and should not require periodic inspection and maintenance;

(iv) it should be mechanically robust in order to withstand damage during installation and service;

(v) it should be capable of fabrication into a variety of forms (round rod, wire, sheet etc.);

(vi) it should have a good electrical conductivity; and

(vii) it should be inexpensive and readily available.

No one material conforms to all these requirements, and although platinum[3] approaches the ideal in many respects, it has a major disadvantage of being very costly. For this reason anodes have been devised in which the mass of platinum used is minimized and the surface area exposed to the electrolyte maximized by using it as a very thin layer on a suitable support, and although materials such as ceramics and plastics have been used for this purpose, they have the serious disadvantage of being non-conductors. Copper and silver supports have also been used, but although they are good conductors they will corrode if exposed to the aqueous environment at any discontinuity or defect in the platinum coating. A tantalum support for platinum was first used by Baum[4] for the electrolytic oxidation of Cl^- ions to chlorate, and in the late 1950s platinized titanium,[5] titanium on which a thin layer of platinum is electrodeposited, became available, and from that time on it has found extensive use in cathodic protection and in other electrolytic processes. More recently, anodes have become available[6] in which the platinum is metallurgically bonded to titanium ("Tibond") or niobium ("Niobond") and the thicker clad layer of platinum compared to the thin electrodeposited layer has the important advantage that it is continuous and free from porosity. These "valve-metal" supports have the advantage that a film of a dielectric oxide will be formed anodically on surfaces exposed at discontinuities, and this film will prevent attack on the metal providing the breakdown potential (~8–12 V in the case of titanium and ~80 V in the case of niobium in a chloride-containing environment such as sea water) is not exceeded.

It is important to draw attention to the following requirements of anodes for cathodic protection to emphasize how these differ from those for anodes for electrolytic processes carried out in a plant. Anodes for cathodic protection will have to withstand arduous conditions (for example, connected to North Sea platforms) for periods of time ranging from 5 to 30 years. They must, therefore, be robust to withstand damage during installation (often under difficult conditions) and during service, when they may be subjected to vibration, impact and abrasion by the environmental conditions prevailing. Since they are often installed in inaccessible places, periodic inspection and maintenance is seldom possible, so that reliability for prolonged periods is essential.

Pb/PbO₂ as an Anode

Lead when anodically polarized in a variety of electrolyte solutions will be oxidized to PbO_2 when the potential of the lead/solution interface has attained a sufficiently positive potential (see Table I for the equilibria and reversible potentials relevant to this discussion), and this is the most important electrochemical characteristic of lead. The PbO_2, which is chemically inert, has the property of semi-conductivity and its electrical resistivity is ~1–4×10^{-4} Ω cm, i.e. about ten times that of lead metal. This means that once a thin film is formed on the surface, further conversion of Pb to PbO_2 will tend to cease and the electrode

Table I. Relevant equilibrium equations.

Equilibrium	Equilibrium Potential (V)
1. $Pb^{2+} + 2e = Pb$	$E = -0{\cdot}126 + 0{\cdot}0295 \log a_{Pb^{2+}}$
2. $PbCl_2 + 2e = Pb + 2\ Cl^-$	$E = -0{\cdot}268 - 0{\cdot}0295 \log a_{Cl^-}$
$K = 1{\cdot}5 \times 10^{-5}$	
3. $PbO_2 + 4\ H^+ + 2e = Pb^{2+} + 2\ H_2O$	$E = 1{\cdot}455 - 0{\cdot}1182\ pH - 0{\cdot}0295 \log a_{Pb^{2+}}$
4. $PbO_2 + 4\ H^+ + 4e = Pb + 2\ H_2O$	$E = 0{\cdot}664 - 0{\cdot}059\ pH$,
5. $PbO_2 + 2\ Cl^- + 4\ H^+ + 2e$	$E = 1{\cdot}592 - 0{\cdot}118\ pH + 0{\cdot}0295 \log a_{Cl^-}$
$\quad = PbCl_2 + 2\ H_2O$	
6. $Cl_2 + 2e = 2\ Cl^-$	$E = 1{\cdot}359 + 0{\cdot}0295\ p_{Cl_2} - 0{\cdot}059 \log a_{Cl^-}$
7. $O_2 + 4\ H^+ + 4e = 2\ H_2O$	$E = 1{\cdot}23 - 0{\cdot}059\ pH + 0{\cdot}0147 \log p_{O_2}$
8. $PbO + 2\ H^+ + 2e = Pb + 2\ H_2O$	$E = 0{\cdot}248 - 0{\cdot}059\ pH$
9. $Pb(OH)_2 + 2\ H^+ + 2e = Pb + 2\ H_2O$	$E = 0{\cdot}242 - 0{\cdot}059\ pH$
10. $Pb_3O_4 + 2\ H_2O + 2\ H^+ + 2e = 3Pb(OH)_2$	$E = 1{\cdot}101 - 0{\cdot}059\ pH$
11. $Pb_3O_4 + 8\ H^+ + 2e = 3\ Pb^{2+} + 4\ H_2O$	$E = 2{\cdot}20 - 0{\cdot}2364\ pH - 0{\cdot}0886 \log a_{Pb^{2+}}$

reaction will be confined to reaction (4); in the case of a Cl^--containing environment, reactions (4) and (5) will occur simultaneously at relative rates that depend on the Cl^- ion activity, pH and anode potential. Thus the system Pb/PbO_2, which may be regarded as PbO_2 on a conducting Pb support from which the PbO_2 is formed, may be used as an inert anode, and possibly its earliest application was in electrorefining and electrowinning. Fink[7] has discussed various types of insoluble anodes used in electrolytic processes, including Pb–4–6% Sb alloys for the electrolyte refining of Cu (to avoid excessive build-up of Cu salts) and Pb–Ag alloys for the electrowinning of Zn and for alkali-chlorine plants. According to Fink, the role of the alloying addition is to enter the oxide surface film and to decrease anode solubility and to increase "catalytic activity," and the entry of Ag_2O into the PbO_2 film was found to have the most influence on these factors. Antimonial lead (Pb–6–8% Sb) has been used from the beginning as an inert anode for the electro-plating of Cr from the CrO_3–H_2SO_4 plating bath, and operates without significant corrosion for prolonged periods, although in certain circumstances an insulating film of $PbCrO_4$ forms over the PbO_2, and has to be removed periodically by scrubbing to restore the electron-conducting properties of the anode. However, Pb–Sn alloys (6–8% Sn) are more widely used at present owing to their superior corrosion resistance.[8]

The predominant anions (SO_4^{2-} and $Cr_2O_7^{2-}$) in many of the solutions considered above are passivating, and the formation of PbO_2 at potentials above the reversible potential for the PbO_2/Pb equilibrium is predictable. However, in chloride-containing solutions the situation is quite different owing to the fact that this ion is inimical to passivation, and results in localized breakdown of the film on a number of metals and alloys with consequent localized pitting. Thus aluminium and many of the stainless steels will pit in chloride solutions at potentials below that of the reversible O_2/OH^- potential in neutral solutions (i.e. ~ 0.8 V), which means that pitting can occur in solutions in which the redox potential is determined by the presence of dissolved atmospheric oxygen. Even the corrosion-resistant high nickel alloys such as the "Inconels" and "Hastalloys" will pit in chloride solutions if their potential is raised to above ~ 1 V by an external source of e.m.f. In the case of the "valve metals" Zr will pit at ~ 2 V and Ti at 8–10 V whereas Nb and Ta will anodize to high potentials (~ 80–100 V) before spark breakdown occurs.[6] It should also be noted that of the precious metals, only platinum, rhodium and iridium are corrosion-resistant when anodically polarized in chloride solutions; gold corrodes rapidly and silver tends to form a non-conducting film of AgCl which prevents electron transfer.

As far as Pb/PbO_2 is concerned, it is quite remarkable that it can be used as an anode in chloride solutions, particularly in view of the fact that, as demonstrated by experience gained over the years with the lead accumulator, Cl^- ions cause breakdown of PbO_2 and corrosion of the Pb substrate.

Lead as an Anode for Cathodic Protection

With the increasing use of impressed current systems for the cathodic protection of structures immersed in sea water, attention was directed to finding alternative materials for graphite, silicon–iron alloys and small electrodes of bulk platinum which had been the original materials for this purpose.[1,2] Graphite has the disadvantage of being fragile, and its electrochemical instability at high current densities means that a large mass of anodes will be required for high current outputs. Silicon–iron (Fe–14% Si) has the disadvantages of fragility, high density and poor corrosion resistance, and its mechanical properties are such that it can be fabricated only by casting; however, when alloyed with Mo or Cr its corrosion resistance in chloride environments is significantly enhanced.[2]

In the mid 1950s, lead alloys were investigated as possible anodes for cathodic protection, and this was probably inspired by the early work of Fink who had discovered the remarkable effect of alloying additions of silver on the corrosion resistance of lead anodes used for the electrolysis of brine. Crennel and Wheeler[1] at the Admiralty Dockyard, Portsmouth, carried out tests on Pb–Ag alloys, and found that Pb–1% Ag was suitable for anodes for the cathodic protection of naval ships, providing the current density did not exceed 100–200 A m^{-2}; at higher current densities an insulating film appeared to form with consequent decrease of the current to a very low value. At about the same time, and in collaboration with Crennel and Wheeler, extensive studies of binary and ternary alloys of lead were carried out by Morgan[1] who found that Pb–6% Sb–1% Ag gave a lower consumption rate and a harder and tougher layer of PbO_2 than either Pb–1% Ag or Pb–6% Sb. This alloy was used very extensively and successfully during the late 1950s and early 1960s, and details of the test procedures have been provided by Morgan.[1] Investigations by Barnard et al.[9] indicated that Pb–2% Ag was the most effective alloy for cathodic protection, and this was used for protecting ships of the Royal Canadian Navy.

In 1958 the effect of introducing small microelectrodes of Pt into pure lead was investigated by Shreir and Weinraub,[10] and it was found that the Pt had a remarkable effect on stabilizing the PbO_2 and preventing its breakdown in chloride solutions. Subsequent practical trials[11] showed that this anode could be used at high current densities, up to 1000 A m^{-2}, and this, coupled with the low cost of the anode (the cost of small microelectrodes of Pt $12 \times 0\cdot5$ mm diam. spaced at about 250 mm interval in a rod anode 38 mm diam. is significantly less than alloying additions of 1–2% Ag), meant considerable economies in the cost of the installation.

Electrochemical Considerations

The anodic behaviour of Pb in Cl$^-$ solutions is complicated by the number of possible competitive reactions, and by the formation of solid corrosion products that may or may not adhere to the surface of the lead. The relevant equilibria are given in Table I, and for simplicity it will be assumed that chloro-complexes, which will be considered subsequently, do not form.

The primary anodic reaction will be the formation of Pb^{2+} ions, and, depending on the concentration of Cl^- ions and volume of solution, this will eventually lead to the formation of crystals of $PbCl_2$; however, the position of formation of $PbCl_2$ for a solution of given Cl^- ion concentration will depend on the current density, and if this is low the Pb^{2+} ions will be able to diffuse and migrate away from the surface of the electrode so that its dissolution will be unimpeded. On the other hand, if the current density is sufficiently high a non-conducting film of $PbCl_2$ will form on the surface of the Pb, and under galvanostatic conditions this will result in an increase in potential owing to the decrease in the true surface area. This increase in potential will be partly due to an increase in the activation overpotential for dissolution, but a much larger contribution will be provided by the resistance overpotential resulting from the large iR drop through the solution permeating the interlocking crystals of $PbCl_2$.

These observations are illustrated (Fig. 1) by the E_H–t curves for Pb anodically polarized at $10 \, mA \, cm^{-2}$ in solutions of different Cl^- concentrations,[12] and it can be seen that whereas the potential remains fairly constant in 0·1 M NaCl it increases significantly at higher concentrations; in 2 M NaCl it can attain 70 V or more. However, in the case of

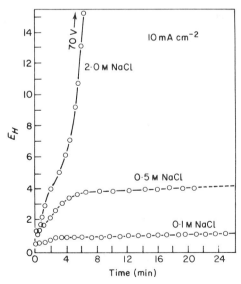

Fig. 1. E_H–time relationships for Pb anodically polarized at $10 \, mA \, cm^{-2}$ in various concentrations of NaCl.[12]

0·5 M NaCl, evolution of Cl_2 and formation of PbO_2 at the periphery of the specimen (a Pb disc mounted in Araldite) showed that the potential of the Pb/solution interface must have been above 1·3 V. In 2·0 M NaCl a very compact layer of $PbCl_2$ formed, and although Cl_2 evolution occurred there was no evidence of PbO_2, which is of some significance in relation to the concentration range of Cl^- ions in which Pb will act as an inert anode.

Briggs and Wynne-Jones[13] showed that when Pb is anodically polarized at low current densities (0·8–15 mA cm^{-2}) in solution containing anions that form sparingly soluble Pb salts there is a period of low overpotential which is followed by a sudden increase in overpotential due to mechanical passivation. The time for passivation, t (s), or the period of low overpotential, is related to the current density (mA cm^{-2}) by the relationship:

$$j^n t = k \tag{6}$$

or

$$n \log j + \log t = \log k \tag{7}$$

where n and k are constants for a given solution and temperature. Figure 2 shows plots of $\log j$ vs $\log t$ for Pb anodically polarized in 0·5 and 2·0 M NaCl, the solutions being saturated with $PbCl_2$ to improve reproducibility. These results confirm the linear relationship between $\log t$ and $\log j$, although the shorter time for passivation in 2·0 M NaCl as compared with 0·5 M NaCl above 4 mA cm^{-2} is anomalous.

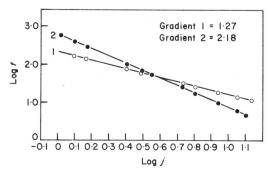

Fig. 2. Linear relationship between $\log j$–$\log t$ (t is the time for passivation) for Pb anodically polarized in 2·0 M NaCl (curve 1) and in 0·5 M NaCl (curve 2).[12]

Figure 3 gives the galvanostatic $E–t$ curves for the anodic polarization of a small Pb–Pt bielectrodes (a 1 cm disc of Pb with a Pt
microelectrode of given diameter in the centre, the disc being mounted
in Araldite so that only one flat surface is exposed) in 0·5 M NaCl. Each
curve is characterized by a rapid increase in potential due to the
formation of $PbCl_2$ on the surface of the Pb followed by a slow decrease
to a steady-state value of ~2 V (*vs* SHE). In the case of the bielectrode
having a microelectrode of only 0·003 in. diameter, the maximum
potential corresponds with that obtained for Pb alone in 0·5 M NaCl
(Fig. 1).

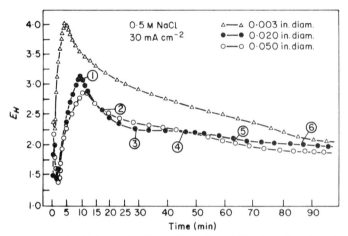

Fig. 3. $E_H–t$ curves for bielectrodes (1 cm diam. disc of Pb with a Pt microelectrode in
the centre) containing microelectrodes of different diameters and anodically polarized
at 30 mA cm^{-2} in 0·5 M NaCl.[12]

Observations of the bielectrode during polarization showed that Cl_2
evolution at the microelectrode and nucleation of PbO_2 on the surface of
the $PbCl_2$ surrounding the microelectrode occurred simultaneously just
before the maximum potential was attained. Continued polarization
resulted in the propagation of the PbO_2 over the surface of the $PbCl_2$
(Fig. 4) and measurements of the area of the PbO_2 at various times
(points 1–6 in Fig. 3) showed that this increased linearly with time.
Similar results to those shown in Fig. 3 have been obtained using a
larger bielectrode (Fig. 5) consisting of a lead rod 2·5 cm × 2·5 cm diam.
with microelectrodes inserted in the centre of the domed surface and at
two points on the curved surface.[14]

Potentiostatic studies[12] in 0·5 M NaCl at different predetermined potentials have demonstrated that a potential of 1·4 V (*vs* SHE) is the minimum for PbO_2 formation and at this potential the PbO_2 is confined to the surface layer of $PbCl_2$ in the immediate vicinity of the microelectrode; at this potential the current is small and constant. At higher potentials the PbO_2 nucleated at the microelectrode propagates over the surface of the lead with a consequent increase in current, which is associated with the conversion of the $PbCl_2$ to PbO_2 (Fig. 6).

These results obtained under carefully controlled laboratory conditions indicate that one role of the Pt microelectrode is to provide a stable surface at which PbO_2 can nucleate readily without being undermined by subsequent formation of $PbCl_2$; this is followed by slow conversion of the $PbCl_2$ surrounding the Pt with a consequent increase in current at

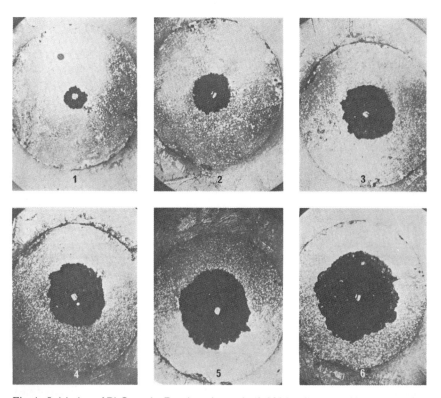

Fig. 4. Initiation of PbO_2 at the Pt microelectrode (0·020 in. diam.) and its propagation over the surface of the previously formed $PbCl_2$. The numbers correspond to the circled numbers in Fig. 3.[12]

constant potential or a decrease in potential at constant current. Its
other role is to limit $PbCl_2$ formation, as indicated by the increase in the
potential to a maximum followed by a decrease.

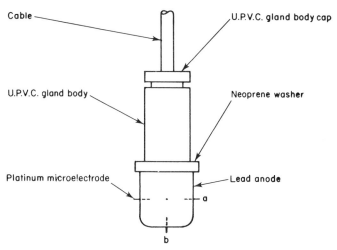

Fig. 5. Bielectrode consisting of Pb or a Pb alloy $2 \cdot 5$ cm $\times 2 \cdot 5$ cm diam. showing
possible locations of Pt microelectrodes at positions *a* and *b*.[14]

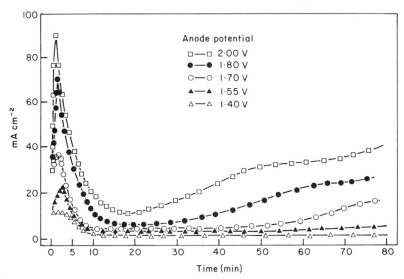

Fig. 6. Potentiostatic *j-t* curves for a Pb–Pt bielectrode anodically polarized in
$0 \cdot 5$ M NaCl.[12]

A simple experiment of some significance is to form a stable PbO_2 film on the bielectrode by anodically polarizing it in 0·5 M NaCl, and to then remove the Pt microelectrode.[14] This is illustrated in Fig. 7, in

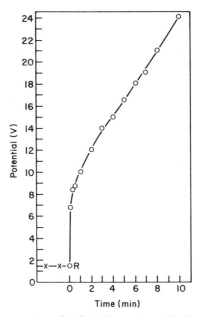

Fig. 7. Potential–time curve for a Pb–Pt bielectrode anodically polarized in 0·5 M NaCl at 300 A m^{-2} for 24 h, before removing the microelectrode (point *R*). [14]

which the bielectrode was polarized for 24 h at 300 A m^{-2}, and attained a steady potential of 1·6 V (*vs* SCE); removal of the Pt microelectrode at point *R* caused an immediate increase in potential. Wheeler[14,15] has commented on this sudden increase in potential when the micro-electrode is removed, and has postulated that the sole function of the Pt is to "maintain a low-resistance between the shell of lead dioxide and the underlying lead". Although this explanation is attractive in its simpli-city, the assumption that the PbO_2 is not in contact with the Pb is not tenable, as will be discussed subsequently.

The converse experiment is to anodically polarize lead in NaCl solution until formation of $PbCl_2$ has resulted in an increase in potential to 15–20 V and then to insert a Pt microelectrode and to continue polarization (Fig. 8). As might be anticipated, the introduction of the microelectrode, at which Cl_2 is evolved with kinetic ease, causes an immediate decrease in potential to about 6–10 V followed by a more gradual decrease the potential falling to ∼3 V in 60 min. This decrease is accompanied by nucleation of PbO_2 at the microelectrode and its propagation over the surface of the $PbCl_2$.

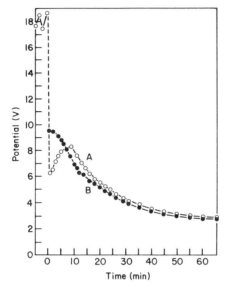

Fig. 8. Effect of inserting a Pt microelectrode into a Pb anode after polarizing it in 0·5 M NaCl at 300 A m^{-2}.[14]

Practical Trials

Laboratory studies can only provide limited information on the behaviour of the Pb/PbO_2–Pt bielectrode in service, and a major difficulty in prolonged laboratory studies is the changing nature of the electrolyte. This applies particularly to studies involving chlorine evolution, since this, together with the OH$^-$ ions generated at the cathode, will result in a solution of higher pH and redox potential than the initial sodium chloride solution, and a solution of this nature is more conducive to PbO_2 formation than a sodium chloride solution.

For this reason practical trials on bielectrodes consisting of different Pb alloys (Pb–1% Ag containing alloying additions of Sb, Te and Bi) have been carried out in high-velocity sea water in a condenser water box at a power station.[11] The bielectrode used for this purpose was similar to that shown in Fig. 5, and Fig. 9 shows the appearance of the bielectrodes at the termination of the test. Most of the bielectrodes were polarized at 500 A m^{-2} for 1 year, but one bielectrode (Pb–1% Ag) was polarized at 2000 A m^{-2} for 3 months. After completion of the test, the consumption of the lead in forming adherent PbO_2 and non-adherent PbO_2 and other corrosion products was determined from the cross-section of the Pb remaining and from weight loss. These tests showed that at 500 A m^{-2} the consumption of the lead ranged from 0·004 to 0·003 lb A^{-1} year^{-1}, but at 2000 A m^{-2} this increased to 0·063 lb A^{-1} year^{-1}; the latter corresponds to an ionic current of 0·19% if it is assumed that Pb is oxidized to PbO_2. During these tests the current remained constant indicating that breakdown of the PbO_2 with subsequent formation of insulating corrosion products did not occur even at 2000 A m^{-2}. Although these results show that the Pb is consumed during polarization, the high consumption rates obtained do not appear to apply in practice over more prolonged periods of polarization (10–15 years), and it is possible that the rate of corrosion is rapid in the first few months, and that it then decreases with time to a very low value. In connection with the mechanism of the Pb/PbO_2–Pt bielectrode attention should be drawn to the surface appearance of the PbO_2 (Fig. 9), which can be seen to be covered with a number of small

Fig. 9. Pb/PbO_2–Pt bielectrodes after testing in high velocity sea water.[11]

being spread along the primary grain dendrites. The extent of the attack depends on the grain orientation; with low binary antimony alloys there is a lack of eutectic continuity and the micrographs show discontinuities in the spheroidal segregation.[46] From 1% to 6% Sb the attack is still largely intercrystalline but with subgrain attack, the grain boundary corrodes more deeply than the channels traversing the grains. These subgrain paths are formed by preferential corrosion of the inter-dendritic network and the antimony-rich areas. As the antimony concentration is increased, the attack is more uniformly distributed along the interdendrite network and the antimony-segregated areas. At higher antimony concentrations, 9% and 10%, there is penetration of the antimony-rich areas of segregated antimony and eutectic. Thus, although the total corrosion is greater, as measured by weight loss, it is normally less destructive due to the moderated grain boundary penetration, with secondary attack in the body of the grain.

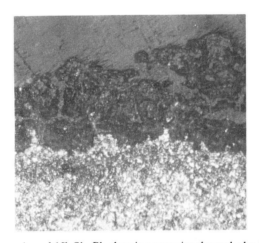

Fig. 12. Corrosion of 6% Sb–Pb showing porosity through the film. × 100.

The binary antimony alloys were shown by Lander[138] to corrode rapidly at the beginning of battery cycling but slowed down later. Corrosion at constant current increased slightly with increased anti-mony concentration;[127,206] however, at constant potential there was a large increase in corrosion with increased antimony concentration.[207] This can give rise to difficulties in comparing lead corrosion data on a quantitative basis without quoting the current density or the operating potential (*see* Fig. 13).

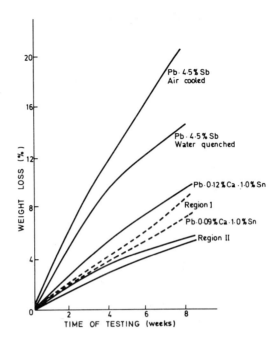

Fig. 13a.

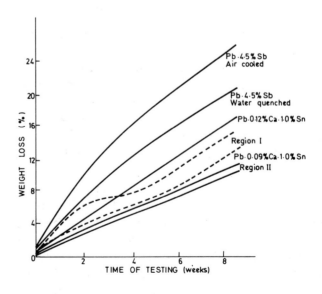

Fig. 13b.

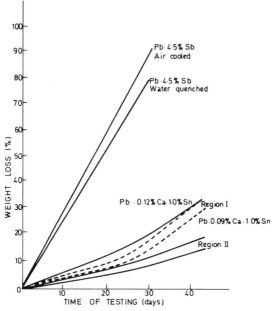

Fig. 13c.

Fig. 13. (*Above and facing page*). Effect of test conditions, alloy composition and casting on anodic weight loss.[266] (a) Constant current density $= 0.42$ mA cm^{-2}. (b) Constant current density $= 0.84$ mA cm^{-2}. (c) 2·5 V constant potential. Region I for the Pb–0·09% Ca Alloy has lower casting temperature than Region II.

Heat treatment of cast antimonial alloys is one method of improving their strength[228] since the distribution in a low concentration matrix is controlled by the quench rate. Micrographs and corrosion data show that heat treatment of a 5% antimonial alloy alters the shape of the eutectic particles brought about by diffusion of lead to the primary dendrite where it reprecipitates epitaxially. Since this cannot happen at the grain boundary there is more extensive intergranular penetration than would be predicted from the chemical analysis of the alloy composition.

Decreasing the antimony content to below about 5%, depending on casting conditions, produces a coarse dendritic structure with a lower strength and a susceptibility to casting defects, shrinkage shown as hot tears or hot cracks and increased brittleness. The cracks appear along the grain boundary due to the solidification in the coarse dendritic structure (Fig. 14). Further cracking of the grid structure due to the combined effect of stress and corrosion may occur during operation of the battery.[221] The grid has to support the weight of the active material and accommodate the stress due to expansion of the lead dioxide matrix, this stress being concentrated at impurity inclusions, casting defects such as porosity and large voids or microporosity as outlined above. Grains with a large boundary area give rise to planes of weakness,

Fig. 14. Shrinkage cracks in the lug of a Pb–6% Sb alloy grid.

which aid crack propagation, therefore grain refining elements are added to improve the mechanical strength as indicated by increases in hardness, relative elongation and tensile strength. Thus, as the antimony content is decreased, the mechanical deterioration is partly compensated by the introduction of small amounts of arsenic and tin.[199,261,290] Increased corrosion resistance is attributed to favourable structural effects,[288] spheroidization of the eutectic being promoted by copper and arsenic. An arsenic:copper ratio of 2 : 1 is recommended[289] since precipitation of the Cu_3As phase results in a uniform corrosion-resistant structure. There are indications that arsenic increases the substrate corrosion but decreases the amount of dischargeable lead dioxide corrosion product;[219] the mechanism is not understood but could be analogous to the discharge inhibition by antimony.

The problem of castability has been quantified (Fig. 15) on a comparative basis by use of a specially designed mould,[22] and further work to standardize castability data to known conditions of mould temperature, specimen sizes etc. would be useful.

Automatic grid casting, pasting and setting of the paste in ovens restricts the practical options available for improving antimonial alloy technology, since a cast Brinell hardness of at least 11–12 BH is required. Although some improvement can be achieved by heat treatment or rapid cooling after casting, this could be mitigated by the decrease in strength due to age-hardening at the temperature of the paste-setting oven. As indicated previously, the hardness increases with increase in antimony concentration, but the loss of strength, due to age hardening, following quenching of low-concentration antimony alloys, also increases with increase in antimony. Thus, although the precipitation of supersaturated antimony as platelets in the lead matrix increases with antimony concentration, this is somewhat offset by precipitation within the eutectic.

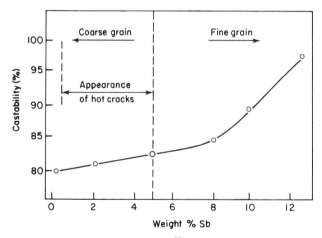

Fig. 15. Castability of lead–antimony alloys.[22] Casting temperature = 600°C, mould temperature 150–175°C.

Grain refinement of antimonial alloys is an alternative method of increasing their strength, and the elements silver, sulphur, tellurium and selenium have been examined and tested in this context.[5,63,64,268,287] The use of selenium has found favour over the past few years[22,28,173] as a means of decreasing the antimony content to between 2 and 3%, but retaining the strength within acceptable casting properties. The ternary phase diagram is not available, but the binary (Fig. 16) indicates an intermetallic, PbSe. The grain size of the ternary alloy is significantly decreased compared to the binary (Fig. 17), with a fine equiax microstructure; thus the dendritic structure is eliminated, together with some of the problems of microsegregation and casting faults. The aged hardness is about 17 BH.

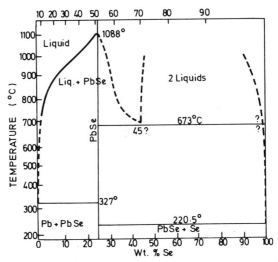

Fig. 16. Equilibrium diagram of the Pb–Se binary system.[218]

Early use of this type of alloy was in the negative grid,[18] which minimized negative grid corrosion in relation to the antimony-poisoning problem (Fig. 9). The obvious advantages, however, of low self-discharge and good charge characteristics, which result from decreased antimony concentration, have extended the alloy application to automotive as well as railway diesel-starting and carriage lighting applications.[22,173] Typical alloying combinations are 1·5–4·0% antimony, 0·005–0·1% selenium, 0·05–2% tin, 0·025–0·2% arsenic, with or without 0·025–0·1% silver.[22,173,292–294,298,299] The corrosion resistance and penetration are similar to the conventional 6% antimony alloys; however, the paste adhesion is more typical of a 10–12% binary[191] and this may be due to selective leaching of the selenium, as predicted by its Pourbaix Diagram,[189] with increased corrosion (region 2 of Fig. 7).

The beneficial effect of silver is well-documented[83,85,92,161,185,200,264,265] and values of up to 1%[162] and 0·15%[75] have been quoted. The silver appears to prevent shedding and thereby minimizes the corrosion;[75]

however, more recent investigations[155] suggest that the silver surrounds the antimony phase spheroids and prevents selective dissolution; a detailed investigation using two levels of silver, 0·1% and 1·0% Ag, lead to the possibility that the silver delays pore closure thereby allowing the antimony to dissolve and aid nucleation,[4] perhaps in a manner similar to selenium. There is also the evidence that silver lowers the oxygen overvoltage[140] and the mechanisms may involve adsorption as indicated in the previous section.

Fig. 17. Surface of a Pb–2·5% Sb–0·1% Se–0·5% Sn alloy after stripping of corrosion product layer. ×2000.

Cadmium additions to the lead–antimony system aimed at a pseudo-eutectic at 1–2% Sb are claimed[17] to reduce casting defects, improve corrosion properties, decrease shelf life and increase overcharge life. Cadmium appears[3] to be more effective in the early stages of corrosion, it decreases the length of the lead sulphate plateau and appears to act as a nucleation centre. In the ternary alloy there appears to be no benefit.[4]

Tin is normally added to the alloy to improve fluidity, particularly in alloys based on the alkaline-earth elements. The solubility of tin in lead is about 1·3% with eutectic composition at 38·1% (Fig. 18).

The alloy crystallizes into dendrites[136] during casting and corrosion occurs by intergranular attack and preferential dissolution of eutectic segregations.[191] The grain structure is of a cellular form.[243] Lander suggested that an additional element is necessary to impart sufficient strength to lead–tin alloys and 0·1–0·5% tin is shown to improve corrosion resistance but reduce ageing[154] of antimonial alloys, although it has been reported that tin has no effect on the tensile strength of antimonial alloys.[185] The corrosion of tin ternary and quaternary alloys shows an interaction depending on the other elements.[180] A 5% Sn and 0·075% Se addition has been suggested to reduce self discharge.[287]

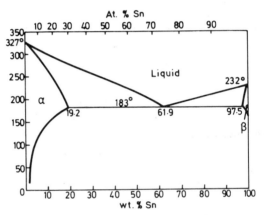

Fig. 18. Equilibrium diagram of the Pb–Sn binary system.[218]

Lead–calcium and related alloys

These alloys form the other main group of interest generally classified as non-antimonial, and are found in standby and post office application with increasing use in automotive batteries, i.e. sealed and maintenance-free.[102,113,213,214,278] Lead–calcium alloys represent the classical precipitation-hardened binary alloy and there are various papers[2,33,106,109,114,165,240,259] which have been published describing their properties, their advantage for the lead–acid battery being in the low rate of self discharge or minimal shelf loss of capacity. The alloys conform to the concept of alloying lead with electronegative elements[102] and were introduced by Schmacher and Philips[214] in 1935 for the Bell Telephone Laboratories. The alloys have found application for the standby and telephone use but capacity loss due to positive active material softening was a fundamental problem for submarine use,[220] and there are some difficulties in recharge after a deep[113] or complete[41] discharge, particularly with constant voltage equipment. The corrosion rates of these alloys is usually low and battery lives of 30 years have been predicted[261] for standby service; excessive grid growth and inter-granular corrosion can cause premature failure, attributable to poor casting conditions. The mechanisms of active material softening and occasional recharge difficulties have not been fully explained; the corrosion product film is physically different[41] from that on antimonial alloys, there being less α-PbO_2[40,43,48,71] and a greater concentration of lead sulphate in the film.[259] Thus the lack of corrosion penetration into the alloy due to pore closure or the absence of suitable adsorption to modify the lead dioxide morphology may be significant. The film shows less tendency to crack into layers, compared to antimonial alloys,[41] but disintegrates as a soft powder.[221]

The metallurgical properties of lead–calcium, lead–barium and lead–strontium alloys are similar and should be considered together, although the calcium alloys have received greater attention than the other alloy systems. The inherent softness of the binary alloys has led to further additions, the most frequently reported being tin. The binary lead–calcium system shows age-hardening due to grain refinement and precipitation of Pb_3Ca.

The equilibrium diagram (Fig. 19) shows a maximum solubility of 0·1 wt % calcium in lead near the peritectic temperature of

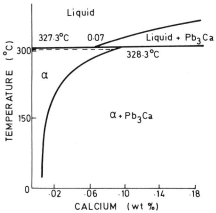

Fig. 19. Equilibrium diagram of the Pb–Ca binary system.[218]

$328 \cdot 3°C^{103,116,218}$ and a solubility of $0 \cdot 01$ wt % at 25°C. Alloys containing greater than $0 \cdot 07\%$ calcium freeze to produce primary Pb_3Ca crystals, and annealing just below the peritectic temperature causes re-solution of the Pb_3Ca in alloys of between $0 \cdot 07$ and $0 \cdot 10\%$ calcium. Primary Pb_3Ca in a cast alloy occurs as cubes or star-shaped dendrites in the lead matrix and precipitation in the solid state produces Pb_3Ca inclusions. Single-phase alloys have a large crystal size with serrated grain boundaries; the alloy solidifying homogeneously as a coarse-grained structure is usually associated with a low secondary creep rate.[266] Thus alloys with a calcium content of $0 \cdot 1\%$ were reported to have a lower rate of grid growth than standard antimonial lead.[240] Higher content alloys are characterized by a fine-grain heterogeneous structure which is probably age-hardened to a small degree giving a low creep strength.[116] Casting conditions influence the structure and the eventual corrosion as indicated by Mao et al.[152] Gravity casting produces large smooth grains with preferential intergranular corrosion, the result of induced stresses at the grain boundary, whereas pressure casting gives rise to fine grains and serrated boundaries in both annealed and unannealed specimens; there is a more uniform precipitation of Pb_3Ca giving rise to a less severe and more uniform attack. The problem of shrinkage and large voids penetrating to the grid surface (Fig. 20) can be mitigated by control of the mould temperature.

The binary calcium alloys increase in hardness with time, grain refinement taking place over a few hours or days depending on both the casting and quenching conditions, as well as the calcium content.[52,98,116,171,205,213,241] Microstructures of the calcium alloy system[50,52,153,171,179] show that the grain refinement increases with increase in calcium content, but also the rate of solidification influences the grain size.[52] The primary dendrites within the grains comprise a fibrous morphology which breaks down into stubby-armed branched dendrites,[52,98,116] this cellular type of structure having been observed in other lead alloy systems by Winegard et al.[62,167,168,254] Annealing of the structure influences the grain growth and redistribution of precipitates,[242] leaving the grain boundary relatively free of active precipitates.[153] Thus, the presence of these precipitates within the grain can be said to promote matrix or intragranular corrosion at the expense of localized grain boundary attack. Impurities increase the strength and castability of the binary alloys;[51] the strength also increases with increasing cooling rate,[242] precipitation taking place from a greater degree of supersaturation. Corrosion behaviour is sensitive to precipitate distribution within the matrix,[153] as shown in Fig. 7, for corrosion at 2·8 V (i.e. 75 mV above an open circuit of 1185 mV after 60 days at 180°F in 1·210 sp.gr. sulphuric acid).

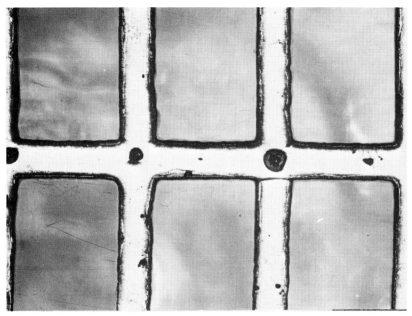

Fig. 20. Voids at the intersection of a cast Pb–Ca–Sn automotive grid.

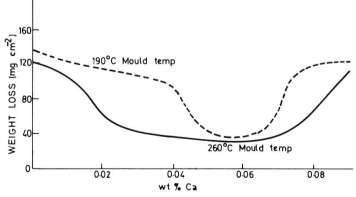

Fig. 21. Effect of casting temperature and calcium content on the anodic weight loss of Pb–Ca alloys after 60 days at 2·8 V.[155]

A low mould temperature introduces an irregular grain and in practice the grain size will vary depending on the grid and wire thickness, but nevertheless the effect of a fine grain size can be seen in the range 0·05 to 0·07% calcium. At less than 0·04% there is little matrix precipitation and the corrosion is intergranular, while above 0·04% the attack becomes intragranular due to the high volume fraction of the Pb_3Ca contained in the lead-rich matrix. Casting conditions are more critical with calcium alloys, due to the higher freezing temperature and difficulties in mould filling; there is a need for improved control of the mould temperature to prevent pores, voids and a honeycomb type of coring arising. The grids are normally gravity cast— pressure-cast positive grids appear to be prone to abnormal growth and preferential grain boundary corrosion[117,153] the reasons for which have been outlined above and are the result of an abnormal structure caused by rapid cooling. Improved control and design should alleviate many of these problems.

The inherent softness of the calcium binary has led to alloy additions for increased strength, tin being the most frequently used element. Intermetallic precipitates of the type Sn_3Ca or Pb_xSn_yCa may be formed, but this has not yet been confirmed by the publication of a ternary phase diagram. The beneficial effect of tin is related to the calcium content of the alloy, the higher levels (0·1%) both with and without tin showing intragranular corrosion;[94] thus at 0·5% tin, there is an optimum calcium content,[177] and increasing the tin content increases the susceptibility to intergranular corrosion, which suggests an interactive effect between these alloying additions.[150] Tin increases the grain size,[51] but impurities such as Bi, Sb, Cu and Ag have little effect on the properties related to grain size and hydrogen evolution at the cathode. Nevertheless, even at the typical concentration of 1·0% tin,[52] the alloy components are soft with an as-cast Brinell hardness of about 11 BH and an aged hardness (10–15 days) of 12–14 BH. With further ageing the hardness increases to 15–16 BH, which is not significantly different to a standard antimonial grid; however, there is the production problem of handling grids in automatic casting and pasting machinery. Tin improves the longer-time hardness,[177] but appears to inhibit the onset of hardening by about 1 h,[171] there being little effect on the microstructure, hardness and castability between 0·5% and 1·0% Sn.[52] Further additions have been considered,[281] and an alloy including aluminium[286] has been proposed: 0·08% Ca, 1·5% Sn and 0·05% Al. A practical problem with casting is the oxidation of calcium in the melt, and methods of minimizing this effect include an inert gas blanket.[52] Also, the recycling of antimonial grids with calcium grids causes difficulty due to the removal of both these alloying elements as a dross containing Ca_3Sb_2. As with the binary calcium alloy, the grain size is dependent on the mould size and casting conditions (Fig. 22).

Lead–barium and lead–strontium alloys have not been as extensively studied as the lead–calcium systems, presumably because the metallurgy is somewhat similar but with an increased cost for the elements and a more reactive alloy, high-content barium alloys decomposing in air. The phase diagram (Fig. 23) for lead–barium shows a eutectic of 4·5% barium at 293°C, but is otherwise similar to the lead–calcium diagram. The intermetallic Pb_3 Ba crystallized from the hyper-eutectic alloys with a maximum solubility of 0·5 wt % barium at 327°C[99] and about 0·02 wt % at 25°C.[212]

Early studies of the binary alloy were concerned with the fatigue properties[255] and the lack of hardness necessary for a bearing metal.[99] The age-hardening characteristics[147] and the corrosion resistance[260] have been reported. Alloys containing 0·1–0·3% barium consist mainly of barium in solid solution, the increasing volume of secondary phase present increasing with concentration. This causes precipitation within the grains, and the binary alloy is considered unsuitable for battery grids. Parr et al.[177] suggested that the barium content should not exceed 0·1%, otherwise the alloy becomes susceptible to intergranular attack, but below 0·07% the strength is adversely affected. Laboratory tests on the ternary system Pb–Ba–Sn indicated that the alloys should have good creep and the corrosion resistance is characterized by the formation of a tenacious oxide layer, which, owing to its interdendritic attachment to the parent alloy, should promote good contact between paste and grid. However, full-scale trials showed the alloy to have poor castability with a high susceptibility to casting defects (hot cracks).

Little published data exist on lead–strontium alloys, although Globe Union have recently indicated[90] an interest in this system for use in a maintenance-free battery, suggesting that it has better properties than calcium.[4] Early references were concerned with its use as a hardener, at concentrations up to 10%, or as a mixture with barium and or calcium for use in bearing alloys.[274–284] References for the use of strontium alloys in battery technology are even more sparse, with only three patent claims granted.[276–280] Cranfield and Kaiser[276] proposed an alloy of

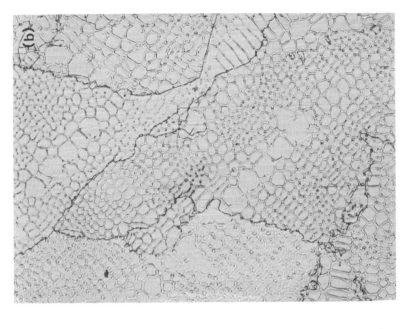

Fig. 22. Grain size in a cast automotive grid prepared from a Pb–Ca–Sn alloy. ×100. (a) Grid wire, (b) side frame.

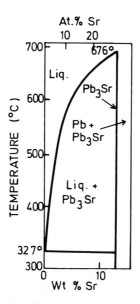

Fig. 23. Equilibrium diagram of the Pb–Sr binary alloy.[218]

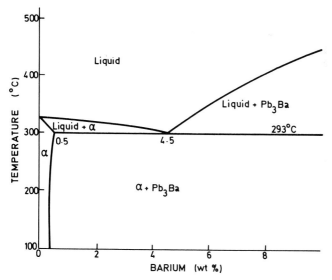

Fig. 24. Equilibrium diagram of the Pb–Ba binary alloy.[218]

0·2–0·6% strontium and 0·2–2·0% tin, with or without the addition of barium or calcium; this was followed[280] by a grid alloy of composition 0·4–0·6% strontium with 0·1–0·5% tellurium. Bouton and Schumacher[280] claimed good strength and freedom from corrosion and embrittlement from an alloy of 0·01–0·15% calcium, 0·005 to 0·05% barium and 0·005–0·2% strontium, which, from our present-day knowledge of calcium alloys, would appear to be a more feasible concentration range. The binary phase diagram (Fig. 24) shows similar characteristics to the lead–calcium with the liquidus rising steeply from the lead side and a limiting solid solubility. There are no published ternary diagrams for the calcium–strontium–barium systems and no corrosion data in the literature for the low-concentration strontium and barium alloys. By analogy with the calcium system one can anticipate low creep resistance with possible grid growth during cycling, but any advantage over the corresponding lead–tin–calcium or even barium alloys would have to be weighed against the increased cost.

Lead–lithium alloys

The use of lithium as an alloying element in maintenance-free batteries was extensively examined by Mao et al. during the late 1960s and early 70s.[151–158] Lithium in the free state has chemical properties similar to calcium, but its metallurgy resembles antimony;[103] thus there was a possibility that the alloys may possess the desirable properties of both the antimony and calcium alloys.

The binary phase diagram[103] shows a retrograde solidus with a solvus that decreases from 0·10% at 237°C to 0·015% lithium at room temperature. Only below this lower figure are the alloys single-phased, and increasing the lithium concentration will result in the precipitation of lithium–lead intermetallics in the grain boundaries, with possible eutectic coring, grain refinement, overageing and embrittlement. The addition of tin refines the grain size and increases grain boundary precipitation, but there is no published diagram to aid phase identification; nevertheless, the ternary alloys in excess of 0·5% tin have adequate hardness and strength to eliminate the necessity for the heat treatment and quenching required for the binary alloys. In all cases there are problems in maintaining an adequate lithium content in the molten alloy.

Corrosion, particularly at high anodic rates, limits the useful composition to less than 0·03% lithium[158] with a claimed optimum range of 0·02–0·03%.[154] The attack is typically intergranular, similar to that reported for alloys containing greater than 0·001% lithium immersed in 50% sulphuric acid.[92] Tin appears to minimize grain boundary precipitation and the decrease in grain size decreases the corrosion penetration resulting in a more general type of attack. The deep intergranular corrosion may result in a product layer of great apparent thickness, which in fact contains uncorroded grains still attached to the alloy. This is especially so at higher potentials where the enhanced migration of lithium is reputed to improve the cycle life[153] with the increased porosity of the corrosion layer accommodating the stresses during cycling, thereby preventing cracking or loss of product.[71] These lithium-based alloys have not replaced the more traditional battery alloys even for maintenance-free application, presumably due to the manufacturing problems and the intergranular corrosion. Nevertheless, they do indicate the approach required to develop an improved alloy system.

Summary

This section has highlighted some of the problem areas responsible for the apparent lack of development of new alloys. This can be seen as the result of the interaction or compromise between metallurgical requirements and the necessity of maintaining a suitable corrosion product film. The problems of handling components during fabrication, combined with the relatively poor creep resistance of lead alloys during service determines the minimum alloying levels required to provide acceptable strength and hardness. The difficulties to be overcome during production are as important as the electrochemical corrosion problems; these include shrinkage (the result of segregation), precipitation hardening, adequate control of required level of alloying elements within the melt, prevention of oxidation of molten reactive elements, and minimizing impurity levels when using recycled lead. Casting technology has had to improve as the concentration of the alloying additions decreased, and the importance of grain-refining elements and intermetallic compounds of the electronegative elements is now more widely recognized. These metallurgical factors also determine the type of corrosion, whether general or intergranular, but there is still a need to understand more fully the complex relationship between the protective properties of the corrosion products and their longer-term adhesion to the alloy substrate as well as their cohesion to the active material of a battery during cycling.

Conclusions

The aim of this review has been to indicate both the scientific principles leading to our present knowledge of the electrochemical corrosion of lead and the technological constraints which limit the use of many alloys in sulphuric acid. Although the literature survey has been wide-ranging, it does not claim to be complete since it is drawn mainly from investigations related to the lead–acid battery. Nevertheless, it should provide a basis for the understanding of corrosion of lead in other systems, since similar alloy developments are mirrored in electro-winning processes. Much of the recent impetus for the battery technologist to minimize the use of expensive or undesirable alloying elements has been the result of economic and marketing pressures; also, improvements in the design and corrosion life of components have decreased the amount of lead required to support the active material. The more recent indications suggest a good basis for future material utilization and the development of newer alloys, but there is a danger in comparing the results of different workers reported for various alloys without information on the microstructure.

   This comparison of alloys is a perennial problem and is due in part to the various anodic test conditions used. The ranking may change depending on the potential range used, and this highlights the inadequacy of tests at constant current. The increasing use of voltage control systems on automotive, traction and standby batteries had also accentuated some problems associated with the corrosion film formation and morphology. The electrical resistance, adhesion and diffusion properties of the film influence not only the rate of corrosion but rate of charge acceptance and shedding of material. Thus the lead dioxide films produced on the support lugs of stationary batteries either fall away harmlessly as a loose powder or the internal stresses built up over the years in a continuous film can give rise to a sudden dislodgment of large flakes. These examples illustrate the interaction between alloy composition and operating voltage, and again emphasize the increasing use of potentiostatic control in corrosion testing. Selection of the appropriate test potential and the use of a planned experiment allows the interaction of alloying elements to be investigated and quantified on a statistical basis.[149] In this context, note should also be made of the recent papers[164,259] concerned with corrosion and accelerated testing where dimensional changes, corrosion penetration and temperature effects are used to provide corrosion equations capable of predicting the life times of battery grids.

These later observations and analysis suggest that the development of lead alloys for use under anodic conditions is decreasingly a matter of subjective experience and is now based on scientific evidence; hopefully, this should allow these materials to be engineered to their known design limits.

References

1. A. A. Abdul Azim and M. M. Anwar, *Corrosion Sci.* **9**, 245 (1969).
2. A. A. Abdul Azim and K. M. El-Sobki, *Corrosion Sci.* **12**, 371 (1972).
3. A. A. Abdul Azim, K. M. El-Sobki and A. A. Khedr, *Corrosion Sci.* **16**, 209 (1976).
4. A. A. Abdul Azim, K. M. El-Sobki and A. A. Khedr, *Corrosion Sci.* **17**, 415 (1977).
5. N. Kh. Abrikosov *et al.*, *Neorg. Materialy* **5**, 741 (1969).
6. A. Augub, A. I. Rusin and M. A. Dasoyan, *Zashchitr. Metal Elecktrokhim. Akad. Nauk. SSSR.* 328 (1965).
7. I. A. Aguf and M. A. Dasoyan, "Technology of Storage Battery Production", TsINTI elektropromyshlennosti i priborostroeniya, Vol. **50** (1960).
8. T. N. Andersen and D. L. Adamson, *Metal. Trans.* **5**, 1345 (1974).
9. R. T. Angstadt, C. J. Venuto and P. Ruetschi, *J. Electrochem. Soc.* **109**, 177 (1962).
10. L. I. Antropov, S. Ya. Popov, T. I. Pochekaeva and N. N. Romenskaya, *Trudy Sovesh. Elektrokhim. Akad. Nauk. SSSR Otd. Khim. Nauk.* **1950**, 539 (1953).
11. G. Archdale and J. A. Harrison, *J. Electroanal. Chem.* **34**, 21 (1972).
12. G. Archdale and J. A. Harrison, *J. Electroanal. Chem.* **39**, 357 (1972).
13. G. Archdale and J. A. Harrison, *J. Electroanal. Chem.* **47**, 93 (1973).
14. G. Archdale and J. A. Harrison *in* "Power Sources 4" (D. H. Collins, ed.), p. 309. Oriel Press, Newcastle-upon-Tyne (1973).
15. I. I. Astakhov, E. S. Vaisberg and B. N. Kabanov, *Dokl. Akad. Nauk. SSSR*, **154**, No. 6, 1414 (1964).

16. I. A. Bagotzkaya and A. N. Frumkin, *Dokl. Akad. Nauk. SSSR* **92**, 979 (1953).
17. N. E. Bagshaw, "Proceedings of the 3rd International Conference on Lead 1968", p. 209. Pergamon (1969).
18. N. E. Bagshaw and T. A. Hughes *in* "Batteries 2" (D. H. Collins, ed.), Pergamon, New York (1965).
19. S. C. Barnes and R. T. Mathieson *in* "Batteries 2" (D. H. Collins, ed.), Pergamon, New York (1965).
20. W. H. Beck, R. Lind and W. F. K. Wynne-Jones, *Trans. Far. Soc.* **50**, 147 (1954).
21. W. H. Beck and W. F. K. Wynne-Jones, *Trans. Far. Soc.* **52**, 1260 (1956).
22. D. Berndt and S. C. Nijhewan, *J. Power Sources* **1**, 3 (1976).
23. D. Berndt and E. Voss, "Batteries 4" (D. H. Collins, ed.), Pergamon, New York (1969).
24. H. Bode and J. Euler, *Electrochim. Acta* **11**, 1211 (1966).
25. H. Bode and E. Voss, *Electrochim. Acta* **1**, 318 (1959).
26. H. Bode and E. Voss, *Electrochim. Acta* **6**, 13 (1962).
27. H. Borchers and S. C. Nijhewan, *Metalloberfläche* **29**, 465 (1975).
28. H. Borchers, S. C. Nijhewan and W. Schwarfenberger, *Metalloberfläche* **28**, 863 (1974).
29. Fr. Brenthel, *Z. Metallkde* **22**, 23 (1930).
30. M. P. J. Brennan, B. N. Stirrup and N. A. Hampson, *J. Appl. Electrochem.* **4**, 49 (1974).
31. J. Z. Briggs, *Metals and Alloys* **9**, 49 (1938).
32. C. H. Brubaker, *J. Amer. Chem. Soc.* **76**, 4269 (1954).
33. H. Buckel and H. Hannemann, *Z. Metalkde* **32**, 120 (1940).
34. J. Burbank *in* "Batteries" (D. H. Collins, ed.), p. 43. Pergamon, New York (1963).
35. J. Burbank, *J. Electrochem. Soc.* **103**, 87 (1956).
36. J. Burbank, *J. Electrochem. Soc.* **104**, 693 (1957).
37. J. Burbank, *J. Electrochem. Soc.* **106**, 369 (1959).
38. J. Burbank, *J. Electrochem. Soc.* **111**, 765 (1964).
39. J. Burbank, *J. Electrochem. Soc.* **111**, 1112 (1964).
40. J. Burbank, *J. Electrochem. Soc.* **113**, 10 (1966).
41. J. Burbank, A. C. Simon and E. Willihnganz *in* "Advances in Electrochemistry and Electrochemical Engineering" (P. Delahay, ed.), Vol. 8, p. 157. Wiley Interscience, London and New York (1971).
42. J. Burbank *in* "Power Sources 1" (D. H. Collins, ed.), p. 147. Pergamon, New York (1967). *See also* J. Burbank, "The Plate Materials of the Lead–Acid Cell, Part 3– Anodic Oxidation of Tetragonal PbO." Naval Research Laboratory Report 6613, Dec. 29 (1967).
43. J. Burbank *in* "Power Sources 3" (D. H. Collins, ed.), p. 13. Oriel Press, Newcastle-upon-Tyne (1971).
44. J. Burbank and E. J. Ritchie, *J. Electrochem. Soc.* **116**, 125 (1969).
45. J. Burbank, *J. Electrochem. Soc.* **118**, 525 (1971).

46. J. Burbank and A. C. Simon, *J. Electrochem. Soc.* **100**, 11 (1953).
47. P. Butler, *The Engineer*, 23.6.1977.
48. W. O. Butler, C. J. Venuto and D. V. Wisler, *J. Electrochem. Soc.* **117**, 1339 (1970).
49. A. G. Cannone, D. O. Feder and R. V. Biagettii, *Bell System Tech. J.* **49**, 1279 (1970).
50. T. W. Caldwell and U. S. Sokolov *in* "Power Sources 5" (D. H. Collins, ed.), p. 5. Academic Press, London and New York (1975).
51. T. W. Caldwell and U. S. Sokolov, *J. Electrochem. Soc.* **123**, 972 (1976).
52. T. W. Caldwell, U. S. Sokolov and L. M. Bocciarelli, *J. Electrochem. Soc.* **123**, 1265 (1976).
53. J. P. Carr, J. R. Calvert and N. A. Hampson, *J. Electroanal. Chem.* **34**, 425 (1972).
54. J. P. Carr and N. A. Hampson, *Chem. Rev.* **72**, 679 (1972).
55. J. P. Carr, N. A. Hampson and R. Taylor, *J. Electroanal. Chem.* **33**, 109 (1971).
56. E. J. Casey and K. N. Campney, *J. Electrochem. Soc.* **102**, 219 (1955).
57. S. M. Caulder, J. S. Murday and A. C. Simon, *J. Electrochem. Soc.* **120**, 1515 (1973).
58. S. M. Caulder and A. C. Simon, *J. Electrochem. Soc.* **121**, 1546 (1974).
59. G. A. Chadwick *in* "Fractional Solidification" (M. Zief and W. R. Wilcox, eds), p. 113. Marcel Dekker, New York (1967).
60. T. Chiku, *J. Electrochem. Soc.* **115**, 982 (1968).
61. T. Chiku and N. Nakajima, *J. Electrochem. Soc.* **118**, 1395 (1971).
62. G. S. Cole and W. C. Winegard, *J. Inst. Metal.* **92**, 322 (1963).
63. T. R. Crompton and G. Elitenbroek, *J. Electrochem. Soc.* **119**, 655 (1972).
64. M. Dasoyan, *Dokl. Akad. Nauk. SSSR* **107**, 863–866 (1956); *Proc. Acad. Sci. USSR, Chem. Technol. Sect.* **107**, 37 (1956).
65. M. Dasoyan, *Nachr. Elecktro-Ind (Russ)* 2 (1957).
66. J. L. Dawson, M. I. Gillibrand and J. Wilkinson *in* "Power Sources 3" (D. H. Collins, ed.), p. 1. Oriel Press, Newcastle-upon-Tyne (1970).
67. J. L. Dawson, J. Wilkinson and M. I. Gillibrand, *J. Inorg. Nucl. Chem.* **32**, 501 (1970).
68. P. Delahay, M. Pourbaix and P. Van Rysselberghe, *J. Electrochem. Soc.* **98**, 57 (1951).
69. C. Drotschmann, *Batteries* **17**, 472 (1963).
70. C. Drotschmann, Blei Akkumulatoren, pp. 94–99. Verlag Chemie, GMBH, Weinheim (1951); *Batteries* **19**, 851 (1966); *Batteries* **20**, 876, 899 (1966).
71. D. L. Douglas and G. W. Mao, *in* "Power Sources 4" (D. H. Collins, ed.), p. 561. Oriel Press, Newcastle upon Tyne (1973).
72. K. Ekler, *Can. J. Chem.* **42**, 1355 (1964).
73. W. Feitknecht, *Z. Elektrochem.* **62**, 795 (1958).
74. W. Feitknecht and A. Gaumann, *J. Chim. Phys.* **49**, 135 (1952).
75. C. G. Fink and A. J. Dornblatt, *Trans. Electrochem. Soc.* **79**, 269 (1941).

76. C. G. Fink and L. C. Pan, *Trans. Electrochem. Soc.* **46**, 349 (1924).
77. A. N. Fleming, J. A. Harrison and J. Thompson in "Power Sources 5" (D. H. Collins, ed.), p. 1. Academic Press London and New York (1975).
78. M. Fleischmann and M. Liler, *Trans. Far. Soc.* **54**, 1370 (1958).
79. M. Fleischmann and H. R. Thirsk, *Electrochim. Acta* **1**, 146 (1959).
80. M. Fleischmann and H. R. Thirsk, *Electrochim. Acta* **2**, 22 (1960).
81. M. Fleischmann and H. R. Thirsk, "Advances in Electrochemistry and Electrochemical Engineering" (P. Delahay, ed.). John Wiley, London (1963).
82. S. Feliu, L. Galan and J. A. Gonzalez, Final Report No. LE-130 Int. Lead and Zinc, Res. Org. N.Y. (1970).
83. S. Feliu, L. Galan and J. A. Gonzalez, *Werkst. Korros.* **23**, 554 (1972).
84. S. Feliu and M. Morcillo, *Corrosion Sci.* **15**, 593 (1975).
85. S. Feliu and M. Morcillo, *Electrochim. Acta* **21**, 1035 (1976).
86. D. A. Frey and H. E. Weaver, *J. Electrochem. Soc.* **107**, 930 (1960).
87. A. N. Frumkin and N. A. Aladzalova, *Zh. Fiz. Khim.* **18**, 493 (1944).
88. M. I. Gillibrand and G. R. Lomax, *Electrochim Acta* **11**, 281 (1966).
89. J. H. Gladstone and A. Tribe, *Nature* **26**, 342 (1882); **26**, 602 (1882); **27**, 583 (1883).
90. Globe Union, American Metal Market, 25 Jan. (1977).
91. J. Goebal, *Z. Metallkde* **14**, 357, 380 (1922).
92. J. A. Gonzalez, J. J. Royeula and S. Feliu, Final Report Project NoLE-70, ILZRO (1967).
93. J. A. Gonzalez, J. J. Royeula and S. Feliu, *Rev. Metal* **7**, 105 (1971).
94. J. A. Gonzalez, J. J. Royeula and S. Feliu, *Werkst. Korros.* **22**, 758 (1975).
95. H. Grafen and D. Kuron, *Werkst. Korros.* **20**, 749 (1970).
96. H. Grafen and D. Kuron, *Werkst. Korros.* **21**, 3 (1970).
97. R. Grauer, P. Wehr and H. J. Engell, *Werkst. Korros.* **20**, 94 (1969).
98. J. N. Greenwood and C. W. Orr, *Proc. Aust. Min. Metall.* **112**, 287 (1938).
99. G. Grube and A. Dietrich, *Z. Elektrochem.* **44**, 755 (1938).
100. N. A. Hampson and J. P. Carr, *Chem. Rev.* **72**, 679 (1972).
101. H. E. Haring and K. G. Compton, *Trans. Electrochem. Soc.* **68**, 283 (1935).
102. H. F. Haring and U. B. Thomas, *Trans. Electrochem. Soc.* **68**, 293 (1935).
103. M. Hansen and K. Anderbo, "Constitution of Binary Alloys". McGraw-Hill (1958).
104. H. S. Harned and W. J. Hamer, *J. Amer. Chem. Soc.* **57**, 27 (1935).
105. J. A. Harrison and H. R. Thirsk *in* "Fundamentals of Metal Deposition", Advances in Electroanalytical Chemistry (A. J. Bard, ed.), Vol. 5 Dekker, New York (1971).
106. O. Heckler, W. Hofmann and H. Hanemann, *Z. Metallkde.* **30**, 12 (1938).
107. W. Herrmann, W. Ilge and G. H. Proepstl, *Proc. 2nd U.N. Int. Conf. Geneva* "Peaceful Uses of Atomic Energy", **19**, 272. Pergamon, New York (1958).
108. W. Herrmann and G. Proepstl, *Z. Elektrochem.* **61**, 1154 (1957).
109. G. O. Hiers, *Mining Metal.* **17**, 43 (1936).

110. G. O. Hiers, "Metals Handbook". Am. Soc. Metals, Ohio, U.S.A. (1948).
111. H. F. Hintermann and C. J. Venuto, *J. Electrochem. Soc.* **115**, 10 (1968).
112. S. Hisano, *Kogyo Kogaku Zasshi* **62**, 376 (1959).
113. E. Hoehne, *Z. Matallkde.* **30**, 52 (1938).
114. E. Hoehne, *Arch. Metallkde.* **2**, 311 (1948).
115. E. Hoehne and H. D. Graf, *Metallwirtschaft* **21**, 218 (1939).
116. W. Hofmann, "Blei und Bleilegierungen". Springer-Verlag, Berlin (1962). English edition "Lead and Lead Alloys". Springer-Verlag, Berlin (1970).
117. A. M. Howard and E. Willihnganz, *J. Electrochem. Technol.* **6**, 370 (1968).
118. S. Ikari and S. Yoshizawa, *J. Electrochem. Soc. Jap.* Overseas edn. **28**, E192 (1960).
119. B. N. Kabanov, I. G. Kiseleva and D. I. Leikis, *Dokl. Akad. Nauk. SSSR* **99**, 805 (1954).
120. B. N. Kabanov, D. I. Leikis and E. I. Krepakova, *Dokl. Akad. Nauk. SSSR* **98**, 989 (1954).
121. B. N. Kabanov, E. S. Weisberg, I. L. Romanova and E. V. Krivolapova, *Electrochim. Acta* **9**, 1197 (1964).
122. T. Katz, *Ann. Chim. 12th ser.* **5**, 5 (1950).
123. V. A. Khitrov, *Izv. Voron. Gos. Ped. Inst.* **29**, 5 (1960).
124. P. J. Killaby, B. J. Taylor and W. C. Winegard, *Amer. Foundaryman* **22**, 59 (1952).
125. G. Z. Kiryakov and V. V. Stender, *J. Appl. Chem. USSR* **25**, 25 (1952); **25**, 33 (1952).
126. A. Kirow, T. Rogatschev and D. Denew, *Metalloberfläche* **26**, 234 (1972).
127. B. V. Krivolapova, *Akad. Nauk. SSSR Otd. Khim. Nauk.* **1950**, 539 (1953).
128. G. Z. Kiryakov and V. V. Stender, *Zh. Prikl. Khim.* **25**, 23 (1952).
129. G. Z. Kiryakov and V. V. Stender, *Zh. Prikl. Khim.* **25**, 1263 (1951).
130. I. G. Kiseleva and B. N. Kabanov, *Dokl. Akad. Nauk.* **122**, 1042 (1958).
131. D. F. A. Koch, *Electrochim. Acta* **1**, 32 (1959).
132. A. E. Koenig, J. U. McEvans and E. C. Larsen, *Trans. Electrochem. Soc.* **79**, 331 (1941).
133. I. M. Kolthoff and B. van't Riet, *J. Phys. Chem.* **63**, 817 (1959).
134. D. Kordes, *Chem. Ing. Tech.* **38**, 638 (1966).
135. E. V. Krivolapova and B. N. Kabanov, *Trudy Sovesh. Elektrokhim. Akad. Nauk. SSSR Otd. Khim. Nauk.* **1950**, 539 (1953); *Chem. Abstr.* **49**, 12161 (1955).
136. J. J. Lander, *J. Electrochem. Soc.* **98**, 213 (1951).
137. J. J. Lander, *J. Electrochem. Soc.* **98**, 220 (1951).
138. J. J. Lander, *J. Electrochem. Soc.* **99**, 467 (1952).
139. J. J. Lander, *J. Electrochem. Soc.* **103**, 1 (1956).
140. J. J. Lander, *J. Electrochem. Soc.* **105**, 289 (1958).
141. F. Lappe, *J. Phys. Chem. Solids* **23**, 1563 (1962).
142. W. M. Latimer, "Oxidation Potentials". Prentice Hall, New York (1953).

143. J. Lecrejewicz and I. Padlo, *Naturwissenschaften* **49**, 373 (1962).
144. K. I. Leikis and E. K. Venstrem, *Dokl. Akad. Nauk. SSSR* **112**, 97 (1957).
145. L. M. Levinzon, I. A. Aguf and M. A. Dasoyan, *Zh. Prikl. Khim.* **39**, 525 (1966).
146. K. J. Linden and C. A. Kennedy, *J. Appl. Phys.* **40**, 2595 (1969).
147. V. S. Lyashenko, *Metallurgia* **13**, 12 (1938).
148. J. McKeown, *J. Inst. Metal.* **60**, 201 (1937).
149. J. McWhinnie, Ph.D. Research, UMIST (1977).
150. M. Maeda, *J. Electrochem. Soc. Jap.* **25**, 197 (1957); Overseas edn. **26**, E21, E183 (1958).
151. G. W. Mao and J. G. Larson, *Metallurgia* **78**, 236 (1968).
152. G. W. Mao, J. G. Larson and P. Rao, *Metallography* **1**, 399 (1969).
153. G. W. Mao, J. G. Larson and P. Rao, *Inst. Metal.* **97**, 343 (1969).
154. G. W. Mao, J. G. Larson and P. Rao, *J. Electrochem. Soc.* **118**, 205c (1971).
155. G. W. Mao, J. G. Larson and P. Rao, *J. Electrochem. Soc.* **120**, 11 (1973).
156. G. W. Mao, T. L. Oswald and B. J. Sobczak *in* "Power Sources 3" (D. H. Collins, ed.), p. 61. Oriel Press, Newcastle-upon-Tyne (1971).
157. G. W. Mao and P. Rao, *Brit. Corrosion J.* **6**, 122 (1971).
158. G. W. Mao, T. L. Wilson and J. G. Larson, *J. Electrochem. Soc.* **117**, 1323 (1970).
159. H. P. Mark, Jr. and W. C. Vosburgh, *J. Electrochem. Soc.* **108**, 615 (1961).
160. J. Marshal, *J. Chem. Soc.* **54**, 771 (1891).
161. D. Marshall and W. Tiedemann, *J. Electrochem. Soc.* **123**, 1849 (1976).
162. V. P. Mashovetts and A. Z. Lyandres, *Zh. Prikl. Khim.* **21**, 347, 441 (1947).
163. N. K. Mikhailova and I. A. Aguf, *Zashchrift. Metal* **10**, 57 (1974).
164. N. J. Maskalick, *J. Electrochem. Soc.* **122**, 19 (1975).
165. P. C. Milner, *Bell Systems Tech. J.* **49**, 1321 (1970).
166. W. Mindt, *J. Electrochem. Soc.* **116**, 1076 (1969).
167. L. R. Morris and W. C. Winegard, *J. Crystal Growth* **5**, 361 (1969).
168. L. R. Morris and W. C. Winegard, *J. Crystal Growth* **6**, 61 (1969).
169. W. J. Muller, *Kolloid-Z.Z. Polym.* **86**, 150 (1939).
170. B. Munasiri, Ph.D. Thesis, UMIST (1977).
171. M. Myers, H. R. Van Handle and C. R. Di Martini, *J. Electrochem. Soc.* **121**, 1526 (1974).
172. K. Nagel, R. Ohse and E. Lange, *Z. Elektrochem.* **61**, 759 (1957).
173. K. W. Nolan, J. L. Hirsch and D. M. Pope, "Proc. 5th Int. Conf. Lead", p. 79. Metal Bulletin (1974).
174. Z. V. Niyazora, A. V. Vakkobov, M. A. Dasoyan, A. Sh. Murodov and E. D. Rozenberg, *Zavad. Lab.* **42**, 584 (1976).
175. R. W. Ohse, *Werkst. Korros.* **11**, 220 (1960).
176. H. S. Panesar *in* "Power Sources 3" (D. H. Collins, ed.), p 79. Oriel Press, Newcastle-upon-Tyne (1971).
177. N. L. Parr, A. Muscott and A. J. Crocker, *J. Inst. Metal.* **87**, 321 (1958).

178. W. T. Pell-Walpole, *Metal Treatment*. **16**, 103 (1949).
179. J. Perkins and G. R. Edwards, *J. Mater. Sci.* **10**, 136 (1975).
180. D. Pavlov, *Ber. Bunsenges. Phys. Chem.* **71**, 398 (1967).
181. D. Pavlov, *Electrochem. Acta* **13**, 2051 (1968).
182. D. Pavlov, M. Boton and M. Stoyanova, *Izu. Inst. Fiz. Bulg. Akad. Nauk.* **5**, 55 (1965).
183. D. Pavlov and N. Iordanov, *J. Electrochem. Soc.* **117**, 1103 (1970).
184. D. Pavlov and R. Popova, *Electrochim. Acta* **15**, 1483 (1970).
185. D. Pavlov, C. N. Poulieff, E. Klaja and N. Iordanov, *J. Electrochem. Soc.* **116**, 316 (1969).
186. D. Pavlov and T. Rogatschev, *Werkst. Korros.* **19**, 677 (1968).
187. J. R. Pierson, *J. Electrochem. Tech.* **5**, 323 (1967).
188. T. S. Plaskett and W. C. Winegard, *Can. J. Phys.* **38**, 1077 (1960).
189. M. Pourbaix, "Atlas of Electrochemical Equilibria in Aqueous Solutions". Cabelcor, Brussels (1966).
190. A. Ragheb, W. Machu and W. H. Boctor, *Werkst. Korros.* **16**, 676 (1965).
191. M. E. Rana, Ph.D. Thesis, UMIST (1975).
192. G. V. Raynor, *Inst. Met. Ann. Eq. Diag.* No. 6.
193. N. F. Razina, *Trudy Chelv. Sovesh. Elektrokhim. Moscow* **1956**, 729 (1959).
194. N. F. Razina, M. T. Kozlovsky and V. V. Stender, *Dokl. Akad. Nauk. SSSR* **111**, 404 (1956).
195. M. Rey, P. Coheur and H. Herbiet, *Trans. Electrochem. Soc.* **73**, 315 (1938).
196. E. J. Ritchie, "Pastes and Grids for the Lead–Acid Battery", Eagle-Picher Industries, Progress Report No. 2, Combined ILZRO Research Contracts LE-82 and LE-84, Dec. (1967).
197. E. J. Ritchie and J. Burbank, *J. Electrochem. Soc.* **117**, 299 (1970).
198. D. H. Roberts, N. A. Ratcliff and J. E. Hughes, *Powder Metall.* **10**, 132 (1962).
199. T. Rogatchev, St. Ruevski and D. Pavlov, *J. Appl. Electrochem.* **6**, 33 (1976).
200. T. Rogatschev, W. Karolena and D. Pavlov, *Metalloberfläche* **24**, 11 (1970).
201. P. Ruetschi, *J. Electrochem. Soc.* **120**, 331 (1973).
202. P. Ruetschi and R. T. Angstadt, *J. Electrochem Soc.* **103**, 202 (1956).
203. P. Ruetschi and R. T. Angstadt, *J. Electrochem. Soc.* **105**, 555 (1958).
204. P. Ruetschi and R. T. Angstadt, *J. Electrochem. Soc.* **111**, 1323 (1964).
205. P. Ruetschi, R. T. Angstadt and B. D. Cahan, *J. Electrochem. Soc.* **106**, 547 (1959).
206. P. Ruetschi and B. D. Cahan, *J. Electrochem. Soc.* **104**, 406 (1957).
207. P. Ruetschi and B. D. Cahan, *J. Electrochem. Soc.* **105**, 369 (1958).
208. P. Ruetschi, J. Sklarchuck and R. T. Angstadt, *Electrochim. Acta.* **8**, 333 (1963).
209. A. V. Sapozhnikov, *Zashchrift. Metal.* **9**, 625 (1973); *Prot. Met. USSR* **9**, 559 (1973).

210. E. Sato and T. Shiina, *J. Electrochem. Soc. Jap.* **32**, 148 (1964).
211. E. Sato and S. Takagi, *J. Electrochem. Soc. Jap.* **32**, 146 (1964).
212. E. Schmid, *Z. Metallkde.* **35**, 85 (1943).
213. E. E. Schumacher and G. M. Bouton, *Metals Alloys* **1**, 405 (1930).
214. E. E. Schumacher and G. S. Philips, *Trans. Electrochem. Soc.* **68**, 309 (1935).
215. J. S. Sellan, British Patent 3987 (1881).
216. T. F. Sharpe, *J. Electrochem. Soc.* **122**, 845 (1975).
217. T. F. Sharpe, *J. Electrochem. Soc.* **124**, 168 (1977).
218. C. J. Smithells, "Metals Reference Book" 5th edn. Butterworth (1976).
219. A. C. Simon *in* "Batteries 2" (D. H. Collins, ed.), p. 65. Pergamon, New York (1965).
220. A. C. Simon, *in* "Power Sources 2" (D. H. Collins, ed.), Pergamon, New York (1969).
221. A. C. Simon, *J. Electrochem. Soc.* **114**, 1 (1967).
222. A. Simon, *Electrochem. Technol.* **3**, 307 (1965).
223. A. C. Simon and S. M. Caulder, *J. Electrochem. Soc.* **118**, 659 (1971).
224. A. C. Simon and S. M. Caulder *in* "Power Sources 5" (D. H. Collins, ed.), p. 109. Academic Press, London and New York (1974).
225. A. C. Simon, S. M. Caulder and E. J. Ritchie, *J. Electrochem. Soc.* **117**, 1264 (1970).
226. A. C. Simon, S. M. Caulder and J. T. Stemmle, *J. Electrochem. Soc.* **122**, 461 (1975).
227. A. C. Simon and E. L. Jones, *J. Electrochem. Soc.* **100**, 1 (1953).
228. A. C. Simon and E. L. Jones, *J. Electrochem. Soc.* **101**, 536 (1954).
229. A. C. Simon and E. L. Jones, *J. Electrochem. Soc.* **104**, 133, 536 (1957).
230. A. C. Simon and E. L. Jones, *J. Electrochem. Soc.* **109**, 760 (1962).
231. A. C. Simon, C. P. Wales and S. M. Caulder, *J. Electrochem. Soc.* **117**, 987 (1970).
232. W. Singleton and B. Jones, *J. Inst. Metal.* **51**, 71 (1933).
233. J. Strange, *Electrochim. Acta* **19**, 111 (1974).
234. D. E. Swets, *J. Electrochem. Soc.* **120**, 925 (1973).
235. E. Tarter and K. Ekler, *Can. J. Chem.* **47**, 2191 (1969).
236. H. R. Thirsk and J. A. Harrison, "A Guide to the Study of Electrode Kinetics". Academic Press, London and New York (1972).
237. H. R. Thirsk and W. F. L. Wynne-Jones, *J. Chem. Phys.* **49**, C131 (1952).
238. U. B. Thomas, *Bell Lab. Rec.* **16**, 12 (1937).
239. U. B. Thomas, *Trans. Electrochem. Soc.* **94**, 42 (1948).
240. U. B. Thomas, F. T. Forster and H. E. Haring, *Trans. Electrochem. Soc.* **92**, 313 (1947).
241. S. S. Tolkachev, *Vest. Leningrad. Univ.* **11**, 152 (1958).
242. A. B. Townsend, USAEC Report No. Y-1307, Union Carbide Nulc. Co., Oak Ridge (1960).
243. K. N. Tu, *Met. Trans.* **3**, 2769 (1972).
244. S. Tudor, A. Weisstuch and S. H. Davang, *Electrochem. Technol.* **3**, 90 (1965).

245. S. Tudor, A. Weisstuch and S. H. Davang, *Electrochem. Technol.* **4**, 406 (1966).
246. N. B. Vaughn, *J. Inst. Metal.* **61**, 35 (1937).
247. E. M. L. Valeriote and L. D. Gallop *in* "Power Sources 5" (D. H. Collins, ed.), p. 55. Academic Press, London and New York (1975).
248. E. M. L. Valeriote and L. D. Gallop, *J. Electrochem. Soc.* **124**, 370 (1977).
249. E. M. L. Valeriote and L. D. Gallop, *J. Electrochem. Soc.* **124**, 380 (1977).
250. G. W. Vinal, *J. Res. Nat. Bur. Stand.* **25**, 417 (1940).
251. G. W. Vinal, "Storage Batteries", 4th edn. Wiley, New York (1955).
252. G. W. Vinal and D. N. Craig, *J. Res. Nat. Bur. Stand.* **14**, 449 (1935).
253. E. J. Wade, "Secondary Batteries", p. 14. The Electrician Printing and Pub. Co. (1902).
254. D. Walton, W. A. Tiller, J. W. Rutter and W. C. Winegard, *Trans. AIME* **203**, 1023 (1955).
255. H. Waterhouse, Research Report No. 440, Brit. Non-Ferrous Metal, Res. Ass. (1937).
256. J. L. Weininger, *J. Electrochem. Soc.* **105**, 577 (1958).
257. J. L. Weininger, *J. Electrochem. Soc.* **121**, 1454 (1974).
258. J. L. Weininger and C. R. Morelock, *J. Electrochem. Soc.* **122**, 1161 (1975).
259. J. L. Weininger and E. G. Siwek, *J. Electrochem. Soc.* **123**, 602 (1976).
260. K. Wickert, *Korros. Metallsch.* **18**, 357 (1942).
261. E. Willihnganz, *Electrochem. Technol.* **6**, 388 (1968).
262. E. Willihnganz, *J. Electrochem. Soc.* **92**, 281 (1947).
263. T. L. Wilson and G. W. Mao, *Met. Trans.* **1**, 2631 (1970).
264. M. Yamaura, M. Kohno, M. Yamane and H. Nakashima, ILZRO Project LE 197, Annual Report, March 20 (1973). YUASA Battery Co. Ltd., Japan.
265. M. Yamaura, M. Kohno, M. Yamane and H. Nakashima, ILZRO Project LE 197, Annual Report, March 20 (1974). YUASA Battery Co. Ltd., Japan.
266. J. A. Young, International Battery Council Meeting (1973).
267. A. C. Zachlin, *Trans. Electrochem. Soc.* **92**, 21 (1947).
268. A. C. Zachlin, *Trans. Electrochem. Soc.* **92**, 259 (1947).
269. H. E. Zahn, *Precis. Metal Mold.* **12**, 68 (1954).
270. A. I. Zaslavaki, J. D. Kondrashov and S. S. Tolkachev, *Dokl. Akad. Nauk. SSSR* **75**, 559 (1950).
271. E. Zehender and W. Herrmann, *Bosch Techn. Ber.* **1**, 126 (1965).
272. E. Zehender, W. Herrmann and H. Leibssle, *Electrochim. Acta* **9**, 55 (1964).
273. V. P. Zlomanov *et al.*, *Met. Trans.* **2**, 121, (1971).
274. German Patent 301,380 (1921).
275. French Patent 772,826 (1934).
276. U.S. Patent 2,013,487 (1935).
277. U.S. Patent 2,040,078 (1936).
278. U.S. Patent 2,042,840 (1936).

279. Australian Patent 110,423 (1936).
280. U.S. Patent 2,170,650 (1939).
281. U.S. Patent 2,142,835 (1939).
282. Australian Patent 114,745 (1941).
283. British Patent 548,775 (1942).
284. British Patent 550,485 (1943).
285. U.S. Patent 2,678,340 (1953).
286. British Patent 712,718 (1954).
287. Canadian Patent 503,663 (1954).
288. D.A.S. Patent 1,097,695 (1958).
289. D.A.S. Patent 1,239,102 (1964).
290. U.S. Patent 2,678,341 (1964).
291. British Patent 1,329,974 (1973).
292. U.S. Patent 3,912,537 (1973).
293. German Patent 2,412,322 (1974).
294. U.S. Patent 3,990,893 (1974).
295. U.S. Patent 3,881,953 (1975).
296. British Patent 1,425,554 (1976).
297. British Patent 1,454,401 (1976).
298. British Patent 1,461,260 (1977).
299. British Patent 1,461,587 (1977).

13 Corrosion of Lead and Alloys in Sulphuric Acid: Detailed Survey and Special Conditions

A. T. KUHN

Department of Dental Materials, Eastman Institute of Dental Surgery, London

By far the largest body of literature on the corrosion of lead and its alloys relates to sulphuric acid solutions, and this material, which stretches over half a century, is confused, largely apparently self-contradictory and has never before, as far as we are aware, been thoroughly reviewed. Much of the confusion arises because the work stems from two completely different types of interest. The battery manufacturers conduct their research, as do outside scientists with battery-related interests, in sulphuric acid of the highest purity, and in a relatively restricted temperature and acid-concentration band. The "anode-oriented" work deriving from the electrowinning of metal interests often relates to sulphuric acid heavily contaminated with Cl^-, F^-, Co^{2+}, Fe^{2+} and many other inorganic and organic constituents. This is the first source of apparent discrepancies. The second, and probably more serious one, stems from the test method and test regime from which corrosion data are derived. The "process-anode" work considers mainly corrosion at high anodic potentials where oxygen is evolved. The occasional publication considers what happens when process anodes are "at rest" but mainly the work described relates to anodes at or close to 2·0 V *vs* R.H.E. Much battery work, on the other hand, describes "positive plate lifetimes". To obtain this data, the plates are "cycled" through a charge–discharge regime, possibly with a "rest" period as well. In consequence, we can say firstly that the anode is only briefly raised to oxygen evolution potential, and secondly, that the overall corrosion data obtained in most of such work constitute some sort of "mean rate" averaged over a history during which the electrode has been briefly raised to the oxygen evolution on the one hand, having spent substantial time in the partly "sulphated" condition on the other. Under the latter conditions, corrosion is much faster than when the anode is PbO_2-coated. Finally, some battery corrosion data are extrapolated from grid expansion measurements, and the relevance of this to anodic corrosion of a massive metal is tenuous, to say the least. In short, it seems best to resist the temptation to link together the two types of data described above, and some authors such as Andersen,[1] for example, specifically mention contradictions arising when one attempts to do so. Nevertheless, certain alloys, notably the Pb–Ag ones, are recognized by both battery technologists and process technologists, as having superior properties. We are left, then, in the unhappy position of not being able to rely on cross-correlation of data, while knowing that it does sometimes tie together.

A similar problem arises when we consider the process-anode data. On one side, we have a corpus of papers in which various alloys have been compared one against the other, in a given solution. On the other, we have work in which a given alloy has been tested in various acid strengths, at different temperatures and in particular at different current densities and with different impurities present. Can we relate results freely across all these situations? Largely yes, one feels. However, there is no certainty, and the aim of this section must be to give a guide to trends which will serve to lead workers in the field most quickly to the optimal anodes for their particular conditions.

The importance of a further parameter has been raised by Sunderland[2] relating to grain size of the alloys used in various studies reported in the literature. Within a given compositional specification, such as 0·5% Ag, we may find alloys with a wide range of grain sizes. Indeed very small amounts of third component, introduced either deliberately or accidentally, may act as grain-size refiners. Lead alloys tend to "age" at disparate rates. Certain of the Pb–Ca alloys used in the battery industry can be seen to "age" in the 2–3 minutes immediately after they are cast. Other alloys take longer, possibly many years, to reach some stable state. One cannot easily think of work in which these variables are reported, and yet Sunderland has found that they exercise a profound influence on the corrosion rates. We must therefore consider the data in all these ways, both in terms of solution effect, and alloy composition and texture.

Effect of Other Constituents in Solution, Solution Effects

Sulphuric acid at extremes of temperature or concentration

Increasing acidity hastens corrosion of lead and its alloys except perhaps at the most extreme dilutions (0.005 M or less) when unusual effects can occur. Andersen[1] shows how corrosion of the ternary Pb–Sb–As alloy increases steadily as acid concentration goes from 50 g l⁻¹ to 275 g l⁻¹. Azim and Anwar[3] have studied Pb–Sn alloys in acid of 440 g l⁻¹. Grafen and Kuron[21] have also studied corrosion in strong acids (to 5 M), while Izidinov[4] has looked at strong acids especially at low temperatures ($-20°C$). Under extreme conditions (high potentials) "per" compounds are formed and he reports that the anodes disintegrate at a visible rate. Azim and El-Sobki[5] have studied Pb and Pb–Cd in 10 N H_2SO_4 and found that the alloy has the lower corrosion resistance.

There are many other papers in this field, and ignoring, for the purposes of this work, those papers which simply relate to free corrosion of lead without any electrochemical measurement, we may cite the work of Schmitt,[45] who gives equations for the prediction of corrosion rates from 25–165°C at a given concentration such as 70%, or similarly over a range of concentrations at a given temperature. Gonzalez investigates acid up to 90% at 100°C using both Pb and its alloys,[46] as does Sapozhnikov[47] for Sb alloys. An electron-optical study of lead corrosion and its products in 90% acid at 80°C is reported by Grauer.[48] The 2–4 μA cm⁻² current density in the passive region falls in line with the findings of other authors such as Grafen.[21] Fullea,[49] in his studies at 100°C of a wide range of alloys, suggests that at least 8 days are required to evaluate the true worth of an alloy, and we would endorse this suggestion. Dunaev[50] cites the following order of merit in 10 N acid: Pb–1% Ag, Pb–4·5% Sb, Pb–2·5% Sn, all at 150 mA cm⁻². Sandybaeva[51] suggests a threshold effect at 10 N acid strength. Above this, he believes corrosion rapidly becomes more severe as the solubility of the $PbSO_4$ in the medium increases. A similar conclusion is quoted for the Sb- or Sn-containing alloys.[54] Elsewhere, Grafen discusses the merits of Pb–Cu–Cd–Sn quaternary alloys[52] and Pb–Pd.[53] Holstein[55] investigates refined Pb in 50–80% acid, while Gonzalez makes the point that the special virtues of the 1% Ag alloy decrease as current density increases, for example, in 30–40% acid.[56] An early but excellent paper by Lander[58] studies corrosion up to 50°C in 40% acid, although mainly in the sulphate region. Mikhailova[59] considers the fracture of anodic films arising during deformation in 10 N acid at room temperature for Pb and Pb–6·6% Sb and ternary alloys with Ag or As. Gonzalez returns to earlier work in 50% and 90% acids using both lead and its alloys[60] and an ILZRO-sponsored survey by the same author considers 185 lead-based alloys.[61] The effect of temperature on corrosion in 1 N acid of pure and technical grade Pb was followed (with potential measurements) by Khitrov,[62] and Holstein also investigated high temperatures (to boiling) of 70% acid.[63] Ekler too[64] studied potentials in up to 20 N acid. Maja[66] has studied overvoltages, a.c. impedance and other parameters of Pb–Sb alloys in 1–15 N acid from 10–45°C.

Sulphuric Acid with Other Species Present

Mixtures with chromic acid and perchloric acid (these being present in large proportions) are treated on pp. 402 and 405.

Fluoride ions

Zhurin and Solov'ev[6] have studied the effect of concentration of F^- ions, from 0 to 800 mg l^{-1}, on the corrosion of Pb and Pb–1% Ag. Addition of 50–100 mg l^{-1} F^- actually causes a decrease in corrosion rate of some 20–30%. Further increase in concentration of F^- causes an approximately two-fold increase in the corrosion rate of the silver alloy. In other cases, the effect is slight and may be beneficial or detrimental. While the corrosion increases with Pb–Ag alloys, the carry-over of Pb into the Zn decreases as F^- increases over the same range. The authors also measure film thickness and porosity as well as phase composition. Somewhat surprisingly, the film is reported to be mainly α-PbO_2, even in these acid conditions. Increasing concentration of F^- ions lowers the fraction of PbO_2 in the α-form. For the Pb–Ag alloy, the corrosion minimum is coincident with formation of the thinnest (7·7 μm) and least porous (34%) film on the surface. Corrosion data were obtained by weight loss after stripping corrosion products with a solution of 160 g l^{-1} NaOH and 200 g l^{-1} sugar. The comments on F^- ions in the section on chromic acid (p. 405 *et seq.*) should be noted as being contradictory in their effect, and accelerating corrosion.

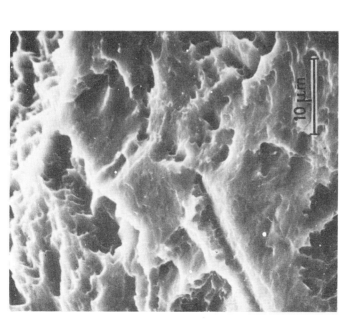

Fig. 8. Detail from Fig. 7, but with 60° camera angle in place of the 30° angle otherwise used throughout this series. Note the resulting prominence of the PbO$_2$ particles. ×20 000.

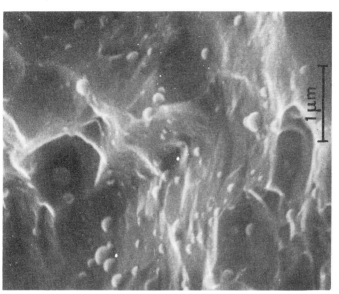

Fig. 7. S.E.M. of chemically roughened Ti treated as above (Fig. 6) then 1 h in boiling 15% oxalic acid. ×2000.

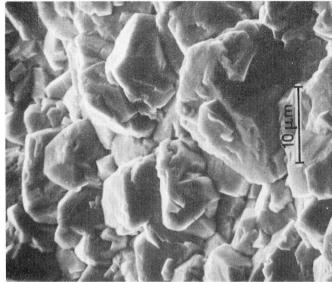

Fig. 10. S.E.M. of Ti anode after 10 min PbO_2 deposition. The final form of the surface corresponds to tetragonal rutile structured deposits. $\times 2000$.

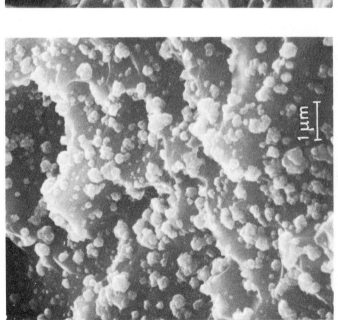

Fig. 9. S.E.M. photo of Ti pretreated with Ti(IV) as described on p. 290 and taken 2 s after commencement of PbO_2 deposition shortly before closing together of the PbO_2 film. $\times 10\ 000$.

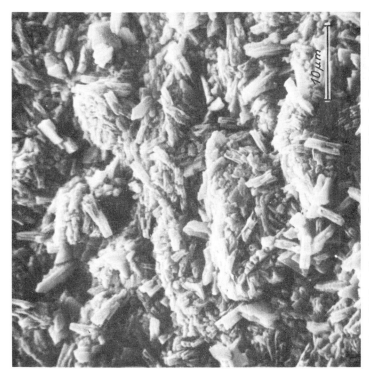

Fig. 11. S.E.M. as Fig. 10 but following "stabilization" of the PbO₂ by means of cyclic voltammetric treatment (1 M H₂SO₄, 30 cycles at 50 mV s⁻¹ with limits set between +0·750 and +2·300 V vs R.H.E.). ×2000.

Fig. 12. S.E.M. of a PbO₂/Ti composite after *c.* 4000 h benzene oxidation (20% H₂SO₄, 40°C, 4 A dm⁻²). ×2000.

that corrosion is broadly related in Tafel fashion to the potential of the anode, and that effects of Cl^- concentration are important mainly in as much as whether or not they "depolarize" the anode. Work has continued on this aspect of the problem and there are indications that the rate of corrosion (as shown by Lartey to be more or less constant over a 2-week period) subsequently declines to reach a lower steady-state value. Though these are only preliminary findings and subject to confirmation, it is worth noting that similar effects have been found in long-term sea-water corrosion tests of platinized titanium or similar precious metal-coated Ti anodes. An explanation was offered there in terms of a high initial loss rate due to poorly adherent metal grains on the outer surface of the anode, which exposed a more strongly adherent layer beneath.

The corrosion products of PbO_2 anodes in laboratory tests appear to be mainly a powdery suspension of PbO_2 which may subsequently settle out at the bottom of the reaction vessel. This poses a dilemma in that it is difficult to explain what is in effect a corrosion not involving any net chemical change in electrochemical (or indeed, chemical) terms. One alternative explanation can be attributed to purely mechanical attrition of the PbO_2 due, for example, to the scouring effect of bubbles of gas formed at the electrode. If this is so, it should be possible to vary the mechanical properties of the deposited dioxide so as to achieve a mechanically more resistant structure. The means by which this might be done are obvious from the foregoing pages. There is, however, a less orthodox explanation which is as follows. It is not widely appreciated that the evolution of a gas at an electrode surface causes substantial potential fluctuations at the electrode–electrolyte interface. With a large electrode, such fluctuations will *appear* to be averaged out, but there is no doubt that they do occur, and can be of the order of 0·5 V. The effect of these fluctuations is to reduce the overvoltage of the system momentarily. It can therefore be postulated that the corrosion protection afforded by the anodically formed PbO_2 is periodically lost when a portion of the electrode is covered by a bubble. In such a case, the potential might fall back to the "sulphate" region. More rapid dissolution would then take place. The dissolved sulphate would, however, be reoxidized by contact with the anode, to PbO_2, although under these conditions much of it would not deposit onto the anode in a mechanically satisfactory fashion. The result would then correspond with what is seen—a build-up of suspended PbO_2 in the solution. Work is presently under way on several fronts, aimed at the elucidation of this question.

References

1. J. Cotton and I. Dugdale, 3rd International Symposium on Batteries, Bournemouth, 1962, pp. 170–183. Pergamon (1963). Also British Patent 869,618.
2. P. Faber, *in* "Power Sources 4" (D. H. Collins, ed.), p. 525 and discussion. Oriel Press, Newcastle on Tyne (1973).
3. F. Beck, *Ber. Bunsenges.* **79**, 233 (1975).
4. B. N. Kabanov, *Electrochim. Acta* **9**, 1197 (1964).
5. A. T. Kuhn (ed.), "Industrial Electrochemical Processes". Elsevier, Amsterdam (1971).
6. R. Thangappan and S. Nachippan, Indian Patent 105731 (1967).
7. R. Huss and D. Wabner, *Metalloberfläche* **8**, 305 (1974).
8. K. C. Narasimham and H. V. Udupa, *J. Electrochem. Soc. Jap.* **29**, 137 (1961).
9. O. De Nora, British Patent 1,192,344.
10. F. Barrett, private communication.
11. D. Wabner and H. P. Fritz, *Z. Naturforsch.* **31B**, 39 and 45 (1976).
12. N. Hampson and C. Bushrod, *Brit. Corrosion J.* **6**, 129 (1971); *Trans. Inst. Metal Finish* **48**, 131 (1970).
13. K. C. Narasimham and H. V. K. Udupa, *Electrochim. Acta* **15**, 1619 (1970); **16**, 1301 (1971).
14. J. F. Smith, *Trans. Inst. Metal Finish* **53**, 83 (1975). Also British Patent 1,340,914.
15. U.S. Bureau of Mines R.I. 8111 "Lead Dioxide-Plated Ti Anodes . . ." L. W. Higley, W. M. Dressel (1976).
16. X. Y. Ghosh, *Electrochim. Acta* **14**, 161 (1969).
17. V. A. Volgina and E. A. Nechaev, *Sov. Electrochem.* **9**, 984 (1973); **9**, 1717 (1973).
18. K. C. Narasimham and H. K. Udupa, "Current Sciences" **30**, 139 (1961).
19. J. C. Grigger, H. C. Miller and F. D. Loomis, *J. Electrochem. Soc.* **105**, 100 (1958). *See also* U.S. Patents 2,872,405 and 2,945,790.
20. M. Fleischmann and M. Liler, *Trans. Far. Soc.* **54**, 1370 (1958).
21. W. Mindt, *J. Electrochem. Soc.* **116**, 1076 (1969).
22. M. Fleischman and H. R. Thirsk, *Trans. Far. Soc.* **51**, 71 (1955).
23. M. Fleischman and H. R. Thirsk, *Electrochim. Acta* **2**, 22 (1960).
24. H. Laitinen and N. H. Watkins, *J. Electrochem. Soc.* **123**, 804 (1976).
25. P. Ruetschi, R. T. Angstadt and B. Cahan, *J. Electrochem. Soc.* **106**, 547 (1959).
26. R. Lartey, Ph.D. thesis, Salford University (1976).
27. J. P. Carr and N. Hampson, *Chem Rev.* **72**, 679 (1972).

28. I. M. Issa, M. S. Abdelal and A. A. El Miligy, *J. Appl. Electrochem.* **5**, 271 (1975).
29. K. C. Narasimham, S. Sundarajan and H. K. Udupa, *Bull. Acad. Polon., Ser. Sci.* **13**, 619 (1965).
30. A. B. Gancy, *J. Electrochem. Soc.* **116**, 1496 (1969).
31. R. L. Clarke, private communication.
32. K. C. Narasimham and H. Udupa, *J. Appl. Electrochem.* **6**, 189 (1976).
33. K. C. Narasimham and H. Udupa, *Can. J. Chem.* **53**, 3327 (1975).
34. V. Lazarev, *Izv. Vyszh. Ucheb. Zaved. Khim.* **18**, 1336 (1975).
35. Japanese Patent 69,13,370
36. German Off Patent 2,200,901
37. USSR Patent 456,054
38. Japanese Patent 72,38742
39. Japanese Patent 72,18,844
40. USSR Patent 495,714
41. D. Wabner, Habil Thesis T. U. Munich (1976).
42. V. S. Krikorov *Izv. Akad. Nauk. SSSR Neorg. Mater.* **11**, 461 (1975).
43. T. Randle, private communication.
44. F. Barrett (Morgett Electrochemicals Ltd.), private communication.
45. N. T. Thomas and K. Nobe, *J. Electrochem. Soc.* **119**, 1450 (1972).
46. I. B. Goldberg and E. P. Parry, *J. Electroanal. Chem.* 54 (1974) 427.
47. K. Hauffe, H. Gruenwald, *Z. Elektrochem.* **56**, 937 (1952).
48. K. C. Narasimham and H. V. Udupa, *J. Electrochem. Soc.* **123**, 1294 (1976).
49. S. G. Canagaratna and N. A. Hampson, *Surface Technol.* **5**, 163 (1977).
50. W. Palmaer, *Z. Elektrochem.* **29**, 415 (1923).
51. A. I. Rusin, *J. Appl. Chem. USSR* **43**, 2656 (1970).
52. D. Bergner, *Chem. Ing. Technol.* **47**, 137 (1975).
53. R. Lartey and A. T. Kuhn, *Corrosion—NACE* **32**, 73 (1977).

12 Corrosion of Lead and Alloys in Sulphuric Acid: Introduction and Basic Principles

J. L. DAWSON

Corrosion and Protection Centre, University of Manchester, Institute of Science and Technology

Introduction

The three main uses of lead as a corrosion-resistant material in contact with sulphuric acid can be identified[116] as: (*i*) the construction of equipment for the Chemical Industry; (*ii*) as anodes in electrolytic processes of electrowinning and electrorefining; and (*iii*) as the support grids in the lead storage battery. The largest single use of both primary and recycled lead is in the lead–acid battery where the economic and technological pressures have ensured continued alloy development as reflected in the electrochemical literature. These influences have been mirrored by similar investigations with the introduction of new lead alloy anodes for use in electrolytic cells.[8] The present survey will largely review the observations of research workers concerned with alloy use in the lead–acid battery and publication of fundamental studies of electrochemical corrosion and alloy development.

The corrosion resistance of an alloy is dependent on its metallurgy, the physical and mechanical properties, as shown in its structure and chemistry, as well as the electrochemical factors which need to be considered when it reacts with the corrosive environment; the influence and concentration of impurities in the service environment are also important but these have often not been fully investigated. Lead relies for its protection on the formation of a passive film of lead sulphate or lead dioxide, therefore the morphology of the film is crucial in the determination of the operating life of a component. The morphological changes and the mode of corrosive attack involve a complex interaction of a number of variables: operating potential, alloy composition and grain size, casting defects and segregation, adsorption of electroactive species, and mechanical erosion effects. It is accepted in the present review that laboratory tests involving only a galvanostatic weight loss determination are often too insensitive and may even be misleading, particularly if one wishes to gain a fundamental understanding of the corrosion processes to predict long-term effects. An interesting corollary is that although the lead–acid battery has been widely used for many years with publication and verification[20,21,88,89,204,251,252] of the following thermodynamic data:

$$Pb + SO_4^{2-} \underset{charge}{\overset{discharge}{\rightleftharpoons}} PbSO_4 + 2e$$

$$E^0 = -0.356 + 0.0295 \log a_{SO_4^{2-}}$$

$$PbO_2 + SO_4^{2-} + 4\,H^+ + 2e \underset{charge}{\overset{discharge}{\rightleftharpoons}} PbSO_4 + 2\,H_2O$$

$$E_\beta^0 = 1.6871 - 0.1182\,pH + 0.0295 \log a_{SO_4^{2-}}$$

$$E_\alpha^0 = 1.6971 - 0.1182\,pH + 0.0295 \log a_{SO_4^{2-}}$$

it is only recently[170,191] that the electrochemical mechanisms associated with these major phase changes have been fully elucidated. This may seem surprising since lead sulphate and both polymorphic forms[54] of lead dioxide are found in the active material[25] of the lead–acid cell and as corrosion products on lead;[36,77,206] the β-form is the low pressure form with a tetragonal rutile type structure[143,241,270] produced from lead sulphate during battery cycling, whilst the α-form, not found in nature, has the orthorhombic structure of columbite.[270] The apparent lag in our knowledge has had to await the result of developments in instrumentation and electrocrystallization theory.[81,236] Potentiostatic investigations and corrosion tests have become more widespread and allow statistically designed experiments to be used in alloy assessment[149] with development of predictions from life test data.[164,259]

There is a substantial catalogue of diagnostic information on the type of corrosion and corrosion products produced by lead and its alloys.[74,85,112,153,181,183,229] However, the selection of a particular alloy for an application is based not only on the anticipated corrosion resistance but also on past experience and economic factors, including the ease of fabrication. The general trend has been towards decreased alloying additions, perhaps a reflection on the acceptable corrosion properties of pure lead, and the standard in many automotive batteries is now based on a 6% antimony alloy. The requirements for a negligible open-circuit capacity loss, a low or maintenance-free battery with a higher energy density and the use of thinner plates, has been an inducement to the industry to develop even lower antimonial alloys and alternative fabrication techniques. These are based on rolled sheet and expanded metal technology since traditional[233] casting methods have probably reached the minimum grid thickness. The trend in the early 1960s was to investigate lead sheet containing dispersed lead oxide as a strengthening acid; however, the rolling and subsequent expanding produced lines of corrosive attack where the hard lead oxide had been rolled into the soft lead. These dispersion-hardened leads have not found use in the industry.[18,198] The late 1960s and early 1970s have seen the increasing use of low concentration antimonial alloys with selenium as a grain-refining element,[22,291-293,297,298] and there is a continuing interest in alloys containing alkaline earth elements.[47,51,52,90,153,259,266] The use of antimonial lead as the grid alloy in the lead–acid battery is based both on technological considerations and the substantial quantity of recycled lead; this has in part been a factor in the relatively slow introduction of non-antimonial alloys, although problems of charge acceptance, shedding of positive material and strength of the grids produced by the newer fabrication techniques should not be underestimated.

There is a predominance of information in the electrochemical literature on the initial film formation processes without later consideration of the substantial corrosion film growth. Even reported testing of alloys is often limited to hours or a few days, whereas in practice the life of lead components is measured in years; thus there is an excessive reliance on test house observations of simulated service conditions. Accelerated tests involving high potentials, but probably more typically high current densities, or extensive potentiodynamic sweeps from extreme anodic to cathodic regions, may not reflect operating conditions. There is therefore a need for simple laboratory schedules, extending over a few weeks or months, which anticipate longer term effects at constant potential, cycling and open-circuit stand; the tests should also discriminate between changes induced in alloy structure as the result of additions, fabrication, heat treatment etc. This is in apparent contrast to other industries which use or develop corrosion-resistant alloys based on iron, copper, nickel, cobalt and chromium. The reader cannot therefore readily turn to tables of lead alloy corrosion data and obtain a predicted corrosion or penetration rate in mm year^{-1}. The reason is partially to be found in the nature of lead as an element with its low creep resistance since this requires the alloying additions to provide strength and these may adversely affect the corrosion resistance. Also during service life, the alloy in the electrode, or battery plate, will be required to operate at a number of different potentials depending on the length of time the system remains on electrolysis or open-circuit stand; the battery may also operate under float or overcharge conditions as well as discharge. The subject is therefore complex involving electrochemistry, metallurgy and morphological changes. The aim of the following sections is to survey, in the context of electrochemical corrosion science and technology, those factors which are important in our present understanding of the corrosion resistance of lead and its commonly used alloys.

The Thermodynamics of Lead Corrosion and the Composition of the Corrosion Product Film

Thermodynamic predictions

The corrosion behaviour of metals in a particular environment may be predicted from the energy changes associated with the probable reactions. The combined results of electrochemical potential and pH for individual elements have been summarized in the well-known Pourbaix Diagrams.[189] Figure 1 shows the predicted behaviour of lead from the simple aqueous diagram,[68] which in the presence of sulphate ions is modified as in Fig. 2.

Further modifications (Fig. 3) by Ruetschi and Angstadt[203] and Barnes et al.,[19] using the data of Bode and Voss,[25] include the basic sulphates $PbO.PbSO_4$, $3PbO.PbSO_4.H_2O$ and $5PbO.2H_2O$. These thermodynamic diagrams are based on equilibrium conditions and predict spontaneous reactions; the corrosion product composition and environmental changes may prevent or decrease corrosion. An abbreviated list of possible potential-dependent reactions is given in Table I and Fig. 3, based on a combined sulphate and bisulphate ion activity of one; for ease of presentation only those reactions involving compounds which have been detected experimentally are included.

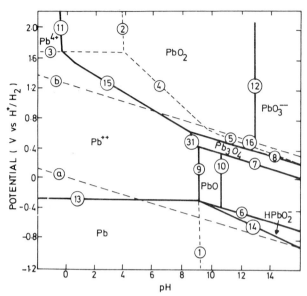

Fig. 1. Potential–pH diagram of the Pb/H$_2$O system at 25°C.[68,189] Lines *a* and *b* indicate the region of thermodynamic stability of water. Thin lines represent an equilibrium between a solid phase and an ion, $a = 10^{-6}$. Thick lines represent equilibrium between two solid phases. Dotted lines represent equilibrium between ions. Circled numbers refer to equations in Table I.

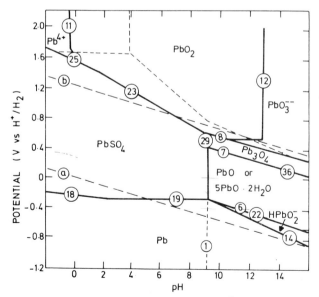

Fig. 2. Simplified potential–pH diagram of Pb/H$_2$O/SO$_4^{2-}$ system at 25°C.[19,68,189,204]

Table I. Thermodynamic data for lead and its corrosion products in aqueous sulphuric acid.

The values below are based on the hydrogen electrode at S.T.P. and pH = 0.

To convert the potentials with respect to a Hg/Hg_2SO_4 electrode or other reference electrode, use the data given by Latimer.[142] Typically, in a battery acid of 4·5 M sulphuric acid, the Hg/Hg_2SO_4 will be c. 575 mV *vs* H_2 *in the same solution* or 600 mV *vs* H_2 at pH = 0.

For the water reactions and equilibrium equations:

(a) $2 H^+ + 2e = H_2$ $\qquad E = -0.0591 \, pH - 0.0295 \log p_{H_2}$

(b) $O_2 + 4 H^+ + 4e = 2 H_2O$ $\qquad E = 1.228 - 0.0591 \, pH + 0.0147 \log p_{O_2}$

Homogeneous Reactions
 Without oxidation

$$HPbO_2^- + 3 H^+ = Pb^{2+} + 2 H_2O \qquad \log \frac{a_{HPbO_2^-}}{a_{Pb^{2+}}} = -28.17 + 3.00 \, pH. \qquad (1)$$

$$PbO_3^{2-} + 6 H^+ = Pb^{4+} + 3 H_2O \qquad \log \frac{a_{PbO_3^{2-}}}{a_{Pb^{4+}}} = -23.13 + 6.00 \, pH \qquad (2)$$

 With oxidation

$$Pb^{4+} + 2e = Pb^{2+} \qquad E = 1.691 + 0.0295 \log \frac{a_{Pb^{4+}}}{a_{Pb^{2+}}} \qquad (3)$$

$$PbO_3^{2-} + 6 H^+ + 2e = Pb^{2+} + 3 H_2O \qquad E = 2.375 - 0.1773 \, pH + 0.0295 \log \frac{a_{PbO_3^{2-}}}{a_{Pb^{2+}}} \quad (4)$$

$$PbO_3^{2-} + 3 H^+ + 2e = HPbO_2^- + H_2O \qquad E = 1.543 - 0.0886 \, pH + 0.0295 \log \frac{a_{PbO_3^{2-}}}{a_{HPbO_2^-}} \qquad (5)$$

$$PbO_2 + 2 H^+ + 2e = Pb + 2 H_2O \qquad E = 0.248 - 0.0591 \, pH \qquad (6)$$

$$Pb_3O_4 + 2 H^+ + 2e = 3 PbO + H_2O \qquad E = 1.076 - 0.0591 \, pH \qquad (7)$$

$$3 PbO_2 + 4 H^+ + 4e = Pb_3O_4 + 2 H_2O \qquad E = 1.122 - 0.0591 \, pH \qquad (8)$$

Heterogeneous Reactions
 Without oxidation

$$PbO_2 + 2 H^+ = Pb + H_2O \qquad \log a_{Pb^{2+}} = 12.67 - 2.00 \, pH \qquad (9)$$

$$HPbO_2^- + H^+ = PbO + H_2O \qquad \log a_{HPbO_2^-} = -15.49 + 1.00 \, pH \qquad (10)$$

$$PbO_2 + 4 H^+ = Pb^{4+} + 2 H_2O \qquad \log a_{Pb^{4+}} = -7.10 - 4.00 \, pH \qquad (11)$$

$$PbO_3^{2-} + 2 H^+ = PbO_2 + H_2O \qquad \log a_{PbO_3^{2-}} = -30.24 + 2.00 \, pH \qquad (12)$$

Table I. (continued)
With oxidation

$$Pb^{2+} + 2e = Pb \qquad E = -0 \cdot 126 + 0 \cdot 0295 \log a_{Pb^{2+}} \qquad (13)$$

$$HPbO_2^- + 3 H^+ + 2e = Pb + 2 H_2O \qquad E = 0 \cdot 706 - 0 \cdot 0886 \, pH + 0 \cdot 0295 \log a_{HPbO_2^-} \qquad (14)$$

$$PbO_2 + 4 H^+ + 2e = Pb^{2+} + 2 H_2O \qquad E = 1 \cdot 482 - 0 \cdot 1182 \, pH - 0 \cdot 0295 \log a_{Pb^{4+}} \qquad (15)$$

$$PbO_2 + H^+ + 2e = HPbO_2^- \qquad E = 0 \cdot 649 - 0 \cdot 0295 \, pH - 0 \cdot 0295 \log a_{HPbO_2^-} \qquad (16)$$

Limits of domains of predominance of soluble lead ions

$Pb^{2+}/HPbO_2^-$	$pH = 9 \cdot 34$	(1)
Pb^{4+}/PbO_3^{2-}	$pH = 3 \cdot 84$	(2)
Pb^{2+}/Pb^{4+}	$E = 1 \cdot 694$	(3)
Pb^{2+}/PbO_3^{2-}	$E = 2 \cdot 375 - 0 \cdot 1773 \, pH$	(4)
$HPbO_2^-/PbO_3^{2-}$	$E = 1 \cdot 547 - 0 \cdot 0886 \, pH$	(5)

When the sulphate ion is present, additional equilibrium conditions have to be taken into account. The data below are based on a sulphate activity of 1.
For the sulphate and bisulphate ions without oxidation

$$SO_4^{2-} + H^+ = HSO_4^- \qquad \log \frac{a_{HSO_4^-}}{a_{SO_4^{2-}}} = -1 \cdot 92 + pH \qquad (17)$$

Heterogeneous Reactions

$$PbSO_4 + H^+ + 2e = Pb + HSO_4^- \qquad E = -0 \cdot 302 - 0 \cdot 0295 \, pH - 0 \cdot 0295 \log a_{HSO_4^-} \qquad (18)$$

$$PbSO_4 + 2e = Pb + SO_4^{2-} \qquad E = -0 \cdot 356 - 0 \cdot 0295 \log a_{SO_4^{2-}} \qquad (19)$$

$$PbO.PbSO_4 + 4e + 2 H^+ = 2 Pb + SO_4^{2-} + H_2O$$
$$E = -0 \cdot 113 - 0 \cdot 0295 \, pH - 0 \cdot 0148 \log a_{SO_4^{2-}} \qquad (20)$$

$$3 \, PbO.PbSO_4.H_2O + 8e + 6 H^+ = 4 Pb + SO_4^{2-} + 4 H_2O$$
$$E = 0 \cdot 030 - 0 \cdot 044 \, pH - 0 \cdot 0074 \log a_{SO_4^{2-}} \qquad (21)$$

$$PbO + 2e + 2 H^+ = Pb + H_2O \qquad E = 0 \cdot 248 - 0 \cdot 0591 \, pH \qquad (22)$$

$$\beta\text{-}PbO_2 + SO_4^{2-} + 2e + 4 H^+ = PbSO_4 + 2 H_2O$$
$$E = 1 \cdot 718 - 0 \cdot 1182 \, pH + 0 \cdot 0295 \log a_{SO_4^{2-}} \qquad (23)$$

$$\alpha\text{-}PbO_2 + SO_4^{2-} + 2e + 4 H^+ = PbSO_4 + 2 H_2O$$
$$E = 1 \cdot 700 - 0 \cdot 1182 \, pH + 0 \cdot 0295 \log a_{SO_4^{2-}} \qquad (24)$$

$$\beta\text{-}PbO_2 + HSO_4^- + 3 H^+ + 2e = PbSO_4 + 2 H_2O$$
$$E = 1 \cdot 659 - 0 \cdot 0886 \, pH + 0 \cdot 0295 \log a_{HSO_4^-} \qquad (25)$$

$$\alpha\text{-}PbO_2 + HSO_4^- + 3 H^+ + 2e = PbSO_4 + 2 H_2O$$
$$E = 1 \cdot 669 - 0 \cdot 0886 \, pH + 0 \cdot 0295 \log a_{HSO_4^-} \qquad (26)$$

(*continued*)

Table I. (continued)

$2 PbO_2 + SO_4^{2-} + 4e + 6 H^+ = PbO.PbSO_4 + 3 H_2O$

$$E = 1\cdot468 - 0\cdot0886 \text{ pH} + 0\cdot0148 \log a_{SO_4^{2-}} \quad (27)$$

$4 PbO_2 + SO_4^{2-} + 8e + 10 H^+ = 3 PbO.PbSO_4.H_2O + 4 H_2O$

$$E + 1\cdot325 - 0\cdot0739 \text{ pH} + 0\cdot0074 \log a_{SO_4^{2-}} \quad (28)$$

$PbO_2 + 2e + 2 H^+ = PbO + H_2O \qquad E = 1\cdot107 - 0\cdot0591 \text{ pH} \quad (29)$

$3 PbO_2 + 4e + 4 H^+ = Pb_3O_4 + 2 H_2O \qquad E = 1\cdot122 - 0\cdot0591 \text{ pH} \quad (30)$

$Pb_3O_4 + 2e + 2 H^+ = 3 PbO + H_2O \qquad E = 1\cdot076 - 0\cdot0591 \text{ pH} \quad (31)$

$4 Pb_3O_4 + 3 SO_4^{2-} + 8e + 14 H^+ = 3(3 PbO.PbSO_4.H_2O) + 4 H_2O$

$$E = 1\cdot730 - 0\cdot1034 \text{ pH} + 0\cdot0074 \log a_{SO_4^{2-}} \quad (32)$$

$5 PbO.2 H_2O + 10 H^+ + 10e = 5 Pb + 7 H_2O \qquad E = 0\cdot260 - 0\cdot0591 \text{ pH} \quad (33)$

$5 PbO_2 + 10 H^+ + 10e = 5 PbO.2 H_2O + 3 H_2O \qquad E = 1\cdot070 - 0\cdot0591 \text{ pH} \quad (34)$

$2 Pb_3O_4 + 3 SO_4^{2-} + 10 H^+ + 4e = 3(PbO.PbSO_4) + 5 H_2O$

$$E = 2\cdot010 - 0\cdot148 \text{ pH} + 0\cdot0443 \log a_{SO_4^{2-}} \quad (35)$$

$5 Pb_3O_4 + 10 H^+ + H_2O + 10e = 3(5 PbO.2 H_2O) \qquad E = 0\cdot960 - 0\cdot0591 \text{ pH} \quad (36)$

$Pb_2O_3 + SO_4^{2-} + 4 H^+ + 2e = PbO.PbSO_4 + 2 H_2O$

$$E = 1\cdot750 - 0\cdot1182 \text{ pH} + 0\cdot0295 \log a_{SO_4^{2-}} \quad (37)$$

Limits of domains of stability of two phases without oxidation

$4 PbO + SO_4^{2-} + 2 H^+ = 3 PbO.PbSO_4.H_2O \qquad \text{pH} = 14\cdot6 + \tfrac{1}{2} \log a_{SO_4^{2-}} \quad (38)$

$3 PbO.PbSO_4.H_2O + SO_4^{2-} + 2 H^+ = 2(PbO.PbSO_4) + 2 H_2O$

$$\text{pH} = 9\cdot6 + \tfrac{1}{2} \log a_{SO_4^{2-}} \quad (39)$$

$PbO.PbSO_4 + SO_4^{2-} + 2 H^+ = 2 PbSO_4 + H_2O \qquad \text{pH} = 8\cdot4 + \tfrac{1}{2} \log a_{SO_4^{2-}} \quad (40)$

A more realistic model of the corrosion product layer formed on lead has been proposed by both Ruetschi[201,204] and Burbank[37,41] as shown in Figs 4, 5 and 6. The corrosion models are consistent with the previous thermodynamic diagrams (Figs 2 and 3), but with the compounds largely confirmed by experimental data obtained from X-ray and electron diffraction studies on specimens corroded at different potentials and pH. As well as indicating the composition of the corrosion product film, the data also account for the redox reactions occurring during film formation and discharge of lead dioxide. Electrochemically these may be observed as either current maxima during a potentiodynamic sweep (i.e. cyclic voltammetry[176,216,217]) or as potential plateaux[184] during constant current experiments. Care must be taken not to ascribe a potential plateau observed during discharge of lead dioxide to a corrosion product produced during initial anodization; thus some basic sulphates, which are a discharge reaction product, could be present in an excess when compared to the initial amount produced by corrosion.

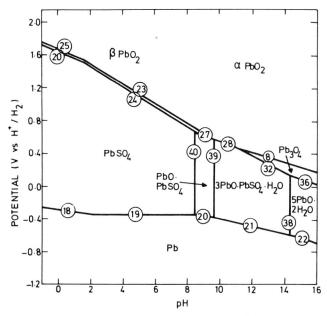

Fig. 3. Solid phases of the $Pb/H_2O/SO_4^{2-}$ system at 25°C.[204] The diagram shows the basic sulphates and assumes $a_{HSO_3^-} + a_{SO_4^-} = 1$.

Stoichiometry of lead oxides and sulphates

The structure and composition of the oxides of lead have received more attention in relation to their production and use in the formation of battery-active material than as corrosion products.[35,38,42,48,55,60,197] The yellow or orthorhombic litharge and the red or tetragonal massicot product layer are found as a distorted lead oxide with an orthorhombic pseudo-tetragonal structure, the stoichiometric formula suggested being Pb_7O_{11}. Collectively the oxides are known as $PbO_t.xPbO_{1.33-1.51}$ for Pb_5O_8, and $PbO_{1.47-1.57}$ for Pb_2O_3. These oxides are formed under the initial sulphate layer, possibly by the reaction of diffusing oxygen species[184] or water,[41,136,201] the attack being either by direct inclusion in the lead lattice[184] or by releasing Pb^{2+}. The tetragonal lead oxide reacts with oxygen species to produce the various intermediate non-stoi-

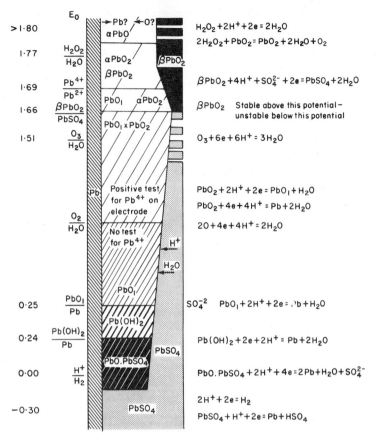

Fig. 4. The first schematic representation by Burbank of lead corrosion films.[37,41] Relative positions of the passivation films and the expected formation potentials are shown.

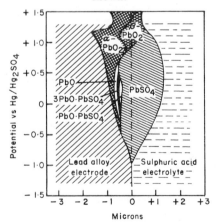

Fig. 5. Representation of the multi-phase corrosion layer by Ruetschi.[204]

chiometric compounds, indicated above, by inclusion of further oxygen in the oxide lattice. The stoichiometry therefore changes with potential and oxygen diffusion[121] from a tetragonal to an orthorhombic structure, the reaction with an oxidation product of water producing $PbO_t.xPbO_2$.

The presence of lead oxides and basic lead sulphates, which are thermodynamically unstable at a low pH is due to a restricted ionic diffusion gradient through the corrosion product layer.[41,204] A more recent analysis[201] suggests that the lead sulphate film may also behave as a perm-selective membrane (see p. 46). $5PbO.2H_2O$,[185,208] PbO (orthorhombic),[48,185] $PbO.PbSO_4$,[35,185] and $3PbO.PbSO_4.H_2O$[48,185] have been detected under the sulphate film. Increased potential results in further reactions involving oxygen species and water to produce α-PbO_2 from $PbO_t.xPbO_2$, basic lead sulphates and the underlying lead.[37,39,118,187,201] β-PbO_2 is produced from the basic lead sulphates and lead sulphate at lower pH.[37,41,54,118,190,183,201,204]

β-PbO_2 covers a wide range of compositions with a possible[122] lower phase limit of $PbO_{1.88}$; the product irrespective of the method of

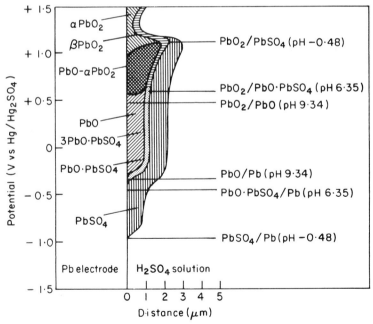

Fig. 6. Films formed during 24 h at constant potential in 4·2 M H_2SO_4 with indication of pH gradient.

preparation is deficient in active oxygen, an n-type semi-conductor, and the composition range lies within the limits:

$$PbO_{1.80-1.99}(OH)_{0.07-0.27}$$

Resistivity and Hall effect measurements[239] indicate that the current carriers are electrons with 1.8×10^{-2} conduction electrons per molecule of β-PbO$_2$, suggesting that interstitial Pb^{4+} ions provide the free electrons. The outer layer of β-PbO$_2$ is attached to the α-PbO$_2$, the latter being produced by solid phase oxidation of lead, the result of oxygen diffusion into the metal. α-PbO$_2$ composition varies[270] as PbO$_{1.93-2.02}$ and is slightly more conductive than the β-form,[6,86] the reported[7] resistivities being 10^{-3} and 4×10^{-3} Ω cm, respectively. The stoichiometry of the oxides and dioxides influence the kinetics of the corrosion processes and the morphology of the corrosion product; some of these aspects will be considered in the next section.

Summary

The composition and potential dependence of the corrosion product layer has been extensively studied by a number of workers using various analytical techniques, and the data have been presented in terms of the thermodynamic equilibrium diagrams and corrosion models (Figs 1–6). However, the major limitation of this approach is that the figures do not contain kinetic information and therefore cannot be used to predict corrosion rates. Nevertheless, a knowledge of the equilibrium conditions required for the formation of the compounds found in the corrosion product layer is a necessary background for the understanding of the corrosion kinetics.

Electrochemical Corrosion Kinetics

The majority of work reported on the anodic corrosion of lead has been obtained using simple constant current apparatus, the galvanostatic technique,[15,72,73,74,102,131,135,151,169,172,206,207] which involves measuring the potential during the time of the experiment. Provided the current remains steady, then the length of the time–potential plateau enables the amount of corrosion product to be computed by Faraday's Laws.

Potentiostatic techniques provide more detailed information on the rate-controlling mechanisms and the current–time transients obtained are now easily interpreted on an electrocrystallization kinetic basis.[81,236] It is important to realize that electrode reactions are heterogeneous processes with potential-dependent rate constants; the potential characterizes the interfacial conditions and the net reaction is proportional to the current density.

The use of constant potential measurements has not been widespread in corrosion studies on lead,[35,36,37,136,139,204] although the technique offers many advantages in both theoretical investigations and practical alloy development. This is surprising in view of the rapid developments made possible in the field of stainless steels and nickel-based alloys by the use of relatively cheap potentiostats. These advantages are well-illustrated in the much quoted data of Lander[136] obtained during his studies on lead alloy and reported in the 1950s for different acid concentrations. Initial current–time transients were observed, but not analysed; even so, the corrosion current and weight loss results were compared. The influence of potential and corrosion product film composition can be seen in the data for 30% sulphuric acid with two maxima, the first at $980 \, mV$ and a second smaller peak at $1075 \, mV$ vs Hg/Hg_2SO_4. However, the corrosion conversion to a constant basis of $1 \cdot 18 \, mg \, cm^{-2}$ tended to distort the data; a preferred presentation of the quasi-steady state corrosion current, i.e. after 24 h, and obtained with improved potentiostatic control, was given by Ruetschi and Angstadt[204] and only a single peak at $1100 \, mV$ was observed. Recent and more detailed investigations on lead, lead–antimony[191] and lead–calcium alloys[149] have helped to resolve these apparent differences. The potentiostatic experiments were carried out at 50-mV intervals, and a minimum was observed at $900 \, mV$ before the peak (Fig. 7). The polarization curve also displayed a break in the linear region at about $-200 \, mV$.

Detailed analysis[191,149] of the four current–time transients which precede the steady-state conditions allow the electrocrystallization processes associated with film formation to be stated with greater confidence. At all potentials an outer layer of lead sulphate is first formed; further oxidation of the underlying metal produces tetragonal lead oxide, basic lead sulphates and α-lead dioxide, and at higher potentials the outer sulphate layers are converted to β-lead dioxide. The kinetic data have been complemented by electron microscopy examination of the corrosion product layer and the conclusions, with regard to crystal size, film thickness and morphology, were similar to previous workers;[73,150,185,203,206,211] a summary of the film formation and growth process is presented below.

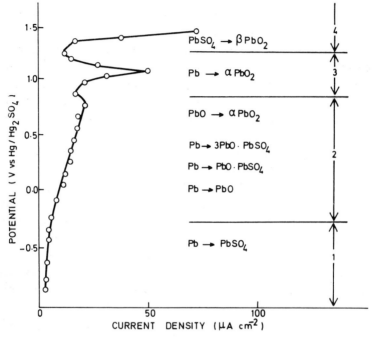

Fig. 7. Steady-state corrosion of lead after 4 h in 4·5 M H_2SO_4.[191]

Anodic dissolution of lead should be considered in four distinct potential regions (Fig. 7). The first, typically from -700 mV to -200 mV *vs* Hg/Hg_2SO_4, involves the formation of lead sulphate by the classic dissolution–precipitation mechanism,[11,12,13,14,191] j *vs* $t^{-1/2}$; there is also evidence of a solid state nuclear process at the higher potentials.[12] From -200 mV to $+695$ mV, the second region, the rate of corrosion increases with potential; the film thickens by formation of tetragonal lead oxide and basic lead sulphates both underneath and within the pores between the initial crystalline lead sulphate film;[73,183,191] this has already been discussed in the previous section in terms of the Pourbaix diagrams and models (Figs 1–6). The crystal size, typically 0·1–10 μm, of the lead sulphate decreases with increase in potential and increase in antimony.[15,43,45,61,73,97,180,181,183,191,257]

The film thickness and its morphology also determines the pH of the interior and since analysis shows the presence of lead oxide a pH of 9·34 is anticipated. A diffusion potential,[201] based on this interior pH with an external bulk acid concentration of pH = 0, has been calculated at 535 mV, $\Delta \psi_d = RT/F \ln(a_{H^+})_1/(a_{H^+})_2$. On this basis Ruetschi has introduced the concept of the lead sulphate behaving as a membrane permeable to H^+ or OH^- but essentially non-permeable to SO_4^{2-}, HSO_4^- and Pb^{2+}. This is not unreasonable since although lead sulphate is considered as being inert, it does possess some conductivity,[81,207] in the order of 10^{-9} to 3×10^{-10} Ω^{-1} cm^{-1}, which may in part account for its dissolution during the oxidation to lead dioxide.* The second region in Fig. 7 with its restricted acid diffusion also involves the limited formation of α-PbO_2 presumably from the oxide $PbO.xPbO_2$ or basic lead sulphate. The upper limit of the second region, observed by Lander[136] as his first maximum, is followed by a slight decrease in the corrosion rate between 850 mV and 965 mV. This may indicate a maximum in the amount of α-PbO_2 present in the low-conductivity oxide–sulphate layer. Traditionally, these first two potential regions have been considered as passive, due to the protective lead sulphate, as indicated by the almost vertical potential–current curve; however, the changing composition in the lead oxide region is more typical of a pre-passive film formation. These observations are not as apparent if cyclic voltammetry is used as the sole electrochemical technique.[176,216,217]

True passivity, involving the formation of an electronically conducting film of α-PbO_2 produced directly from lead, occurs in the third and fourth potential regions above 965 mV. This reaction has been documented by other workers and is normally seen as a peak between 965 mV and 1165 mV, depending on acid concentration; this corresponds to the second and smaller maximum observed by Lander. Above 1165 mV, β-PbO_2 is produced by oxidation of lead sulphate and basic lead sulphates; typically[149,191,248,249] a transient of j vs $t^{3/2}$ is observed for the conversion of lead sulphate to lead dioxide, indicative[236] of an electrocrystallization mechanism of solution diffusion control with a three-dimensional nucleation and growth process. This final solid phase change is followed by oxygen evolution in the transpassive region and the corrosion increases rapidly with potential (Fig. 8).

* See p. 46.

Lander's early results[136–140] clearly demonstrate the importance of
weight loss determinations in alloy evaluation, especially in the oxygen
evolution region where it is impossible to compute the corrosion rate
from Faraday's Laws. It is also essential to examine the specimens for
intergranular penetration which may be more catastrophic than the
simple weight loss would indicate. Corrosion rate *vs* time relationships
also provide information on the corrosion mechanism provided the test
is of adequate duration; in this context the reader should again note that
the majority of the electrochemical and weight loss data available on the
corrosion of lead have been obtained on initial or, at best, only after a few
hours exposure. Extension of a constant voltage test by means of simple
potentiostats with virtually no drift is now a practical option enabling a
realistic assessment[149] over at least a few weeks or preferably a few
months exposure. Extrapolation of the data to predict anticipated
service life can then be made within known confidence limits.

Logarithmic equations of weight loss *vs* time usually represent the
initial stages of reaction or a low activation process, whereas a parabolic
rate indicates that the corrosion is either controlled by diffusion
through the film or that the growth is limited by the conductivity of the

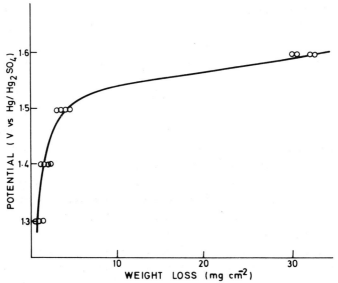

Fig. 8. Effect of potential on the corrosion rate of a Pb, 0·05% Sr, 0·5% Sn alloy after 4
weeks in 5 M H_2SO_4.[149]

electrolyte in the pores and electronic conductivity of the film. A linear relationship would be expected only if the reaction rate at the metal surface was controlling and diffusion played little part in the process, i.e. a constant rate of growth. A number of researchers have noted exponential type rates[259] and this is consistent with the observations of Russian workers[16,121] who clearly demonstrated that anodic corrosion of lead is by the diffusion of atomic oxygen through the layer of lead dioxide and not by an oxidation reaction at the bottom of possible pores. Their experiments used lead foil sandwiched between two electrochemical cells, and it was noted that polarization of one side of the lead in the first cell resulted in potential changes of the lead surface exposed in the second cell. The absence of pores through the lead was confirmed by three independent electrochemical tests and it was also suggested that the rate of combination of adsorbed atomic oxygen into molecules was a slow step in the oxygen discharge reaction. This may be significant since the oxygen overpotential at the lead dioxide–electrolyte interface could presumably affect the corrosion rate of the underlying metal.

This section on the kinetics of the corrosion of lead illustrates a number of factors which are important in assessing corrosion rates. Lead when used as a cladding material will corrode in a manner predicted by corrosion theory, as expressed in an Evans type of polarization diagram, and since the driving force for the reaction will be provided by the naturally occurring cathodic reactions of hydrogen evolution and oxygen reduction, then the film formation and growth will take place in the first two potential regions discussed above and shown in Fig. 7. The reactions are the same as for the self discharge of a negative battery plate,[102] and the initial film formation rate will be controlled by the levels of impurities present on the surface which will decrease the hydrogen overpotential. In the case of antimonial alloys the potential may rise to the redox potential of the Sb/SbO^+ couple, and antimony will then dissolve in the presence of oxygen in the sulphuric acid. With non-antimonial alloys, the potential plateau of the Sb/SbO^+ couple is not observed and a corrosion potential in the second potential region is adopted in aerated acid.

Under applied potential conditions, as in electrolysis or overcharge of a battery, the important factor is to retain a coherent film of α-PbO_2. In the absence of cracks or pores through this layer, the corrosion rate is apparently controlled by a solid-state diffusion process. The composition of the alloy determines the structure of the dioxide substrate and it is significant to note that antimonial alloys tend to give a linear rate of corrosion under constant voltage conditions whereas calcium alloys exhibit a logarithmic or parabolic rate. In the latter case it has not been established if the reaction rate-determining step is diffusion through the film or diffusion in the metal, but obviously metallurgical factors such as grain size, grain boundary composition and the related problem of corrosion resistance of eutectic compositions and intermetallic compounds become increasingly important in controlling the life of the component. The metallurgical factors will be considered later in relation to specific alloys.

Morphological and Chemical Influences on the Protective Properties of the Anodic Film

The composition and morphology of the corrosion product film, the ratios of lead sulphate, α- and β-lead dioxide, obviously play a crucial but as yet a rather ill-defined role in protecting the underlying metal. Although α-PbO_2 is produced by deposition from alkaline baths[205] and also by the formation at a high internal pH during the positive plate formation from tetrabasic sulphate[25,230] and residual lead,[25] its presence in the corrosion product layer appears to be promoted by high acid concentrations, increased anodic polarization and lower temperatures.[41] α-PbO_2 is the more stable phase in 4·62 M H_2SO_4 at 50°C and above, whereas the β-form is more stable under standard conditions as indicated by the differences in the reversible potentials shown in Table I. Leaching of selected phases from the alloy will give rise to a roughened surface with porosity into the metal grain boundary or even dissolution within the grain, which is clearly observed by sectioning the samples followed by microscopic examination. In extreme cases dislodgement of the entire grain is possible as shown by the embrittlement of lead–magnesium alloys.[116] Electrolyte penetration of the porous interfacial structure with a consequent pH change has been proposed as an explanation for the presence of α-PbO_2, and yet the mechanisms involved in the long-term corrosion of the substrate are more typical of a solid state process as indicated previously. The interaction of solid state diffusion of oxygen on dissolving alloyed metals has not been fully investigated.

Mechanical erosion of the outer dioxide layer can also occur during extended oxygen evolution—this involves fracture of the structure[170] and loss of particles. Exposure of underlying α-PbO_2 could result in a change of oxygen overpotential, typical values[205] for the Tafel slopes being $0 \cdot 14$ mV decade^{-1} for β-PbO_2 and $0 \cdot 07$ mV decade^{-1} for α-PbO_2; this change could be significant in determining the rate of oxygen diffusion in the dioxide and alloy matrix.

A further factor that requires clarification is the chemical role of certain elements and their influence in the modification of the lead dioxide film; of particular interest are antimony,[30,45,66] cobalt[10,129,130,131,250] and silver,[75,139,162,182,195] either as an addition to the electrolyte or as a soluble species produced during corrosion of the appropriate alloy. These elements decrease the oxygen overpotential and appear to be effective in reducing anodic corrosion. The mechanisms involved in these processes are not understood, as indicated in the review of the lead–acid cell by Burbank et al.[41] The possibility of the ions acting as a redox catalyst to provide an alternative path for oxygen evolution has been considered:[10,139,194]

$$2\,H_2O + 2\,Ag^{2+} = H_2O_2 + 2\,Ag^+ + 2\,H^+$$

or

$$2\,H_2O + 2\,Co^{3+} = H_2O_2 + 2\,Co^{2+} + 2\,H^+$$

followed by $2\,H_2O_2 = O_2 + 2\,H_2O$. An alternative route[131] for cobalt is

$$4\,CoOH^{2+} = 4\,Co^{2+} + O_2 + 2\,H_2O.$$

It is significant that antimony[66] and cobalt[10] in the battery acid both undergo an adsorption–desorption with cycling, the ions being adsorbed[66] or incorporated[197] into the lead dioxide during charge and then released on discharge. Cobalt, although apparently decreasing the corrosion, has the disadvantage of oxidizing certain battery separators. In the case of antimony, the adsorption–desorption process forms part of the solution route occurring during "antimony poisoning" of the battery where during service the antimony is corroded from the positive grid. The antimony (V) anions[67] may diffuse through or round the separators, where they are reduced to antimony(III), and are then deposited as antimony metal on the sponge lead of the negative plate (Fig. 9).

Antimony has a lower hydrogen overpotential than lead and its presence promotes hydrogen evolution with consequent water loss and self discharge on storage. Negative grid corrosion, either by the production of stibine during overcharge or by selective dissolution of the antimony eutectic and reprecipitation of the antimony during open-circuit stand or discharge (a mechanism analogous to the dezincification of brass) also causes antimony poisoning or self-discharge problems. However antimonial alloys when used in the positive grid do appear to have an ability to both retain the adhesion of the corrosion product film during cycling and maintain the electrochemical reactivity of the lead dioxide plate material. This accounts, in part, for the apparent conflict in the use of antimonial alloys, particularly those proposals to add antimony to the positive paste of batteries in order to improve performance and minimize shedding[291] compared to other proposals which remove antimony from the grid surface prior to pasting in order to minimize antimony poisoning.[63]

The corrosion product layer on lead–calcium alloys is more dense than that found on antimonial alloys, perhaps because of the minimal penetration of corrosion into the metal substrate.[219] However, studies have shown an excess of sulphate in the dioxide layer on calcium alloys and this may be indicative that the calcium is assisting the nucleation of lead sulphate; this again is perhaps a question for further investigation.

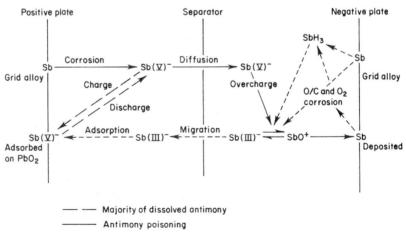

$$\text{------- Majority of dissolved antimony}$$
$$\text{——— Antimony poisoning}$$

Fig. 9. Antimony transference in the lead–acid battery.[66]

It is also noteworthy that barium sulphate is never added to the positive plate of the battery, in contrast to its wide use as a nucleation additive to the negative plate.

Corrosion of the antimony from the alloy tends to provide space for the lead dioxide film with less internal stress occurring at the 11% Sb eutectic composition,[221] also the surface roughness increases up to the eutectic[55,53] and there is an increase in film conductivity with increase of antimony concentration.[55] Increased porosity due to leaching of the antimony has been considered above and Burbank has suggested the possibility of lattice modification resulting from incorporation of antimony(V) into β-PbO_2.[197] There is now substantial evidence[66,191] that antimony(V) is adsorbed from solution onto the surface and inhibits[55,42] the growth of the lead sulphate (where the reaction requires a further 20 mV polarization[191] to initiate nucleation) and discharge of β-PbO_2 (where some 75 mV extra polarization is involved). Since the presence of dissolved antimony also decreases the overpotential for oxygen evolution on lead dioxide and modifies the discharge reaction, presumably the data reflects the competition at interfacial sites between the adsorption of sulphate ions, antimony ions and water molecules.

Antimony modifies the observed crystal structure and morphology of both lead sulphate and β-lead dioxide;[30] the crystal growth is retarded, and adhesion of active material and corrosion product are increased by improvements or retention of the non-stoichiometry. Thus X-ray analysis shows line broadening due to smaller crystallite size or lattice strain, deformed structures giving a stronger intercrystalline bridge; differential thermal analyis indicates that antimony slows down the gradual conversion with cycling of the electrochemically reactive and non-stoichiometric lead dioxide to the more stoichiometric, but less reactive, form. Some of these aspects of the morphology of lead dioxide have been considered by Caulder and Simon et al.[57,58,223-226] during investigations on automotive battery plate materials. Scanning electron microscopy has shown that the emergence of the "coralloid" stucture is indicative of the approach of the end of the useful life of the plate.

A detailed survey of the problems of retention of positive plate activity is not the aim of the present section, even though there are similarities between the problems of adhesion of corrosion product and the structural changes observed in the active material. In fact it should be noted that the corrosion product may not have the same morphology and physical characteristics as the battery compound.

Open-circuit stand of an electrode after anodic polarization can result in morphological changes and corrosion analogous to the self discharge of the positive plate of the lead–acid battery, α-PbO$_2$ being converted to tetragonal lead oxide.[41,137,139] In the case of charge–discharge cycling to simulate the service conditions, care must be exercised that the test does not overemphasize any particular aspect. Discharge of lead dioxide produces lead sulphate and basic lead sulphates depending on the potentials involved.[175,180,204] Potentiostatic discharge at low overpotentials, 50 mV, gave two current transients,[170] corresponding to the formation of lead sulphate (j vs $t^{-1/2}$), and the monobasic sulphate (j vs t^2). Recharge current transients on a number of different lead alloys under similar potentiostatic investigation[149] also show two peaks. With continued shallow-discharge cycling the formation of lead sulphate is favoured whereas deep-discharge cycling favours the formation of the basic sulphate.[170] At even lower potentials the discharge could result in the production of tribasic sulphate or even lead if the reaction was allowed to go completely cathodic. Thus, in corrosion testing of alloys, it is important to specify both the potential regions of interest and the time spent at any particular potential so that the morphological changes occurring in the corrosion product layer mirror the anticipated service conditions.

Lead Alloys

The composition of alloys used for the fabrication of electrodes in electrolysis cells and as the support grid of the lead–acid battery have apparently remained relatively unchanged over the years in spite of extensive investigations with wide-ranging compositions. The most commonly used alloying additions are antimony, tin, arsenic and calcium; a major consideration being to increase the strength and resistance to creep. It is interesting to note that the most significant trend over the past 25 years has been the decreasing concentration of the alloying elements, an indication that compositions approaching that of pure lead may offer advantages, provided the mechanical and localized corrosion difficulties can be overcome. The fundamental factors involved in the electrochemical film formation processes have been outlined above but it is also important to place these in the context of the alloy composition as this can appreciably modify the nature of the film.

Lead is one of the more malleable and ductile metals with a melting point of 327°C; major considerations in the development of a practical alloy include mechanical strength, corrosion resistance, character of the corrosion product and cost. Cost is self-explanatory since economic factors would rule out many prospective alloys regardless of their properties. Castability determines the integrity of the component, including minimizing defects due to shrinkage, and resistance to distortion, particularly where high production rates are required from automatic grid casting machines with a low scrap level. Strength is required to facilitate handling during processing as well as to retain the electrode structure during service. Early alloys were often limited by the casting technology, and the temperature difference between the liquidus and solidus is still an important practical consideration. Most lead alloys undergo some form of age hardening due to precipitation within the grain or at the grain boundary; alloying additions are therefore designed to promote the required precipitate of an intermetallic of known composition, or to modify the grain size. This aspect of the alloy strength is important in battery fabrication since the grid often has to be stored for a few days prior to pasting to allow the hardness to increase from below 10 BH after casting to above the 17 BH required for automatic pasting. Details on the related metallurgical corrosion aspects of the commonly used alloys are given below but the reader is also recommended to assess, for example, the review on the lead–acid cell,[41] the earlier papers by Simon and Jones,[227–230] and the detailed text-book by Hofman,[116] which covers much of the pre-1960 work on lead alloys.

It is generally realized that many corrosion processes are more destructive than is evident from weight loss and electrochemical measurements.[201] Burbank[36] reported the relation between anodic attack, self-discharge characteristics and microstructure of lead and lead alloys. Corrosion resistance of the alloy is important with regard to mode and extent of attack, e.g. whether the attack is general or concentrated at grain boundaries, general attack being preferred as intergranular attack leads to grain separation and damage greater than would be predicted by Faradaic considerations. The character of the corrosion product is also important from the standpoint of electrochemical passivation and resistance to shedding or erosion, resulting from prolonged oxygen evolution, as discussed previously.

Some trends relating the corrosion process to the alloy micro-structure have been observed,[22,29,46,153,171,242] and these are mainly concerned with the major alloys below. It has been suggested that lead should only be alloyed with electronegative elements,[102] particularly if the element itself may form a passivating layer; in fact, alloys containing calcium, sulphur, silver and tellurium were least attacked[64] and also exhibited grain refinement. The effect of silver, cobalt and antimony in terms of them affecting the chemistry of the corrosion process has been discussed above and all are reported to improve the corrosion resistance.

Tin in combination with other elements appears to produce corrosion-resistant intermetallics, but its historical use at up to 4% Sn has been to improve the fluidity of the metal in the mould. A more usual range is now 0.18–1.25% Sn. The metallurgical factors influencing the use of the alloys will now be reviewed to indicate some of the practical constraints and to anticipate future developments.

Lead–antimony alloys

Lead–antimony alloys were first proposed for lead–acid battery grids in 1881 by Sellon,[253] and at the present time the antimony content ranges from 2 to 12%. The economy of the battery industry is based on a large consumption of recovered lead and it is therefore cheaper to adjust the concentration of antimony in the alloy than to remove it completely; the use of antimonial alloys is also widespread in the electrowinning and electroprocess industries. Antimony improves the castability of lead by reducing its freezing point as well as increasing the strength hardness and resistance to creep;[116,251] an extended freezing range allows the material to flow in the mould, and this is important in the mass production of battery grids. The continuing use of antimony alloys is a compromise between requirements for good castability and favourable properties with acceptable self discharge and cost.

The lead–antimony phase diagram[103,116,218] shows a eutectic composition of 11.1% Sb, with a eutectic temperature of $252 \pm 0.5°C$. A maximum solubility of 3.45% Sb in solid solution at the eutectic, decreasing to 0.1% at room temperature, suggests that lower compositions should contain no eutectic, but in cast structures, even down to 0.01% Sb,[116] coring often produces a eutectic structure. The increased hardness, 10 BH to 16 BH for the 2% and 12% Sb alloys, respectively, and increased tensile strength from 1.5 kg mm^{-2} for pure lead to about 2.7 kg mm^{-2} for the 4–6% alloy, is mainly the result of interstitial precipitation of antimony as supersaturation is relieved. The extended freezing range, from 327°C to 252°C, provides a wide tolerance for the liquidus to solidus in automatic grid casting machinery.

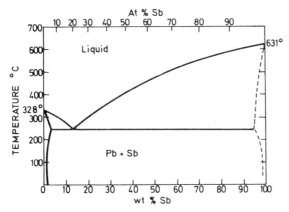

Fig. 10. Equilibrium diagram of the Pb–Sb binary alloy system.[218]

Microstructures of the hyper-eutectic, the eutectic and the hypo-eutectic are well-documented[36,116] as is the influence of anodic corrosion on the type of attack.[36] Since it is in practice impossible to produce a homogeneous alloy of uniform solid solution, even when the amount of antimony is reduced to an insignificant amount,[116] the alloy is segregated on the macro and/or microscale giving rise to shrinkage cracks and voids in the casting. Macrosegregation is of a random nature, depending to a large extent upon the mode of casting, cooling and the shape of the mould. The macrostructure usually referred to is the cored structure which results from the precipitation of successively richer alloy layers upon the primary dendrites. Microsegregation may be defined as segregation which occurs on a microscopic scale within individual grains; thus above the 3·45% Sb maximum solubility, microsegregation is always present.

Inverse segregation, i.e. decreased concentration of the solute towards the centre of the casting with an abnormally high concentration at or near the surface, has been the subject of reviews[178,246] and the inverse segregation of lead–antimony alloys has been reported by

Simon and Jones.[228] For alloys other than fast-chilled, a larger excess of antimony is present at the surface than would be expected from the alloy composition, hence the suggestion for its removal by pickling.[63] A further form of segregation, i.e. gravity segregation, occurs when the densities of the crystals formed are greatly different from that of the liquid from which they are crystallizing. This causes the crystal either to float or sink in the melt. The effect has been illustrated by Killaby *et al.*,[124] by using antimony-124. Autoradiographs of an antimonial lead ingot showed a large concentration of antimony at the top of the ingot with little or no antimony at the bottom.

The alloy structure is therefore formed of primary dendritic lead-rich crystals surrounded by an antimony-rich phase, which may approach a eutectic composition during freezing. It is this liquid which fills any voids formed during shrinkage of the primary crystals, and it should also be noted that segregated antimony may be precipitated as particles[36,116] even at 1% antimony or less. These deleterious effects produced by the segregating characteristics of antimony may be the initial sites for corrosion with selective leaching of the antimony-rich component, causing microporosity at the alloy–corrosion product interface and the linear corrosion kinetics.

Lead and the low-concentration lead–antimony alloys, 0·5% Sb, are shown[46,191] to exhibit intergranular corrosion, the corrosion attack also

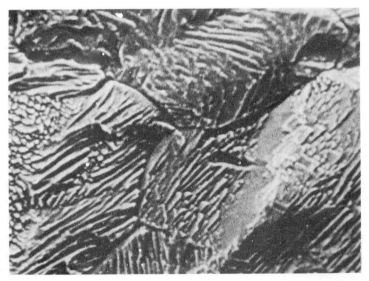

Fig. 11. Corrosion of lead with surface stripped of corrosion product.[191] ×500.

mounds. Examination of the lead beneath these mounds showed that it had been attacked locally, the size and depth of attack corresponding approximately with the size of the mound.

Analysis of the PbO_2 formed on the Pb–1% Ag bielectrodes for Ag content gave values in the range 0·6–0·7%, but whether or not this has any significant influence on the properties (semi-conductivity, plasticity etc.) of the PbO_2 is not clear.

Limitations of the Bielectrode

Nature of environment

The nature of the environment is of major importance in the choice of an anode for cathodic protection systems, and this in turn will determine to a large extent the current output for a given area of surface of the anode. For the protection of steel pipes buried underground, in which diffusion of water is a rate-controlling factor, a very large mass of coke breeze (a groundbed) is used, and although the anode current density cannot be determined with precision, it is obviously very low indeed. On the other hand, anodes for protection of structures in sea water operate in the range 100–1000 A m^{-2}, depending on the nature of the anode and the type of structure to be protected. There are obviously a number of economic and technological advantages in operating anodes at high current densities, although these may be offset sometimes by the disadvantage of uneven current distribution on the structure to be protected.

For an anode to operate at high current densities it is essential that (i) there is an ample supply of anode reactant in order to minimize transport overpotential, (ii) the anode reaction occurs without excessive activation overpotential, and (iii) the anode products can diffuse readily into the bulk water so that the composition of the water in the vicinity of the anode remains unaffected. These conditions apply to anodes used in sea water, since the concentration of Cl^- ions is high and the Cl_2-evolution reaction occurs with only a small overpotential on a number of surfaces, particularly platinum.[16] In fresh water, or in brackish waters of low Cl^- ion concentration, O_2 evolution will be the dominant anode reaction and since this occurs with considerable overpotential (j_0 for O_2 evolution on Pt is $\sim 10^{-12}$ A cm^{-2}) operation at high current densities is less attractive. In sea water, although Cl_2 evolution is the dominant reaction it will be accompanied by O_2 evolution, the relative rate of the latter increasing with current density.[16]

As far as the Pb/PbO_2–Pt bielectrode is concerned, and similar consideration probably applies also to Pb–Ag alloys, the composition of sea water appears to be highly conducive to PbO_2 formation; laboratory studies have shown that about 0·5 M NaCl appears to be the optimum concentration of Cl^- to form PbO_2 (sea water contains ~3·5 w/v % of NaCl, i.e. ~0·6 M), and that PbO_2 is formed more readily in sea water than in 0·5 M NaCl owing to the presence of SO_4^{2-} and HCO_3^- ions which facilitate passivation.

Laboratory studies[12] in solutions containing various concentrations of NaCl have shown that PbO_2 formation is confined to the range 0·1–1·2 M NaCl. Above this concentration the $PbCl_2$ forms a finely crystalline, closely interlocked, adherent film of very high ohmic resistance, which appears to preclude the formation of PbO_2. This may be largely due to the very low activity of Pb^{2+}, but the formation of chloro-complexes must also make a contribution, since it has been observed that wafers of PbO_2 become pitted when anodically polarized in 1·2 M NaCl but are unattacked at lower concentrations.[17] Studies of the anodic oxidation of Pb^{2+} ions to PbO_2 at a Pt electrode in different concentrations of NaCl have shown that the coulombic efficiency at potentials below the reversible Cl^-/Cl_2 potential is 100%, providing the concentration of NaCl is <1·2 M. At and above this concentration there is a decrease in efficiency, which has been explained by complex formation:

$$Pb_{aq}^{2+} + Cl_{ads} + 3\ Cl^- - e\ \rightarrow\ PbCl_4\ \rightarrow\ Na_2PbCl_6\ \rightarrow\ 2\ Na^+ + PbCl_6^{2-}$$
(8)

Similarly, the observation that the Pb metal will not form PbO_2 at and above 1·2 M NaCl is explained by the possibility that it immediately reacts with the NaCl and with Cl_2 to form $PbCl_6^{2-}$.

$$PbO_2 + Cl_2 + 2\ NaCl\ \rightarrow\ Na_2PbCl_6 + 2\ H_2O$$
(9)

It has been shown that in low concentrations of Cl^- ions, $PbCl_2$ does not crystallize on the surface of the lead (Fig. 1) and the reaction Pb → Pb^{2+} proceeds at a low rate, which in time could lead to excessive corrosion of the lead. This provides an oversimplified explanation of the fact that the Pb/PbO_2–Pt bielectrode is not suitable for use in waters of low salinity; the precise resistivity below which the anode normally operates satisfactorily has not been determined but it is probably of the order of 100 to 250 Ω cm (sea water is $\sim$25 Ω cm). Tudor and Ticker[18] in a series of laboratory tests on Pb anodes in waters of different resistivities (1000–5000 Ω cm) found that in water of 1000 Ω cm the bielectrode failed to form PbO_2, and became severely corroded and coated with $PbCl_2$ when polarized at 10 A m^{-2}. This very low current density was selected to simulate a ship operating with a fixed maximum voltage output entering a water of high resistivity; however, at this very low current density it is doubtful if the bielectrode would form PbO_2 even if polarized in sea water.

The inability to use the bielectrode in waters of high resistivity is a serious limitation, and has prompted further studies by Helber and Littauer,[19] who preformed (18 h at 500 A m^{-2} in 0·5 M NaCl) a bielectrode to obtain stable PbO_2 film, and then scraped off the PbO_2 from half the surface leaving the remainder in contact with the Pt microelectrode. It was established that the PbO_2 formed on the bare Pb when the bielectrode was subsequently polarized in dilute Cl$^-$ solution or even in fresh water. It was also observed that Pb exposed to the atmosphere for some time was more conducive to PbO_2 formation than the freshly cast metal, and that this beneficial effect could be accentuated by heating the Pb for 1 h at 200°C in a furnace containing sulphur so that the surface of the lead became filmed with PbO and PbS. After this treatment, rapid formation of PbO_2 occurred when the bielectrode was polarized in tap water (1060 Ω cm) or in 0·0001 M NaCl (8960 Ω cm).

Helber and Littauer consider that in very dilute solutions of Cl$^-$ ions, (0·001 M), the passivation produced by the formation of $PbCl_2$ in more concentrated solutions is replaced by sluggish formation of PbO by a reaction of the type:

$$Pb + H_2O \rightarrow PbO + 2 H^+ + 2e \tag{10}$$

which is insufficiently insulating to provide the prepassivating film that is the precursor to PbO_2 formation.

These studies indicate that the bielectrode should be capable of being polarized in high-resistivity waters without corroding, providing it is preformed in 0·5 M NaCl or given a film of PbS + PbO; however, there is little evidence that it has been used in these environments for cathodic protection systems.

Pressure

Hollandsworth and Littauer,[20] on the basis of observation made in service, have investigated the effect of hydrostatic pressure on the corrosion of Pb–6% Sb–1% Ag and the Pb–Pt bielectrode. Laboratory studies in artificial sea water in a pressure vessel showed that severe corrosion occurred at pressures corresponding to a depth of immersion of 30 m or more. The rate of formation of PbO_2 increased significantly (by a factor of 41 at a pressure equivalent to a depth of 352 m) at depth compared to surface operation, and the thick layer of PbO_2 developed cracks and fissures and became separated from the underlying Pb, although in the case of the bielectrode, failure was somewhat delayed owing to the electrical contact provided by the Pt microelectrode. It would appear that under these circumstances the rate of Cl_2 evolution decreases so that a higher proportion of the charge is consumed in PbO_2 formation.

These results show that Pb alloys and the Pb–Pt bielectrode are unsuitable for use in the sea at depths greater than 30 m.

Comments on the Role of the Pt Microelectrode

The formation of a stable passive layer of PbO_2 on Pb alloys or on Pb plus electron-conducting microelectrodes during anodic polarization in Cl^- solution is a highly complex process, and although considerable work has been carried out in this field, it cannot be said that the mechanism is fully understood. This complexity is due to the number of competitive reactions, of which the formation of Pb^{2+}, $PbCl_2$, PbO_2 and Cl_2 are the most important, and to the fact that during the early stages of the process the PbO_2 formed at the surface is being continuously undermined and detached by the formation of $PbCl_2$. A further complication that arises under practical conditions is the growth of the PbO_2 during prolonged polarization, and to the development of internal stresses that may lead to cracking of the oxide and to separation of the PbO_2 from the underlying Pb.

The ability of alloying additions of Ag to improve the corrosion resistance of Pb when used as an anode has been known for some years, and it is possible that the Ag crystals of the Pb–Ag eutectic (3·5% Ag) perform the same role as the Pt microelectrodes, although other mechanisms have been suggested[21,22] (*see also* Chapter 14). However, the Pb–Pt bielectrode has an important advantage in that the microelectrode can be readily observed and studied, a situation that does not apply to studies of lead alloys.

As far as the Pt microelectrode is concerned, it would appear that there is nothing specific about this metal, and that other inert electron-conducting materials such as iridium, rhodium, platinized titanium, graphite etc. may be used. All that is required is that the microelectrode must be in direct contact with the Pb, and must remain so during the life of the anode. Recent work by Narasimhan and Udupa[23] has shown that pieces of PbO_2, prepared by anodic oxidation of Pb^{2+} at a graphite electrode, can act in the same way as Pt.

On the basis of laboratory studies carried out for short periods under carefully controlled conditions, it is possible to summarize the various stages in the conversion of Pb/Pt to a Pb/PbO_2–Pt bielectrode as follows:

(*i*) Rapid formation of a non-conducting mechanically passivating layer of $PbCl_2$ in which the extent of formation of $PbCl_2$ is limited by Cl_2 evolution at the Pt, i.e. the Pt may be regarded as a "valve" that limits the increase in potential.

(*ii*) Oxidation of Pb^{2+} to PbO_2 and its nucleation at the $Pb/PbCl_2$–Pt interface with simultaneous evolution of Cl_2 at the Pt.

(*iii*) Propagation of PbO_2 nucleated at the Pt over the surface of the $PbCl_2$ formed initially.

(*iv*) Gradual conversion of the underlying $PbCl_2$ to PbO_2, which is in direct contact with the surface of the Pb.

(*v*) Evolution of Cl_2 from both the Pt and the PbO_2, which are now equipotential, with a consequent decrease in PbO_2 formation.

However, these laboratory observations do not provide information on the changes that may take place during prolonged polarization in service, and similar considerations also apply to prolonged laboratory tests owing to the changes in composition resulting from the electrode products.

Determinations of the polarized potential and current have been widely used in these studies, and an increase in potential at constant current or a decrease in current at constant potential have been regarded as diagnostic of the formation of a non-conducting layer; the latter also applies in practice during which anode failure manifests itself as a marked decrease in current output, which is associated with anode corrosion. Failure of anodes in service are difficult to investigate but observations of the lead have shown that attack is associated with the formation of lead compounds, presumably $PbCl_2$, $PbCO_3$, $PbSO_4$ etc. admixed with PbO_2.

It is relevant at this stage to consider the proposal made by Wheeler[14,15] that the sudden increase in potential when the Pt is removed from Pb (Fig. 7) proves that its primary function is to make electrical contact between the Pb and the PbO_2. This may well be the case in the early stages of polarization, but after a short period of time the PbO_2 and the Pb will be in contact. This is illustrated[12] by the sections of the Pb/PbO_2–Pt bielectrodes shown in Fig. 10, and can be demonstrated by the following experiment.[14] A bielectrode of the type shown in Fig. 5, but with only one microelectrode at position a is anodically polarized for 24 h in $0·5$ M NaCl, and the bielectrode is then raised so that the microelectrode is above the level of the solution. On removing the microelectrode from the Pb, no increase in potential is observed, showing that the PbO_2 must be in direct contact with the Pb, and the bielectrode can be polarized for a considerable time before the potential starts to increase. It is also apparent that in the case of the Pb–Ag alloys, the alloy and the PbO_2 must be in direct contact, although the area of the Pb alloy/PbO_2 interface in contact may only be a proportion of the total. Thus the immediate increase in potential when the Pt microelectrode is removed must be due to exposure of a small area of bare Pb to the solution, a situation that simulates the breakdown that sometimes occurs in service.

The practical trials carried out by Peplow have shown that after prolonged polarization at high current densities, the PbO_2 can attain a thickness of 1–5 mm and that its surface is covered with small mounds that appear to be associated with attack on the underlying lead. This is considered to be indicative of breakdown of the PbO_2, corrosion of the Pb followed by repair of the PbO_2. Little is known concerning the stresses that develop in PbO_2 under these conditions, but it is apparent that, in view of the growth of the PbO_2, passage of charge is not solely electronic. If the film thickens by ionic transport it will develop stresses and at high rates of thickening of the film, such as occurs at very high current densities, cracking may occur at weak areas. This rapid thickening of the PbO_2 and cracking observed by Hollandsworth and Littauer during short-term laboratory studies of PbO_2 formation at

Fig. 10. Sections of Pb/PbO$_2$–Pt bielectrode polarized in 0·5 M NaCl at 30 mA cm^{-2}. (a) After 2 h, showing lead (centre, white) surrounded by PbCl$_2$ (grey) and PbO$_2$ black; (b) After 24 h, showing Pb (lower half, white) in direct contact with PbO$_2$.[12]

moderate current densities and at elevated hydrostatic pressures confirms this view.

If the PbO_2 is in contact with the Pb the question now arises as to why it appears to become inert and ceases to evolve Cl_2 when a small area of Pb becomes exposed to the solution, and why the potential increases to a high value owing to $PbCl_2$ formation at the Pb exposed to the solution. Unfortunately, there does not appear to be a simple answer to this question, but the fact that both $PbCl_2$ and PbO_2 are formed simultaneously during initial polarization of the Pb–Ag alloys and the Pb–Pt bielectrodes shows that there is a very delicate balance between the conditions that favour continued formation of one or the other of these compounds. An analogous situation arises in the pitting corrosion of a passive metal such as a stainless steel in which localized areas become attacked while the rest of the surface remains protected by the passive film. This phenomenon is still not fully understood, although it is recognized that the geometry of a pit (or a crevice or crack) leads to a situation in which the pH of the anolyte within the pit becomes significantly lower (pH 1–4) than that of the neutral bulk solution. It has been established also that the Cl^- ion concentration within a crevice or pit is far higher than that in the bulk solution. Thus, if breakdown of the PbO_2 film occurs at a weak area in the film, or if the film cracks owing to internal stresses, the exposed Pb surrounded by the overlying PbO_2 could lead to a situation similar to that within a pit or crevice at the surface of a passive metal. An autocatalytic situation could then arise leading to a decrease in pH and increase in Cl^- ion concentration, and to conditions that favour continued formation of $PbCl_2$.

On the other hand, if a Pt microelectrode is in contact with the lead, and assuming that the surface is equipotential, then the potential of the Pb/solution interface will not exceed that of the Pt, i.e. 2·5–3·5 V (*vs* SCE) for the range of current densities used in practice. Under these circumstances some corrosion of the Pb exposed at a discontinuity in the PbO_2 will occur, with the formation of non-conducting corrosion products. However, as the potential is maintained at 2·5–3·5 V by Cl_2 evolution at the Pt, continued formation of $PbCl_2$ will not be possible, and conditions will be such that it will be slowly converted to PbO_2. Possibly the major role of the Pt microelectrode is to act as an internal potentiostat that limits continued formation of $PbCl_2$ and maintains the potential of the Pb/solution interface at a value conducive to the formation of PbO_2. These findings were also endorsed by Royuela *et al.*[25] who examined Pb–Sb–Ag alloys in NaCl both with and without Pt microelectrodes. Their conclusion that, at current densities greater than 60 mA cm^{-2}, no advantage was gained, does not seem to concur with practice.

Practical Considerations

There are now a wide variety of anode materials available for impressed-current cathodic protection systems, and the choice of one in preference to all others is dictated by a number of factors, particularly costs. Platinized Ti and Pt-clad Ti and Nb are the most widely used anodes for protecting structures immersed in sea water, but Pb alloys and the Pb/PbO$_2$–Pt bielectrode still have important applications largely because of their low cost, ease of manufacture and availability.

The Pb–Pt bielectrode normally consists of Pb or Pb alloy rod, 25–50 mm diam. and of length ranging from 1 to 10 m. Microelectrodes of Pt wire (usually containing small percentages of alloying metals such as Ir or Rh to increase hardness), 0·5–0·75 mm diam. and about 13 mm length, are inserted at intervals of about 0·25 m along the length; this is achieved by drilling a hole of diameter slightly smaller than that of the Pt, gently tapping the microelectrode in so that its end projects slightly from the surface of the Pb, and peening the Pb surrounding the Pt to ensure good contact. It may also be used in the form of a billet mounted in a steel casing, but insulated from it by means of Araldite. The anodes may be used at current densities in the range 250–750 A m^{-2}, but as confidence in these anodes has grown there has been a tendency to operate at the highest current density. It is usual to polarize initially at 200–250 A m^{-2} and then to increase to 750 A m^{-2} gradually over a period of 1–2 weeks.

Owing to the complex patent position, the use of the bielectrode has been largely confined to Metal and Pipeline Endurance Ltd. (MAPEL) in the U.K. and to the Lockheed Company in the U.S.A. One of the earliest applications by MAPEL was for the cathodic protection of the cooling water culverts in a power station in Malacca, Malaya, which were installed in 1962 and designed to operate at 250 A m^{-2}. These are still operating satisfactorily after 15 years. Other examples of large structures protected by MAPEL are the jetties in Europort and North Sea gas and oil platforms. The Lockheed Company[24] have used the bielectrode extensively for the protection of bulk carriers, tankers, ocean liners, offshore drilling rigs and oil wells, and possibly the largest installation is for the cathodic protection of the San Francisco Bay Area Rapid Transit Tube. This is a steel tube about 25 m diam. and 4 miles long which is used to carry rail traffic under the Bay. Each anode section is 10 m long, and provides 250 A at 15 V, and a total of 16 anodes are used to give an output of 4000 A.

References

1. J. H. Morgan, "Cathodic Protection". Leonard Hill, London (1959).
2. L. L. Shreir, "Corrosion", Vol. 2, Section 11. Newnes–Butterworths, London (1976).
3. L. L. Shreir, *Plat. Metals Rev.* **21**(4), 110 (1977); **22**(1), 14 (1978).
4. G. Baum, U.S. Patent 1,477,099 August (1922).
5. J. B. Cotton, *Chem. Ind.* **68** (1958).
6. W. R. Jacobs, "Proceedings of Symposium on Cathodic Protection" (Marston Excelsior Ltd), London (1975).
7. C. G. Fink, *in* "Corrosion Handbook" (H. H. Uhlig, ed.), p. 606, J. Wiley, New York (1953); *Ind. Eng. Chem.* **16**, 566 (1924).
8. J. K. Dennis and T. G. Such, "Nickel and Chromium Plating". Newnes–Butterworths, London (1972).
9. K. N. Barnard, G. L. Christie and D. G. Cage, *Corrosion* **15**, 581t (1959).
10. L. L. Shreir and I. Weinraub, *Chem. Ind.* 1326 (1958); L. L. Shreir, *Plat. Metals Rev.* **44**, 3 (1959); L. L. Shreir and Metal & Pipeline Endurance Ltd., British Patent Application 15852/58.
11. D. B. Peplow and L. L. Shreir, *Corrosion Technol.*, April, No. 4, 16 (1964); D. B. Peplow, *Brit. Power Eng.* **3**(5), 61–65 (1961).
12. E. L. Littauer and L. L. Shreir, 1st International Congress on Metallic Corrosion, London, 1961 (Published by Butterworths, 1962).
13. G. W. D. Briggs and W. F. K. Wynne–Jones, *J. Chem. Soc.* **574**, 2966 (1956).
14. L. L. Shreir, *Corrosion* **17**, 118t (1961).
15. W. C. G. Wheeler, *Chem. Ind.* **75** (1959); *see also* Discussion in Ref. 14.
16. E. L. Littauer and L. L. Shreir, *Electrochim. Acta* **11**, 527 (1966).
17. E. L. Littauer and L. L. Shreir, *Electrochim. Acta* **12**, 465 (1967).
18. S. Tudor and A. Ticker, *Mat. Prot.* **53** (1964).
19. H. Helber and E. L. Littauer, *Corrosion Sci.* **10**, 411 (1970).
20. R. P. Hollandsworth and E. L. Littauer, *J. Electrochem. Soc.* **119**, 1521 (1972).

21. E. Sato, *J. Electrochem. Soc. Jap.* **31**, 105 (1963).
22. E. Sato, *Bull. Chem. Soc. Jap.* **39**, 1592 (1966).
23. K. C. Narasimhan and H. V. K. Udupa, *Electrochim. Acta* **22**, 197 (1977); British Patent 1,423,093, Feb. (1974).
24. E. L. Littauer, Private Communication.
25. J. J. Royuela and M. Serra-Ribera, *Metales No Fereos*, 135–148, Sept. (1963).

11   The Lead Dioxide on Inert Basis Anode

H. HAMSAH,* A. T. KUHN‡ and T. H. RANDLE§

* Department of Chemistry and Applied Chemistry, University of
Salford, Lancs
‡ Department of Dental Materials, Eastman Institute of Dental
Surgery, London
§ Department of Chemistry, Swinburne College of Technology,
Hawthorne, Australia

This chapter describes the manufacture, applications and underlying
scientific knowledge—such as it is—of electrodes made of a more or less
inert basis material, coated with a layer of lead dioxide.

Such electrodes have many uses, and are becoming more important.
These uses will be classified in the following pages.

Applications

In batteries

Primary Batteries

The lead primary battery is not widely known, although at the same
time it serves a useful purpose in naval, military and other applications.
The positive plate is lead dioxide coated on an inert basis, often nickel.
The negative is lead, likewise coated on nickel. The electrolyte is
fluoboric acid. The batteries have an indefinite shelf life since the
electrolyte is stored in a sealed ampoule or in similar fashion. When the
batteries are required for operation, the electrolyte is released into the
cell and the battery is ready to provide power after a few seconds (see
also reference to "sea-water batteries" on p. 385).

Secondary batteries

Sulphuric Acid Systems

Some 22% of the weight of an average automotive battery resides in the grids (electrochemically inert) which support the active mass in both positive and negative plates and serve to conduct the electric current out of the cell. There is a long history of attempts to reduce the mass of these grids either by changing their design or by using a lighter material for the purpose. In terms of the positive grid, titanium is the only metal which combines low density with corrosion resistance at an economically realistic price. For this reason a number of patents and publications exist relating to this subject. Without describing this technology, we should be clear that the demands made of such an electrode are more severe than in the case of either a "one-shot" primary battery, as described above, or the "process anodes" described in the section on choice of substrate. For in these cases, current will only flow in one direction and if the layer of lead dioxide is destroyed, as is the case in the primary battery, no further use is required. In the case of the secondary battery, however, one must contemplate the situation where the substrate completely sheds its coating of PbO_2 and is then, on charging, required to rebuild it. The best known paper relating to secondary sulphuric acid batteries is that of Cotton and Dugdale,[1] while a more drastic departure from the traditional "grid with active materials" was described by Faber[2] based on active materials (positive) on titanium "wool". But in both cases, the construction was such that the titanium required a coating of an inert non-corroding conductor such as Pt or Au, so that if the plate became fully sulphated, a conducting surface was still available for the re-oxidation of the sulphate to the PbO_2.

In addition to these extensions of conventional battery technology, there are more radically different concepts such as the secondary batteries using PbO_2 deposited on TiN (e.g. U.S. Patent 3,576,674 or D.O.S. 1,938,409), while Beck[3] has been a notable pioneer in the development of secondary lead/lead dioxide batteries in which the lead salts dissolve in the electrolyte rather than forming insoluble sulphates (see Chapter 4). Wabner[41] mentions that Ti anodes, pretreated in a special manner (see later), were evaluated as positive electrodes in a secondary lead–acid battery and there survived 450 fairly deep discharges, although no precious metal or similar functioning interlayer was employed apart from the somewhat mysterious Ti(IV) pretreatment he describes.

Anodes for electrochemical processes

Among the oldest and most enduring of process anodes is the one made of massive lead or lead alloy. In service, this becomes coated with PbO_2 and the behaviour of this system is described in Chapters 12 and 13.

Such anodes, however, have certain disadvantages. Firstly they are very heavy, since lead especially in its pure form, possesses little mechanical strength. Furthermore, in service, lead tends to "creep" and may bow or otherwise become deformed unless a very massive section is used. Finally, it is known[4] that the corrosion of such anodes takes place to no small extent not at the outer PbO_2 surface, but rather at the interface between the 4-valent PbO_2 and the 2-valent oxide beneath it.

For these reasons, there is a long—and now successful—history of anodes using lead dioxide on some other inert basis metal. The early stages of this work have been discussed.[5] Iron, nickel and platinum have all been used. However, the substrates which lent themselves to serious consideration from a technological point of view were graphite, tantalum, niobium and lastly, of greatest importance, titanium. These anodes which possessed far greater dimensional stability than the lead-based ones have been successfully used in a range of processes including chlorate, bromate manufacture, cathodic protection, electroflotation and sodium hypochlorite electrolysis. For further details, the reader is referred to the book "Industrial Electrochemical Processes".[5] We shall see that in almost all circumstances titanium is the ideal substrate. Nonetheless, it was the last to be adopted because of real or imagined problems of passivation.

The Choice of Substrate

The choice of substrate is simple enough. The material must be at least a moderately good conductor. It must not corrode in the event of a pinhole in the covering layer of PbO_2. Lastly, it must possess good mechanical properties. We can consider two or three substrates which fulfil all except the last of these criteria and which have in fact been used or suggested. These are graphite, iron silicide and magnetite. All are reasonably conducting and corrosion-resistant. However, all of these are heavy, fragile and available only in thick section or cylindrical form and thus poorly compatible with the modern cell designs which use thin flat sections often in the bipolar mode. Where such electrodes are used, therefore, it is either in cells having a cylindrical cathode concentric with an inner cylindrical anode, or in cells operating at low current density where rows of cylindrical anodes confront a flat-plate cathode. In such cells, of course, there is poor current distribution and the "back" or "side" of the anode is poorly utilized.

Given these disadvantages of the above materials, we must now consider the valve metals Ta, Nb and Ti. The first two of these are many times more expensive than Ti and could be used wherever Ti might be used except from an economic point of view.[19] The problems lie in relation to the usefulness of titanium.

It is well known that titanium is corrosion-resistant on account of a thin but very perfect oxide film on its surface. This film is also a good insulator, and it is impossible to pass more than a few mA cm^{-2} at a titanium anode in aqueous medium unless the applied voltage is so high that this film breaks down. For this reason it was appreciated early that the film must be removed before application of the lead dioxide onto the Ti. The controversial question—as we shall see—is whether even for a process anode, some intermediate layer is required to prevent a growth of this TiO_2 beneath the PbO_2 in service, thereby leading to a condition where a seemingly perfect anode is useless on account of the high electrical resistance of the titanium oxide layer.

Manufacture of Electrodes

Almost all methods for the manufacture of electrodes are based on electrode position of the lead dioxide from a bath containing its salts. Dürkes, in German Patent 2,259,821, describes a thermal method for making a lead-coated electrode, the lead being subsequently oxidized to lead dioxide. The patent describes the application of a paste of ferrous or zinc chloride and powdered quartz and lead, followed by a drying and then a firing at 600°C. But apart from this example, all other methods described here are electrochemical. We shall, for reasons which will become apparent, discuss graphite-based anodes separately.

Unless otherwise mentioned, all the coatings described here, derived as they are from acid baths, are preponderantly of the β-PbO$_2$ variety. A much more limited amount of work has been done on the preparation and behaviour of α-PbO$_2$ coatings.

PbO$_2$/graphite anodes

The important earlier patents are D.A.S. 1,182,211 and U.S. 2,945,791 (1960) and D.A.S. 1,496,962 (1966), all from the Pacific Engineering Corp. The second of these specifies addition of a surface-active agent (such as alkyl phenoxy polyoxethylene ethanol type) to reduce gassing, increase throwing power and give a more compact deposit. Nitric acid is also stated to give better throwing power. The stepwise reduction of current density is mentioned in both the first and the second of these patents. All the Pacific patents are characterized by details of "bath management" which are trivial on the laboratory scale, but become crucial when plating on an industrial scale. Typical bath composition from the earlier patents might then be: lead nitrate, 50–400 g l^{-1}; copper nitrate trihydrate, 0–20 g l^{-1}; nickel nitrate trihydrate, 0–10 g l^{-1}; sodium fluoride, 0·5 g l^{-1}; surface active agent, 0·75 g l^{-1}; nitric acid, 0–4 g l^{-1}; and pH = 1·5. This bath is operated at 45–70°C at currents which may be constant (e.g. 25 mA cm^{-2}) or increase or decrease during the plating. The fluoride ions may serve to etch the oxides present on graphite. The patent D.A.S. 1,496,962 somewhat amplifies the question. The nickel is stated to act as a grain-size refiner, the copper to prevent deposition (and thus depletion) of lead on the cathode. This it does presumably by reducing the overvoltage for hydrogen evolution at the cathode. (Wabner[41] has tested PbO$_2$ deposits for the presence of metals such as Cu or Ni and finds no detectable traces.) This patent also claims advantages in the maintenance of the level of dissolved iron below 0·02 g l^{-1}.

Further reading shows that in D.A.S. 1,496,962 (which is equivalent more or less to U.S. Patent 3,463,707), the emphasis is on removal of iron (to less than 0·02 g l^{-1}) and chlorides, and obviating the need to use the relatively costly Ni and Cu salts previously employed. These baths thus contained only sodium fluoride and nitric acid, apart from the lead nitrate at a concentration of 200 g l^{-1}. Thangappan and Nachippan[6] describe the benefits to be obtained by deposition of the PbO$_2$ in a bath filled with fluidized inert particles, i.e. one in which the mass transport of species has been enhanced compared with normal conditions. A much smoother deposit is claimed. Huss and Wabner[7] use paraffin-filled graphite and evacuate it before deposition, to ensure absence of gases, while Narasimham[8] describes electrolytic pretreatment of graphite in alkali with a 24-h "soak" period. Little has been published in the PbO_2–graphite area, although an interesting glass-fibre/PbO_2 composite "sleeve" to fit graphite rods has been described.[38] Graphite is anisotropic in its electrical conductance, and Wabner[41] suggests this may lead to poor current distribution and thus increased corrosion. He also states that the graphite is capable of being "burnt" by the PbO_2, i.e. oxidized, and reports that this can lead with time to formation of a void between the graphite and the inner surface of the PbO_2 (which would presumably in any case be reduced). This constitutes another reason (according to him) why such anodes are comparatively short-lived, although the "Pepcon" commercial anodes are stated to last for 2 years approximately when fitted in hypochlorite cells.

Narasimham and Udupa,[48] in their recent and excellent paper reviewing preparation and uses of these anodes, do not support the views of Wabner. In their work, they describe how anodes in rod (75 cm long, 20 cm diam.) or plate (90 cm × 18 cm × 2 cm) form behave, lasting for as long as 2 years. Although these authors conclude by stating that Ti might be a better substrate, it is quite clear that graphite-based anodes are a wholly workable proposition.

Titanium-based anodes

Although graphite-based anodes are still in use, notably by the Pacific Engineering Corp. for their chlorate and hypochlorite production, the growth is now seen to be in titanium-based electrodes which allow thinner and mechanically superior cell designs. The first mention of these (as distinct from graphite) was in U.S. Patent 3,463,707 or British Patents 1,189,183 (1970) and 1,159,241 or French Patent 1,534,453. In the entire treatment, the question of the formation and deleterious effects of the passive oxide layer must always be borne in mind. As Wabner points out, the concept of an electrode based on a Ti (a strongly reducing metal) substrate coated with 4-valent, highly oxidizing PbO_2 goes against all expectations of success. The thermal coefficient of expansion of Ti (in contrast to graphite) is close to that reported for PbO_2, thus minimizing thermal shock problems.[50]

While pinholes in the outer PbO_2 layer will not, with these substrates, lead to catastrophic corrosion, it will nevertheless allow a passive film of high resistance to spread on the (Ti) substrate surface beneath the PbO_2 layer.

Electrodeposition on Ti—General

The above-mentioned problems of deposition, whether anodic or cathodic, on Ti have given rise to several patents aimed at solving the problem. Thus German Patent 1,170,747 discusses a Cr^{3+} and F^- etch bath as pretreatment in electrodeposition, and U.S. Patent 2,734,837 specifies a pickling in 90°C HCl for 5–15 min with or without a previous HF/HNO_3 etch or a molten NaOH descale bath. U.S. Patent 3,207,679 calls for anodization of the Ti to a "yellow colour" prior to PbO_2 deposition (but with a Pt "flash").

Complete Descriptions for Anode Manufacture

There are dozens of patents describing the best way in which to prepare PbO_2 anodes. There is no way of knowing, in most cases, the true worth of the ideas proposed, for there is almost no example of performance or appearance of an anode so prepared compared with one in which the step was omitted. The best procedure is to detail all the steps, it being understood that several of these may well be omitted.

Preliminary cleaning and treatment. We here consider mechanical abrasion and degreasing. De Nora[9] suggests sandblasting with fine silica sand followed by degreasing with benzene—a solvent virtually proscribed in many cases today. Trichloroethylene is now usually specified. They note that a pre-etch in boiling HCl leads to irregular PbO_2 deposition, with a tendency to cracking. Barrett[10] suggests that the sandblasting is vital to ensure a good "keying-in" of PbO_2 to the Ti substrate, and that choice of correct sandblasting conditions is vital.

Preplating stages. Prior to electrodeposition, almost all procedures specify some means of removing the titanium oxide (although as seen above, its formation is also specified). U.K. Patent 1,373,611 specifies a removal by cathodic reduction of the oxide. In theory this is convenient in that it requires no separate stage but simply a reversal of electrode polarity, although subsequent work shows H_2SO_4 is the best cathodization solution. Most other sources specify an etch in HCl, HF, NaF or oxalic acid. D.O.S. 2,012,506 specifies 10 min at 25°C of a 0·5 M HF/4 M NaF solution, another etch being specified in Belgian Patent 702,806. A completely different idea is proposed by Wabner et al.[11] They suggest that the deeper scratches which can be seen under the microscope after mechanical treatment act as latent fault sites, even though they appear to be evenly covered over by the lead dioxide. After many hours operation, these spring open again. Carborundum inclusions, and hollows with covered patches of oxide, are other sources of failure. These workers etch for 1 h in boiling oxalic acid (15%). After a short time, when the oxide has been etched away, a violent hydrogen evolution is seen, the metal dissolving to form a reddish-brown titanium oxalato(III) complex. Figures 7 and 8 show S.E.M. photographs illustrating the "stormy-sea" effect of this process. In the next stage (to which the authors state they attach the greatest importance), the electrode, after washing, is boiled in Ti(IV) oxalate and oxalic acid (0·2 M and 1·25 M, respectively). The authors state that neither with the naked eye nor with the S.E.M. does this surface show any difference from the one seen after the previous etch. The effect is only seen after plating. The whole idea is taken from D.O.S. 2,306,957 and it is worth quoting what the authors themselves stress,[11] namely, that if the PbO_2 deposit is chemically stripped from a Ti substrate so treated, the "effect" remains and a fresh PbO_2 deposit on that surface constitutes an anode as good as the original one. Whether this argues in favour of some compound or species formed on the surface, or simply reflects the physical state of subdivision of the surface, is not known. We would opt for the latter suggestion. In a somewhat similar vein, D.O.S. 2,023,292 suggests an intermediate firing process of the Ti following the application of an aqueous solution of $CrCl_3$ or similar compounds which are listed there.

Wabner[41] reports many experiments aimed at illustrating the effect that the Ti(IV) pretreatment has on the substrate. For example, even at 60°C, he states that TiO_2 formation is inhibited in air, in contrast to the untreated metal. By means of S.E.M. photographs, he shows how the PbO_2 deposits more evenly on treated Ti instead of forming isolated clumps, and it is suggested that PbO_2 growth into fissures is likewise facilitated, such fissures in their uncovered state being postulated as future electrode-failure sites. Time–potential plots of Ti in the etch bath and in $PbNO_3$ plating baths all reveal differences in behaviour, and Wabner quotes the work of Thomas and Nobe[45] as well as that of Goldberg and Parry[46] in showing that Ti(IV) ions inhibit corrosion of the metal apparently by facilitating passivation. This work is possibly the source of inspiration underlying the Ti(IV) pretreatment concept.

It is impossible to condense here the extensive work of Wabner, much of which has been published. We do not believe that any organic compound could survive prolonged contact with PbO_2 without itself undergoing oxidation and that if the "magic" of the Ti(IV) treatment is to be explained, it lies either in the subsequent decomposition of the oxalato or perchlorate to form a stable inorganic species such as the titanoplumbate, or in some effect on the morphology of the surface. In so far as the performance of such treated anodes can be matched in every way, as far as can be ascertained, by well-prepared Ti alone, we opt for the latter suggestion. D.O.S. 2,344,645, which again raises problems of electrode failure after operation at high anodic potentials or long terms, seeks to avoid this failure (due to oxide passivation on the Ti) by creating an interlayer. Plasma-arc spraying of 0·15-cm thick layers of carbides or borides of Ta or Ti gives a conducting layer which inhibits oxide formation. A grain size of 40–90 μm is suggested. Such a suggestion recalls earlier ideas of Beer based on Ti anodes with a nitride coating (D.A.S. 1,421,047), although in this case the coating was intended to be the outer one in contact with solution.

The electrodeposition process. Lead dioxide may be anodically deposited from a variety of baths. One of the best critical evaluations of these comes from the work of Hampson and Bushrod.[12] Since commercially viable depositions must operate at reasonable current densities, Hampson *et al.* excluded baths based on fluoborates, plumbites and silicofluorides, since at current densities above 5 mA cm^{-2}, these gave highly stressed, flaky and poorly adherent deposits. Nitrate and perchlorate salts of lead alone gave satisfactory deposits. With the latter, solutions of 2 M lead perchlorate and 1 M acid gave good deposits at up to 50 mA cm^{-2}, provided the pH did not fall below 0. Acidity was controlled by circulation of the solution through a bed of PbO. Hampson also reports that lead growths on the cathode were a problem and that this was minimized by using the lowest possible current density on the cathode, i.e. having a larger cathode:anode area ratio. Current efficiencies were approx. 100%. In almost all patents, the nitrate bath is preferred, possibly because of the explosive hazards of perchloric acid. Other workers specify pH 1·5–2·5 as the best condition.

Reporting on the nitrate bath, Hampson showed that the major phenomenon[12] in the nitrate bath was the build-up of nitrite, which, because it can be re-oxidized to nitrate at the anode, leads to loss in current efficiency. A graph shows that deposition efficiency ranges from 100% (nitrate absent) to less than 20% when 6% of nitrite is present. At higher current densities, the nitrite is further reduced to ammonium ions. The nitrite levels were held down by circulation of the electrolyte through Pb_3O_4. Alternatively, H_2O_2 may be added to remove NO_2^-. These authors report no effect of temperature in the range 20–40°C, although commercially, baths are operated at 60°C or higher. The U.S. Bureau of Mines[15] reports cracking when deposition takes place above 70°C.

Narasimham and Udupa[18] have studied the throwing power of the nitrate bath which they state passes through a maximum, but against what is unclear. The effect of nitric acid and surface active agents on throwing power was referred to previously. Oddly very few of the patents cited here refer to the stress which occurs in electrodeposited PbO_2. It is again Hampson and Bushrod[12] who study this, and their results are not easy to summarize. Simple nitrate baths gave compressively stressed deposits.

An equally thorough study of the effect of current density, temperature and organic additives on stress of deposited PbO_2 on nickel and graphite is due to Narasimham and Udupa.[13] In the later paper they explored the use of lead acetate as a buffer to relieve stress. Smith[14] also discusses the effect of deposition conditions on stress, higher stresses occurring in more dilute electrolytes. Current density below 70 mA cm^{-2} had little effect while well-stirred solutions gave stress-force deposits. Organics, notably sodium acetate, acted most effectively as stress relievers.

Indian workers with much experience in this field appear to lay special emphasis on CTAB (cetyl trimethylammonium bromide) for stress relief and a very perfect surface. This has been described,[32] as has its rate of loss[33] by oxidation etc.

Commercial baths operate at 60–70°C. Wabner[41] states that lower bath temperatures result in anodes giving higher overpotentials in subsequent use (for example, for oxygen evolution) and this probably results from the higher overpotentials obtained during the deposition process at lower temperatures and so there is thicker TiO_2 film formation. All contain approx. 200 g l^{-1} lead nitrate. Some have 10 ml conc. HNO_3 l^{-1}, others contain copper and/or nickel salts as described earlier. De Nora (U.K. Patent 1,192,344) operates at 300 g l^{-1} lead nitrate and 3 g l^{-1} copper salts with 0·92 g l^{-1} "Tergitol" (a sodium alkyl sulphate), and uses 250 A dm^{-2} current density. Bath compositions designed to produce correctly stressed anodes (for primary batteries) are discussed in U.K. Patent 1,340,914. It is pointed out that stress can be relieved by heating the finished electrode in air and allowing it to cool slowly. The 360 g l^{-1} nitrate solution is unbuffered at pH 3–4 and no other bath constituents are used except in one example where a two-stage deposit was built up, the inner one as above and the outer one with a lead acetate/sodium nitrate bath. (It should be pointed out that both this patent and Hampson's work cited above relate to depositions on nickel. However, the findings are probably of general applicability.) U.K. Patent 1,373,611 uses a bath with additions of Cu, Ni salts, free nitric acid, sodium fluoride and heptafluobutanol operating at 70°C, and uses a two-step deposition rate, the first 20 mA cm^{-2}, the second 60 mA cm^{-2}.

Among other ideas and suggestions we can include (frequently unspecified benefits) are D.O.S. 2,012,506 (effects of 100 g l^{-1} urea addition), D.O.S. 2,306,957 (the laying down of two discrete PbO_2 layers, the outer one harder than the inner one), the effect of current interruption on PbO_2 structure, and the use of glass beads in the plating bath to avoid pitting due to oxygen bubble entrapment.[15] The importance of stirring or agitating the bath is universally stressed. One paper uses a fluidized (silica) bed to improve quality and uniformity of deposits.[6]

In spite of the widespread use of nitrate baths, advocates exist for other baths, for example, Lazarev[34] finds $H_2SiF_6 + CH_3COOH$ as good as the nitrate bath; a similar bath has been advocated, while sulphamate baths have also been patented.[36]

Miscellaneous work on deposition. Ghosh[16] has studied the effect of various levels of superimposed a.c. on the d.c. used for electrodeposition of the lead dioxide. Certain differences are noted in the case of deposition of α-forms but no significant changes are seen during deposition of the β-form. A series of Soviet papers describe lead dioxide deposition from trilonate (EDTA) baths.[17] However, no indication is given as to the benefits so obtained, although the work is thorough in respect of stress measurement etc.

Narasimham et al.[29] have studied the effect of cathode geometry on the mode of deposition (mainly thickness) of the PbO_2 at the anode. Wabner[41] maintains that deposition using ultrasonic probes results in a 200–300 mV lower overvoltage for oxygen evolution when the electrode is in service.

Glassy PbO₂. Wabner[41] has reported the formation of a "glassy" variation of PbO_2. It is formed by deposition from a conventional nitrate type bath to which has been added some Ti(IV). The precise formula is given as follows: 2·5 ml of $TiCl_4$ are hydrolysed in 300 ml distilled water, neutralized with NH_3 and the hydroxide/oxihydrate so formed is well washed with water. It is then redissolved in HNO_3 (10 ml) and allowed to stand at 20°C overnight. Finally 66 g of lead nitrate are added, the solution is made up to 200 ml (corresponding to 1 N lead nitrate) and clarified with active charcoal. The deposition takes place at 40 mA cm^{-2} and at 20°C.

It was shown that the glassy deposit was β-PbO_2 of the "Plattnerite" form (in fact without any of the traces of α-PbO_2 normally found) and that the grain size was approx. 600 Å in contrast to the 6000 Å of normally prepared deposits. Thus methods such as S.E.M. failed to resolve any structure in such deposits at × 10,000 magnification. These are highly stressed and, after standing in water, or presumably in service, the morphology changes and visible crystalline structure begin to develop. The alkaline "tempering" process described elsewhere also has the same effect after 12 h.

This interesting form of the dioxide may well find certain uses in analytical applications—for example, where its low specific surface area gives the benefits normally associated with polished and smooth metals, and where its apparent lack of long-term stability is less important.

Post-treatment of anodes. Wabner and Fritz[11] describe an electrochemical post-treatment consisting of cyclic voltammetry between +700 mV and +2280 mV. The effects of this are apparent to the naked eye (colour change from light grey to brownish-black) while an S.E.M. examination (as shown in this work) depicts a change from smooth rounded contours of the immediately formed anode to the mass of needles after the formation. A further cycling, this time between the more anodic limits of 1650 mV to 2350 mV is stated to give a more reproducible electrode for kinetic measurements.

We have already described the effects of heating electrodes in air to remove strain in the deposit. A similar proposal is found in D.O.S. 2,306,957 where the coated anode is suspended in weak hot alkali for a period of time.

Regeneration of anodes. As they become more widely used, methods of regenerating older anodes, removing the coating and corrosion products and redepositing fresh PbO_2 will become available. To date, the only reference appears to be a Japanese procedure[39] for boiling the anodes in nitric and oxalic acid mixtures.

The form of the anode. The traditional graphite-based anodes were and still are rods of 2–10 cm diameter. Adhesion of the coating to these was usually good. The advent of metallic substrates posed problems. Attempts to deposit the dioxide on continuous expanses of flat section lead to progressive failure, even when the initial defect is only localized. Expanded mesh has proved to be an ideal form on which to coat PbO_2 while another idea described in a U.S. Bureau of Mines report[15] and in D.O.S. 2,012,506 is to drill holes in flat sheet at regular intervals. The lead dioxide, with its good throwing power, deposits into and behind these holes and anchorage is thereby provided. Another comment made by several workers is the importance of avoiding sharp edges and corners on the form to be coated. If these are present, nodular growths of lead dioxide will form there.

Deposition of α-PbO_2

The foregoing description of anode preparation relate to coatings which consist predominantly of β-phase PbO_2. Little has been written concerning the deposition or behaviour of the α-phase although it is known[25] to exhibit a lower oxygen overvoltage. Ruetschi[25] describes the deposition of α-PbO_2 on Pt. A detailed description of an α-PbO_2

deposition bath is also given by Grigger[19] using alkaline lead tartrate. However, the work reported by Grigger[19] in fact relates to nitrate (e.g. β-phase) deposits. Lartey[26] describes the formation of a very ductile but thin coating of α-PbO$_2$, which is of good adherence and is black, shiny and "slippery", apparently having been deposited in laminae which flake off when a given thickness is exceeded. The Faradaic efficiency of the deposition, even at 5 mA cm^{-2} was c. 30%. Carr and Hampson[27] have also reviewed baths for α-phase deposition while Issa *et al.*[28] have prepared deposits of mixed α- and β-PbO$_2$ (92%, 80% and 60% α-phase) using formaldehyde as a reductant in the deposition bath. The study of Gancy[30] covers both Faradaic efficiency and also stoichiometry, as well as other details. Clarke[31] states that during or after PbO$_2$ deposition from alkaline baths, some sort of luminescence is observed and that deposits exhibit a high degree of preferred orientation. Recently, Soviet workers have patented a bath based on acetone (6–8 g l^{-1}), urea (3–5 g l^{-1}) and lead acetate (7–13 g l^{-1}) in a 30–50 g l^{-1} NaOH solution.[37]

Scientific Studies of Lead Dioxide on Inert Basis Anode

In spite of the many recipes for their preparation, few studies of the completed article exist. Largely for this reason, it is very difficult to evaluate the comparative merits of the various means of preparation.

Metallographic (S.E.M.) studies of surface textures are reported by Wabner[7,11,41] and in the U.S. Bureau of Mines study;[15] X-ray analyses of composition (85% β-phase, 15% α-phase) are cited in Belgian Patent 702,806 and also by Wabner and Fritz.[11] Rusin[51] shows how the $\alpha:\beta$ ratios vary with PbNO$_3$ concentration and current density. A very fine study of the kinetics of deposition[15] of α-PbO$_2$ is reported by Fleischmann and Liler[20] and this is only one of a series of fundamental studies devoted to the electrocrystallization process by Fleischmann.[22,23] A more recent kinetic study of the electrodeposition and cathodic reduction of PbO$_2$ on conductive tin oxide is due to Laitinen and Watkins.[24] The best physical study of (electrical) properties of the PbO$_2$ is probably that of Mindt.[21] Among his more important findings are the increase in resistivity with time (over 12 days) and the effect of loss of oxygen on this. Also the fact that resistivity of the α-phase is greater than that of the β-phase, by an order of magnitude ($\rho_\beta = 10^{-2}$ Ω cm,

$\rho_\alpha = 10^{-3} \, \Omega$ cm). In terms of the composite PbO_2–TiO_2 anode, Wabner[11] points out that β-PbO_2 and TiO_2 are isomorphous (rutile), differing only by 7% in lattice size and that formation of a mixed crystal might be possible. This can be considered both in the context of the Pb–Ti interface and the binding processes there and also in the context of a "modified" $PbO_2 + TiO_2$ anode along the lines mentioned above. However, such an electrode could not, one feels, be made purely by electrodeposition processes.

Wabner[11] also speculates as to the possible existence of other compounds in the system. He mentions lead titanates or titanoplumbates, and the importance of such species, which have been referred to in the literature,[42] remains to be resolved.* He quotes the work of Hauffe et al.[47] who show that the electrical conductance of TiO_2 can be readily increased by several orders of magnitude when certain other oxides are added and this fact again may or may not be relevant to making better composite anodes. A range of TiO_x compositions can also exist with differing electrical properties.[52] Scientifically as opposed to technologically, Wabner has published more than any other worker on the PbO_2/Ti composite and his writings therefore deserve careful analysis. His use of a.c. techniques to investigate processes occurring on PbO_2/Ti anodes has been criticized (see Chapter 9) and Canagaratna and Hampson[49] illustrate some of the pitfalls of this technique.

The basis of his "oxalato" pretreatment is hard to understand. Possibly it does create on the surface of the Ti a monolayer or so of a species which impedes the oxide passivation which might otherwise take place when the anodic deposition occurs. Randle,[43] measuring the potential of the Ti anodes during conventional deposition from a nitrate bath onto simply etched (but otherwise untreated) Ti, notes that when the potential of deposition is greater than 2.0 V vs N.H.E., the resulting electrode is useless or has at best a life of a few hours before failure due to high electrical resistance occurs. Wabner points out that electrodes pretreated with the oxalato method retain their special advantages even when the PbO_2 is stripped off after some considerable time and redeposited. In view of the extremely strong oxidizing properties of PbO_2, it seems inconceivable that any oxalate remains unoxidized after being so coated. Wabner believes that attempts to obtain current–potential data with PbO_2/Ti anodes are useless because

* See the General Bibliography chapter.

of the large (and possibly indeterminate) iR drop across the PbO_2/Ti interface which he finds impossible to separate from the overall iR drop. This may well be so.

The main problem with Wabner's findings is the weight of evidence to contradict them or at least to restrict their validity. In fact, anodes manufactured in the first commercially successful place,[44] based on PbO_2/Ti with no interlayer or special pretreatment such as he describes, have been operating in hundreds of installations in the U.K. and elsewhere (mainly in electroflotation plants) with a proven lifetime of 2 years or longer. In further apparent contradiction to the assertions of Wabner,[44] the anodes made and sold by Messrs Morgett Electrochemicals Ltd. are frequently stored in air for extended periods with no known deleterious effects on their lifetime, instead of leading to the early passivation failure he suggests would take place. It is not possible here to state whether Wabner's assertions stem from experiences with poorly prepared anodes or whether the passivation failures he describes occur sooner in his solutions (2 N H_2SO_4) than they do in the chloride-containing media in which commercial anodes frequently operate. Certainly passivation would occur more easily in the former than in the latter solutions. Wabner's other suggestion that 100 mA cm^{-2} is the maximum permissible current rating on account of temperature rise from ohmic heating in the interlayer region also appears to be dubious. Electrodes have been operated by us and others for 20–30 days in chloride media at 400 mA cm^{-2} without showing the voltage increase he describes. In conclusion, the Ti(IV) treatment does appear to reduce the rate of Ti passivation, but using properly prepared electrodes, such a procedure seems unnecessary. A discussion of corrosion of composite anodes is given on p. 299.

Future of the PbO$_2$/Metal Anode

There seems little doubt that the PbO_2/metal (especially Ti) anode has an important future in the electrochemical process industry. Experience in our own laboratories and elsewhere has shown that a considerable degree of "know-how" is required before an anode capable of operating successfully over a long period can be made. Even now, it is not clear what limits the lifetime of anodes, whether build-up of highly resistive films or simple loss of PbO_2 are the main problems. From the large and diverse "recipes" for their manufacture, it might well be deduced that none of the elaborate "inter-layers" or pre/post-treatments are really necessary. Future work will probably therefore proceed in two directions. These are:

(i) Study of the importance (or otherwise) of interlayers either metallic or of TiN or similar compounds.

(ii) Study of mixed oxide coatings PbO_2–MO_x.

In this respect, the history of the chlorine anode which has evolved from a RuO_2, through a RuO_2–TiO_2 to a RuO_2–TiO_2–SnO_2 electrocatalyst will undoubtedly be borne in mind. However, Mindt[21] argues that high concentrations of "dopants" will be needed to cause any significant effects. In this connection, for example, USSR Patent 495,714[40] describes the preparation of "resistive materials" based on PbO_2, Sb_2O_5 and other oxides. But knowing that, if well made, simple PbO_2/Ti anodes have already a lifetime of 2 or more years, any improvements will have to show marked benefits especially if they are difficult to carry out.

Morphology of the PbO_2 Anode

As should now be obvious, the morphology of the lead dioxide anode can assume a variety of forms when examined under the optical or scanning electron microscope. Dr D. Wabner has kindly provided a selection of micrographs which are shown on the following pages and which represent a typical range of PbO_2 surfaces.

Corrosion of PbO_2/Inert Substrate Anodes

We have seen elsewhere that the mode of corrosion of PbO_2/Pb anodes takes place basically not at the surface but rather at the Pb(II) interlayer between the metal and the fully oxidizing PbO_2. In the composite anode which forms the subject of this chapter, no such intermediate valence state exists (at least not as far as we know) and one would therefore expect such anodes to corrode more slowly, if at all. They do indeed show much higher corrosion resistance than their PbO_2/Pb analogues. A recent paper by Lartey[53] summarizes published data and shows new corrosion information on these composites. The findings of Lartey are

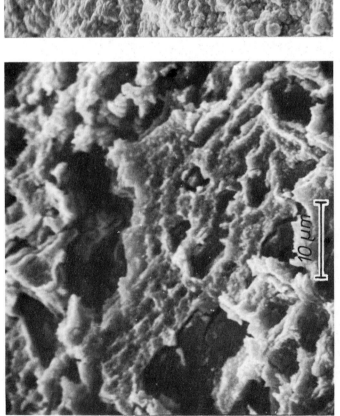

Fig. 1. S.E.M. photo of a Pb anode pickled in 20% HNO_3 and etched in 20% acetic acid with trace of $NaNO_3$. PbO_2 developed following 2 h at 10 mA cm^{-2} in 1 M H_2SO_4. ×2000.

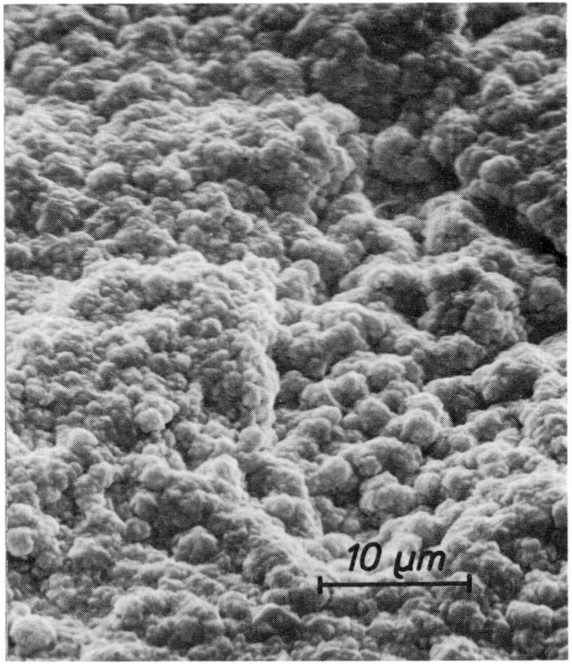

Fig. 2. S.E.M. of an "amorphous" PbO_2 deposit (0·2 M Pb amidosulphate solution deposited at 15 mA cm^{-2}). ×2000.

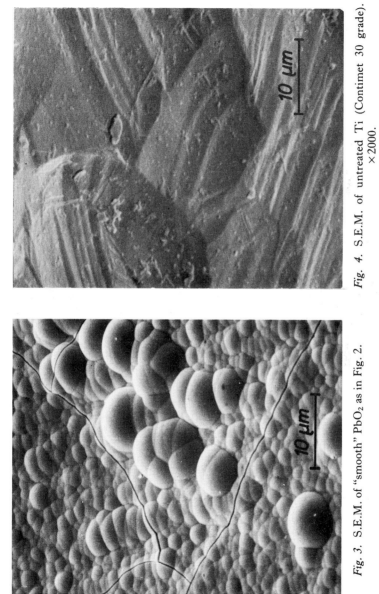

Fig. 3. S.E.M. of "smooth" PbO$_2$ as in Fig. 2.

Fig. 4. S.E.M. of untreated Ti (Contimet 30 grade). ×2000.

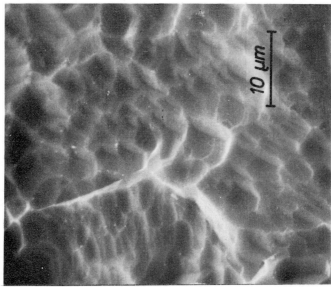

Fig. 6. S.E.M. of a smoothly etched Ti surface (etch was 20% HCl + 20% HCOOH at boiling point, with subsequent smoothing stage in 4% HF + 3% H_2O_2 at 20°C for 1 min). ×2000.

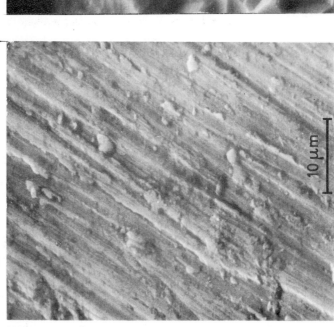

Fig. 5. S.E.M. as above, but after degreasing and polishing with 400 grade corundum paper.

Marinina[26] studied Pb–Ag anodes in an HF–H$_2$SO$_4$ mixture which was designed to leach out niobium or tantalum oxides. A film of PbF$_2$ was found on the anode, and the presence of the minerals (e.g. tantalite) was found to depolarize the anode.

Chloride ions

Andersen[1] states that "Small amounts of Cl$^-$ and NO$_3^-$ in the Chuqui-camata electrolyte made unusable C.P. lead anodes which were serviceable in other conditions". A level of $0 \cdot 1 \text{ g l}^{-1}$ of Cl$^-$ ions is commonly found in electrowinning circuits and must be presumed to have a slightly deleterious effect. Some "process-anode" data are deliberately obtained in solutions of this composition in order to obtain meaningful data. Stender and Seerak[7] suggest that $0 \cdot 15 \text{ g l}^{-1}$ marks a "threshold" concentration. They also study solutions with up to $1 \cdot 5 \text{ g l}^{-1}$ of chloride. Table I is reproduced from their work (see also refs 31, 32, 33).

Table I. Anodic corrosion of lead and lead alloys in the electrolysis of aqueous H$_2$SO$_4$ solutions.

Current Density (A m^{-2})	1000	500	250	100	50	
Anode material	Destruction as calculated for Pb(g(100 A h)$^{-1}$)					Electrolyte
Ordinary lead	0·5483	0·9297	1·1520	0·7148	0·3835	H$_2$SO$_4$—220 g l^{-1}
Ordinary lead	—	1·1763	4·1835	13·8877	13·6065	H$_2$SO$_4$—220 g l^{-1} Cl$^-$—1 g l^{-1}
Ordinary lead	1·9160	1·7550	1·2981	1·2112	—	H$_2$SO$_4$—220 g l^{-1} Cl$^-$—0·1 g l^{-1}
Alloy: Pb + 6% Sb	2·4512	2·3174	1·0769	0·2068	—	H$_2$SO$_4$—220 g l^{-1}
Alloy: Pb + 11% Sb	1·1930	1·9565	1·1875	0·7540	—	H$_2$SO$_4$—220 g l^{-1}
Alloy: Pb + 1% Ag	0·0444	0·0783	0·0945	—	—	H$_2$SO$_4$—220 g l^{-1}
Alloy: Pb + 1·1% Ag	0·1261	0·377	0·945	—	—	H$_2$SO$_4$—220 g l^{-1} Cl$^-$—1·5 g l^{-1}

The same authors in a later work[8] examine a range of alloys in 0, 100 and 500 mg l^{-1} chloride solutions. Apart from one odd figure (which may well be a misprint), it is indeed seen that the corrosion escalates at the highest chloride levels. The best anode is the ternary 0·5% Ag and 1·5% Tl, while data for pure lead are not quoted. Studies at $3 \cdot 5 \text{ g l}^{-1}$ Cl$^-$ and above are reported in Bray and Morral.[9] The authors evolved a Pb–Ca–Te–Tl quaternary alloy which could operate under these conditions.

Metal cations and anions

Certain metallic impurities are found not to affect the rate of anode corrosion. Andersen's paper is specially informative here and Table II is reproduced from his paper.[1,25]

Table II. Chemical species having little or no effect on lead anode corrosion.

Cu^{2+}	0–40 g l^{-1}	Na_2SiO_3	Saturated
Al^{3+}	12 g l^{-1}	Na_2SO_4	Saturated
Mg^{2+}	8 g l^{-1}	Cl^-	0·1 g l^{-1}
Ca^{2+}	Saturated	Animal glue	70 mg l^{-1} day^{-1}

Conditions: 50–275 g l^{-1} H_2SO_4, 28–50°C, 8–40 mA cm^{-2}.

That copper has no effect is stated by Ravindran[10] who also confirms that cobalt reduces corrosion. Neither of these two facts can readily be ascertained from his paper, however, and the rate of corrosion as shown in his Fig. 3 (added Co) appears to be as fast as that shown in his Fig. 2 (no Co). Ravindran's statement that "the effect of acidity . . . is not very pronounced" runs counter to the findings of almost all other workers, and one must thus have reservations as to the accuracy of this paper, which also reports X-ray studies of corrosion products. Other metals, anionic or cationic, however, do affect the rate of corrosion.[25] Once

Table III. Inhibiting effects of Fe, Ni and Co on the corrosion of 10% Sb–0·4% As–Pb in H_2SO_4 solutions.
(Current density = 258 A m^{-2}; $T = 28$°C)

	Corrosion Rate (mm year^{-1})	
Species in Solution	180 g l^{-1} H_2SO_4	50 g l^{-1} H_2SO_4
H_2SO_4 only or H_2SO_4 + $CuSO_4$	2·6	1·3
Fe, 0·5 g l^{-1}	2·2	
Fe, 5 g l^{-1}	1·6	
Fe, 10 g l^{-1}	0·9	0·9
Fe, 10 g l^{-1} (100% Fe^{3+})^a	1·0	
Fe, 10 g l^{-1} (8 g l^{-1} Mg + 12 g l^{-1} Al)	1·9	0·5
Ni, 8 g l^{-1}	0·3	
Ni, 3 g l^{-1}	2·9	
Co, 0·05 g l^{-1}	0·07	

^{a} Fe maintained in the +3 state by separating anode and cathode with a porous diaphragm.

again, Andersen[1] has admirably summarized the situation as shown in Table III. (All of Andersen's results relate to a 0·4% As, 10% Sb ternary alloy.) Andersen points out the very low concentrations at which Co effects a corrosion inhibition, in contrast to the other transition metals examined. Aluminium salts give rise to increase in viscosity and this is suggested as the reason for their apparent mitigation of the protective effect of iron. The same is suggested (to a smaller extent) for magnesium salts. The findings tabulated above are largely not novel, but confirmations of long-established facts. The results of Rey and Coheur[11] are reported in Table IV. The same authors also found that pretreatment of lead anodes in cobalt-containing solutions gave subsequent protection in cobalt-free electrolyte. Quoting one example, a 3-h period of electrolysis in cobalt solution at 300 A m^{-2} gave 250 h protection in subsequent electrolysis, i.e. the corrosion rate was reduced in line with Table IV.

Table IV. Effect of cobalt addition to anode corrosion. Current density $= 3 \text{ A dm}^{-2}$.

Co^{2+} (mg l^{-1})	0	7·5	22·5	68
Anode wt loss (mg)	700	9	−23	−28
PbO_2 (slimes) recovered	910	Trace	0	0

There are numerous other publications giving similar data for Co and, for example, for Ag. Zuborovskii and Stender[12] suggest that silver alloyed with lead is beneficial because it provides Ag^+ ions in solution. (Note that Fink gives a very different explanation for chloride media on p. 391.) They examine the corrosion rate of lead and 1% Ag alloy anodes in sulphuric acid, to which varying amounts of Ag^+ or Co^{2+} have been added. They show that increasing Ag^+ concentration up to 3 mg l^{-1} slows down corrosion. Beyond this there is no further effect and they refer to their own results in which anodes with greater than 1% Ag likewise gave no better protection. Quoting Kiseleva and Kabanov,[13] they state that in presence of 10 mg l^{-1} Ag^+, the lead anode is more stable than the 1% Ag alloy—this is surprising. Again this calls to mind the effect shown in chromic acid solutions (p. 406), but runs counter to the work of Fink on p. 390. As in the work of Coheur and Rey,[11] anodes prepolarized in Ag^+-containing solutions have been found to retain a high degree of subsequent corrosion protection (in this case for 80 h).[14] A solution content of 3 mg l^{-1} is stated to give better protection than that found by adding 1% Ag to the bulk metal.

The same workers discuss Co as an additive, finding that protection increases sharply up to $3 \, mg \, l^{-1}$, but only slowly thereafter. Again confirming Andersen, they mention that Ni has a small effect and Fe not much greater. Mn has no effect at all.

Just how these additives work is a matter of speculation. However, it is a fact that oxygen overvoltage is reduced in both cases (see p. 227), and whether this is a truly catalytic effect, or whether Co^{3+} or Ag^{2+} ions are formed (the former one at least is capable of decomposing water to form the divalent ion again, thereby providing an alternative pathway to oxygen evolution) remains uncertain.

An earlier paper reaching very similar conclusions to both Andersen and Zuborovskii and others is that of Koenig et al.[15] These workers studied anode potentials of pure lead in solutions containing: 0–20 (Co), 0–100 (Fe), 0–200 (Mn), 0–50 (Ni), 0–50 (Cr), and 0–100 (both Na and Mg) $mg \, l^{-1}$. At the same time, they analysed the Zn cathodes formed in the electrolyses for Pb content, thus providing a good index of corrosion at the anode.

Na and Mg exerted no effects; Cr at $50 \, mg \, l^{-1}$ lowered the anode potential by 16 mV. Mn showed no effects, nor did Ni. Fe at $100 \, mg \, l^{-1}$ lowered the anode potential by 65 mV and corrosion (as measured) was reduced to one-third. Cobalt, as found by all other workers, was most dramatic, giving a 117 mV potential drop at $20 \, mg \, l^{-1}$, and reducing corrosion to less than 10% of the control level, though the "post-treatment" benefit of Co decayed somewhat more rapidly than that found by Coheur and Rey. In apparent contradiction to the Mn results cited above, Larsen[16] quotes Hanley[17] where $600 \, mg \, l^{-1}$ Mn was found to have halved the amount of Pb found in the cathode. In the limit, one would expect an MnO_2-coated anode to result when sufficiently high Mn concentrations obtain, and further confirmation that Mn retards corrosion comes from Ravindran.[18] A concentration of $4 \, g \, l^{-1}$ Mn reduces corrosion to approximately 15% of its value in the absence of Mn. Ravindran also reports on the interaction between Mn and Co ions in the same solution, which is obviously complex. Co ions appear to inhibit oxidation and deposition of MnO_2, and hence its corrosion-inhibiting action.

A very full study was carried out by the U.S. Bureau of Mines,[25] which followed on earlier work of Bradt and Oaks[23] and Fink and Kolodney,[24] the latter selecting a 30–50% Sn and 0·3–0·4% Co alloy for Mn electrowinning. Such alloys are, however, too costly to be useful.

The U.S. Bureau of Mines study was based on a solution with 30–35 g l⁻¹ Mn plus 125 g l⁻¹ ammonium sulphate. Some 20 alloys were tested as anodes and rated on the basis of corrosion, rate of oxidation of Mn, cell voltage and current efficiency. With respect to corrosion, Pb–4% Tl, Pb–2% As and Pb–0·2% Te scored better than others. Whether one can extrapolate these results to any other situations than this, in which substantial accretions of MnO_2 grow on the anode, is extremely doubtful, and the 1% Ag alloy, which is normally a good standard by which to judge, came well down the list.

The final section is devoted to a study of microstructure of alloys and the effect on corrosion rates. Selecting the best, an average and the worst anodes in the test, photomicrographs showed that all consisted of primary crystals of Pb surrounded by small amounts of eutectic, with no microstructural differences. On this basis, and in a situation, as we have noted, which should probably not be extrapolated, the authors conclude that there is no effect deriving from this parameter.

Fukushima[27] finds that small amounts of Mn (his solutions went up to 8 g l⁻¹ Mn) lead to less corrosion than larger concentrations, and that the scale so formed is less prone to fall off, thereby exposing the lead or alloy.

Another paper on cobalt additions in relation to corrosion is that of Kripolova *et al.*,[19] and a more recĕnt one again confirming earlier work is due to Gendron.[20] The cobalt work has found at least one practical outlet as reported by Andersen[1] in a plant where it is added to a liquor circuit as corrosion inhibitor.

Zhurin[28] again confirms that Mn up to 3 g l⁻¹ reduces corrosion and that Zn^{2+} ions have no effect. The corrosion rate of the 1% Ag alloy is four times less that of pure Pb. Kharabadze[29] investigates the effect of Mn^{2+} ions in 15–20 N sulphuric acid using a rotating cylinder electrode. Under these extreme conditions, its protective effect appears to be less marked than in more dilute acids. Antonov[30] investigates the effect of Co^{2+} ions. Like other workers, he finds that oxygen overvoltage decreases, as does corrosion and also "charge capacity". This falls in line with the suggestion (p. 227) that a thinner, tighter oxide is formed in the presence of cobalt. Smialowski[31] investigates "time to passivate" and the effect of Co^{2+} ions, while Dunaev and Kiryakov (p. 228) suggest that Tl ions reduce corrosion.

Zuborovskii[41] points out that although Ag^+, Co^{2+} and Fe^{2+} all repress corrosion, when added to the lead as alloy constituents they dissolve out. This would seem to be (i) not wholly true, and (ii) not relevant (*see also* Bryntseva[43] and a paper by Astakhov,[57] who reports X-ray studies of PbO$_2$ films formed in Co^{2+}-containing solution from 25–65°C at various pH values). Behaviour of Pb–Tl alloys (with discussion of others) in acid copper sulphate with 10 g l^{-1} NO$_3^-$ and 1 g l^{-1} Cl$^-$ was reported by Fink and Eldridge.[65] Addition of approx. 3% Tl reduces corrosion three-fold, and further additions reduces it to a sixth of the Pb value.

Organic additives

There is no doubt that many, if not all, organic compounds when present in solution exert a deleterious effect on lead anodes. Thus, while it has[21] been suggested that glue, an additive commonly used in electrowinning, reduces the amount of Pb "picked up" in the cathode, it seems likely that the explanation of this lies in the greater viscosity of glue-containing solutions and thus the greater difficulty of Pb reaching the cathode[22] either in the divalent state or as flocculated PbO$_2$. Whatever the correct explanation, it seems unwise to compare work using glue with that which omits it. Papers relating to this field include Dashiani[34] (various alloys with sulphuric and tartaric acids) and Gratsianski[35] (citric with sulphuric, Pb–In). Corrosion in 60% sulphuric acid is less than that in 10–35%. Kiryakov and Dunaev[36,42] suggest that surfactants can exert a profound effect (Pb, Pb–Tl, Pb–Ag and Pb–Tl–Co alloys). Lukin[37] studied sulphuric acids up to 16 M. Surfactants (or, for example, 15 mg l^{-1} gelatin) affect the corrosion rate when the acid is dilute, but not at higher strengths. In another paper, Kiryakov[38] uses 10^{-3} mol l^{-1} surfactant. This can cause a 300% increase in the rate of Pb corrosion, or a 30% increase for the 1% Ag alloy. The surface oxide is thinner and oxygen overvoltage increases in the presence of surfactants. Investigating 25% acid with 3% organic residues, Ivascanu[39] finds Pb "satisfactory" at 20–60°C. Eggett and Naden[40] discuss problems arising when metals are electrowon following solvent-extraction and again suggest that anode corrosion is aggravated as a result. Accelerated ageing of anodes before use, and Pb–0·1% Ca, Pb–Tl or Pb–Ca–Sn alloys are recommended.

Miscellaneous

Gonzalez[44] reports the effect of Cl^- ions $(0 \cdot 10 \text{ g l}^{-1})$, SO_2 (19 g l^{-1}) and both together. SO_2 reduces the corrosive effect of Cl^- ions. The effect of oxygen is uncertain (see p. 389). There are suggestions that a complex is formed (as with chlorides).

Table V. Summary table of corrosion of alloy anodes in sulphuric acid. PD = potentiodynamic, RT = room temperature, PS = potentiostatic.

Composition	Conditions	Ranking	Remarks	Reference No.
0·07–0·1% Ca	"Battery Float"	1	Grid growth measurement	67
12% Sb		2		67
0·1–0·14% Ca		3		67
Pb	7 N	1	Galv. i–V O_2; to 1·75 V	68
2·2% Sb–0·11% As		2 =		68
w/w.o. annealing				
2·9% Sb	5 g l^{-1}	—	PD	69
10% Sb	2·5 and 5 g l^{-1}	1	PD	69
6% Sb		2	Wet and dry	
3% Sb		3	polish sample	
9% Sn	2·5 and 5 g l^{-1}	1 =	PD	69
6% Sn		1 =		
3% Sn		1 =		
Pb	1·25 sp.gr., 25 days	3 =	Wt loss at 3 mA cm^{-2}	70
6% Sb		4		
0·05% Ca		2		
0·05% Ca–0·1% Ag		3 =		
0·1% Ag		1		
Pb–Ca–Sn	"Battery"		Wt loss	71
(a) Sn = 1·3–1·4%				
+ Ca = 0·04%		1 =		
0·08%		1 =		
0·11%		2		
0·04% Ca		3		
(b) Ca = 0·06–0·07%	"Battery"			
+ Sn = 0·41%		5 =	Wt loss	71
0·9%		3		
1·4%		2 =		
1·7%		1		
2·2%		2 =		
2·7%		4		
3·0%		5 =		
0·07% Ca		6		

(continued)

Table V. (continued)

Composition	Conditions	Ranking	Remarks	Reference No.
1·43% Sn		1 =	71	
0·08% Ca–1·38% Sn		1 =		
Pb		6		71
3% Sb		3		
7% Sb		4		
0·05% Ca		5 =		
0·1% Ca		5 =		
0·1% Ca–1·5% Sn		2		
0·8% Ca–1·3% Sn		1		
0·01% Ca	1·115 sp.gr.,	3 =	PS. Samples	72
0·03% Ca	2·8 V	2	cast at 190°C	
0·06% Ca	microstructure	1	corrode more	
0·08% Ca	data	3 =	than those	
0·09% Ca		4 =	cast at 260°C.	
4% Sn	"Battery	1		73
Pb	cycling"	2		
6% Sb	10–40% acid	3		
Pb	0·84 mA cm^{-2} and	3	Wt loss with	74
12% Sb	0·14 mA cm^{-2}	5	coarse and fine	
6% Sb	56 days	4	structure alloys	
0·46% Ag		1	(ranking for coarse)	
0·1% Ca		2		
Pb	10 N, 3 N, 1 N, 0·1 N	3	Same ranking in	75
0·2% Cd		2	all acid strengths	
0·4% Cd		1 =		
0·6% Cd		1 =		
0·8% Cd		1 =		
Pb	1·115 sp.gr.,	3 =	PS wt loss	76
4·5% Sb	(a) 82 mA cm^{-2}	4		
0·07% Ca	(b) 2·8 V for	2		
0·01% Li	4 weeks	1 =		
0·02% Li		1 =		
0·03% Li		3 =		
Pb	1·21 sp.gr.	5		77
6% Sb	1·05 V *vs*	7		
0·09% Ca	Hg$_2$SO$_4$	6		
0·15% Ca		8		
0·05% Ca		3		
0·05% Sb		4		
4·5% Sn + 0·05% Ca		1		
4·5% Sn + 2% Sb		2		
Pb	1·21 sp.gr.	2	5–25 h	77
4·5% Sn	Range of	1	test	
6% Sb	potentials	3		

Table V. (continued)

Composition	Conditions	Ranking	Remarks	Reference No.
Pb	Cycle	3		77
1% Sn		2		
5% Sn		1		
Pb		3	9 weeks	77
6% Sb	120°F	4		
4% Sn	cycle	1		
5% Sn		2		
			Recorded wt loss $(g\,cm^{-2})$	
Pb		4	0·041	77
2·4% Sn		3 =	0·027	
4·0% Sn	cycle	1 =	0·024	
5·5% Sn	5 weeks	1 =	0·024	
6·5% Sn	RT	2	0·026	
8·5% Sn		3 =	0·028	
6% Sb		5	0·045	
Pb 85–97·5%	Chuquicamata		Optimum is	78
Sb 2·5–14%	leach		94% Pb–6% Sb	
Ag 0–1%	liquor		corrosion c.d. independent $(45–195\,A\,m^{-2})$	
Pb (chemical)	173 A m^{-2}	5	4·2 mm year^{-1}	79
10% Sb, 0·4% As	Cu leach liquor	4 =	2·6	
10% Sn	30–100 days	3	1·8	
5% Sb		4 =	2·6	
2% Sb		4 =	2·6	
0·068% Te		2	1·6	
0·1% Ca		1	0·07	
Pb		11	3250	80
1% Ag		9 =	760	
0·5% Ag		10 =	880	
2% Tl	wt loss	10 =	804	
9% Tl	RT	7	410	
0·5% Ag, 0·2% Tl	500 h	8	620	
0·5% Ag, 0·5% Tl		9 =	760	
0·5% Ag, 1% Tl		3	80	
0·5% Ag, 1·5% Tl		4	90	
0·5% Ag, 2% Tl		2	45	
0·5% Ag, 9% Tl		1	30	
0·5% Ag, 0·2% Te		10 =	820	
0·5% Ag, 0·5% Te		10 =	820	
0·5% Ag, 2% Se		5	170	
0·5% Ag, 2% Se		10 =	810	
0·5% Ag, 0·5% Ca		6	330	

(continued)

Table V. (*continued*)

Composition	Conditions	Ranking	Remarks	Reference No.
4% Sb, 0·5% As, 0·3% Sn, 0·1% Ag		1		
6% Sb, 0·5% As, 0·3% Sn, 0·1% Ag		2	Best alloy is ×4 the worst (life)	
6% Sb, 0·5 As, 0·2% Sn		3		
6% Sb, 0·5% As, 1·7% Sn	S.A.E. overcharge test (3 mA cm^{-2})	4		81
3·5% Sb	4·5 M acid	5		
Pb		8		
6% Sb		6		
7% Sb		7		
0·01% As, 0–8% Sb	6 mA cm^{-2}	1	Optim = 1% Sb, 0·0% As	
0·2% As, 0–10% Sb	overcharge	2 =	= c. ×2 better	82
0·5% As, 0–10% Sb	*RT*, 100 h	2 =	than worst case	
0–2% As		4		
2% Sb, 0–2·5% As		3	Optim = 0% As = ×2	
3% Sb, 0–3% As	As above	2	worst case	82
4% Sb, 0–3% As		1		
Pb		5		
0·1% Ag	3·5 M acid	4		
0·4% Ag	6 mA cm^{-2}	3		83
1% Ag	*RT*	2		
5% Ag		1		
Pb, 0–3% Ag		1 =	Sb has little effect.	
Pb, 6% Sb, 0–3% Ag	As above	1 =	Alloys improve with	83
Pb, 10% Sb, 0–3% Ag		1 =	increased Ag or Tl	
Pb, 0–4% Tl		1 =	content. 4% Tl = 3% Ag	
Pb, 6% Sb, 0–1% Tl		1	Optimum Pb, 6% Sb, 1%	
Pb, 8% Sb, 0–1% Tl	As above	2	Tl, ×2 worse than	
Pb, 11% Sb, 0–1% Tl		3	Pb–4% Tl or above alloys	
Pb, 0·2–0·3% Tl, 0·5% As	As above		Optimum 0·5% As, 0·3% Tl, ×2 worse than 4% Tl	83
			Long-term test with 100s of samples in full-size plant	
0·6% Tl, 0·3% Sb, 0·1% Ca	Zn leach liquor	4	1·26 mg A h^{-1}	
0·7% Tl, 0·5% Ag, 0·1% Ca		1	0·02 mg A h^{-1}	84
0·3% Sb, 0·3% Cu		6	2·63 mg A h^{-1}	

Table V. (continued)

Composition	Conditions	Ranking	Remarks	Reference No.
0·7% Tl, 0·02% Co, 0·3% Sb, 0·1% Ca		5	2·2 mg A h^{-1}	
0·2% Ag, 0·2% Sb		2	0·65 mg A h^{-1}	
1% Ag		3	0·33 mg A h^{-1}	
Pb	500 A m^{-2}	5	Potential measurements	
1% Ag	*RT*	1	also quoted	85
0·05% Co		3		
0·16% Co		2		
0·1% W		4 =		
0·1% Zr		4 =		
0·1% Mo		6		
0–100% Sb		3 =	Optimum at *c*. 4% Sb	
0–100% Bi	70 days	4	Optimum at *c*. 10% Bi	
0–100% Sn	1 M acid	3 =	Optimum at *c*. 5% Sn	86
0–100% Tl	50 mA cm^{-2}	2	Optimum at *c*. 10% Tl	
0–100% Ag		1	Optimum at *c*. 6% Ag	
1% Ag		5	With 100 mg l^{-1}	
1% Ag, 1% Se	1 M acid	4	Cl$^-$	
1% Ag, 2% Tl	400 A m^{-2}	3		80
1% Ag, 1% Ca	*RT*	2		
1% Ag, 0·3% Sn, 0·02% Co	*c*. 500 h	1		
Pb		Fail	Data for	
1% Ag		2 =	0, 100 and 500 mg Cl$^-$	
0·5% Ag, 0·2% Tl		1	given	80
0·5% Ag, 1·5% Tl		2 =		
1% Ag, 0·1% Tl	As above	Fail		
1% Ag, 2% Tl	167 h	Fail		
0·5% Ag, 2% Se		3 =		
0·5% Ag, 0·5% As		3 =		
1% Ag		6		
1% Ag, 0·1% Tl		4 =		
1% Ag, 2% Tl		2	Intermediate time	
1% Ag, 0·1% Ca	As above	5	data also given	80
1% Ag, 1% Ca	100 mg Cl$^-$ 500 h	1		
1% Ag, 0·1% Se		4 =		
1% Ag, 1% Se		3		
1% Ag		5	Ranking after 170 h;	
5% Ba		10	after 1000 h 1% Ag	
1% Ba		8	was best, then	
0·5% Ag, 5% Ba		4	0·5% Ag, 5% Ba.	
1% Ag, 1% Ba		3 =	All others failed.	80
0·3% Ag, 1% Ba	As above	6		
0·5% Ag, 0·5% Ba		7		

(*continued*)

Table V. (*continued*)

Composition	Conditions	Ranking	Remarks	Reference No.
1% Sr		9		
0·5% Ag, 3% Sr		1		
0·5% Ag, 1% Sr		3 =		
1% Ag, 1% Sr		2		

(A series of quaternary alloys of Ba, Sr, As, Sn, Co were also tested. The single alloy with 0·02% Co was best at 850 h, followed by 0·5% Ag 1% Sn. Differences in behaviour were small overall)   80

Composition	Conditions	Ranking	Remarks	Reference No.
Pb	1, 2, 3 M acid	1	Ranking for most	80
1% Ag	0·4; 1; 10 kA m^{-2}	2	severe conditions	
1% Ag, 1% Ca	25–50°C	3	(10 kA, 50°C, 3 M)	
1% Ag, 0·1% Ba, 0·1% As		1	but little difference between most and least severe for each alloy	
5% Sb	4·6 M, 50°C	—	Effect of structure and potential	87
Sb (alloys and 2nd Ag phase in Pb)		—	Very detailed study of solid solutions and multi-phase	88
Ca (0·2–0·8%)	0·1–10 N acid	—	Transient measurements	89
4% Sb, 0–1% Cd	10 N acid 10 days		0·1% Cd is optimum	90
Pb–Tl Pb–Ag	1 M acid *RT*		Full range, optimum = 4% Tl, 40% Ag (binary alloys). Current density also varied	91

References

1. T. N. Andersen and D. L. Adamson, *Metal. Trans.* **5**, 1345 (1974).
2. G. Sunderland, private communication.
3. A. A. Azim and M. M. Anwar, *Corrosion Sci.* **9**, 193 and 245 (1969).
4. S. O. Izidinov, *Sov. Electrochem.* **4**, 579 (1968).
5. A. A. Azim and K. M. El-Sobki, *Corrosion Sci.* **11**, 821 (1971).
6. A. I. Zhurin and E. M. Solov'ev, *Russ. J. Appl. Chem.* **46**, 1613 (1973).
7. V. V. Stender and I. J. Seerak, *Trans. Amer. Electrochem. Soc.* **68**, 510 (1935).
8. G. Z. Kiryakov and V. V. Stender, *J. Appl. Chem. USSR* **24**, 12 (1951).
9. J. L. Bray and F. R. Morral, *Trans. Amer. Electrochem. Soc.* **80**, 55 (1941).
10. K. Ravindran, *Bull. Soc. Chim. Belg.* **83**, 173 (1974).

11. M. Rey, P. Coheur and H. Herbiet, *Trans. Amer. Electrochem. Soc.* **73**, 315 (1938).
12. A. I. Zuborovskii and V. V. Stender, *Prot. Met.* **7**, 43–46 (1971).
13. I. C. Kiseleva and B. N. Kabanov, *Dokl. Akad. Nauk. SSSR* **108**, 864 (1956). *also* **22**, 1042 (1952).
14. E. V. Krivolapova and B. N. Kabanov, *Zh. Prikl. Khim.* **16**, 335 (1941).
15. A. E. Koenig, J. U. McEwan and E. C. Larsen, *Trans. Amer. Electrochem. Soc.* **79**, 331 (1941).
16. E. C. Larsen, *Trans. Amer. Electrochem. Soc.* **79**, 344 (Discussion) (1941).
17. C. Y. Clayton, *Trans. Amer. Inst. Min. Eng.* 275 (1930).
18. K. Ravindran, *Bull. Soc. Chim. Belg.* **84**, 83 (1975).
19. E. P. Kripolova and B. N. Kabanov, *Trudy Sov. Elektrokhim.* 539 (1950).
20. A. S. Gendron, V. A. Ettel and S. Abe, *Can. Metal. Quart.* **14**, 59 (1975).
21. H. Grafen and D. Kuron, *Werkst. Korros.* **20**, 749 (1969).
22. A. E. Koenig and J. U. McEwan, *Trans. Amer. Electrochem. Soc.* **19**, 340 (1941).
23. W. E. Bradt and H. H. Oakes, *Trans. Electrochem. Soc.* **71**, 279 (1937).
24. C. G. Fink and M. Kolodney, *Trans. Electrochem. Soc.* **76**, 401 (1939).
25. B. Schlain and J. D. Prater, U.S. Bureau of Mines Research Report of Investigations 3863 and 3872 (1947).
26. K. I. Marinina, *Chem. Abstr.* **80**, 127468r
27. S. Fukushima, *Chem. Abstr.* **55**, 4195e
28. A. I. Zhurin, *Zh. Prikl. Khim.* **46**, 2099 (1973).
29. N. I. Kharabadze, *Chem. Abstr.* **68**, 65044r
30. S. P. Antonov, *Met. Abstr.* **6**, 34–0304
31. M. Smialowski, *Chem. Abstr.* **47**, 12047g
32. T. N. Klepinin, *Chem. Abstr.* **83**, 87065
33. J. Gonzalez, *Metallurgie Belg.* **4**, 65 (1963).
34. T. S. Dashiani, *Met. Abstr.* **5**, 35–0676
35. N. N. Gratsianskii, *Khim. Zhim.* **34**, 970 (1968).
36. G. Z. Kiryakov, *Chem. Abstr.* **68**, 74598b.
37. E. G. Lukin, *Chem. Abstr.* **79**, 86704.
38. G. Z. Kiryakov, *Chem. Abstr.* **72**, 106501.
39. S. Ivascanu, *Rev. Chim. Buch.* **24**, 693 (1973).
40. G. Eggett and D. Naden, *Chem. Ind.* 389 (1975).
41. A. I. Zuborowskii, *Zaschrift. Metal* **7**, 43 (1971).
42. G. Z. Kiryakov, *Chem. Abstr.* **68**, 74598.
43. V. I. Bryntseva, *Chem. Abstr.* **86**, 62515.
44. J. Gonzalez, *Metallurgie Belg.* **4**, 65 (1963).
45. G. Schmitt, *Werkst. Korros.* **24**, 118 (1973).
46. J. A. Gonzalez, *Corrosion Prot.* **7**, 7 (1976).
47. A. V. Sapozhnikov, *Zaschrift. Metal.* **9**, 332 (1973).
48. R. Grauer, *Werkst Korros.* **20**, 94 (1969).
49. J. Fullea, *Corrosion Prot. (Spain)* **6**, 229 (1975).
50. Yu. D. Dunaev, *Metal. Abstr.* **8**, 35–1719 (1975).
51. A. B. Sandybaeva, *Chem. Abstr.* **81**, 130163.
52. H. Graefen and D. Kuron, *Werkst. Korros.* **21**, 3 (1970).

53. H. Graefen and D. Kuron, 4th Int. Lead Conf. Hamburg, pp. 134–140 (1971).
54. A. B. Sandybaeva and Yu. D. Dunaev, *Chem. Abstr.* **83**, 17620.
55. G. Holstein and E. Pelzel, *Metallurgia.* **14**, 765 (1960).
56. J. A. Gonzalez, *Rev. Metal.* **7**, 105–111 (1971).
57. I. I. Astakhov, *Lead Abstr.* **4**, 2888.
58. J. J. Lander, *J. Electrochem. Soc.* **98**, 213 (1951).
59. N. K. Mikhailova, *Zaschrift. Metal.* **10**, 57 (1974).
60. J. A. Gonzalez, 3rd Int. Conf. Lead, Venice (1968).
61. J. A. Gonzalez, ILZRO Research Report Project LE-70 March (1957).
62. V. A. Khitrov, *Zh. Fiz. Khim.* **37**, 2391 (1963).
63. G. Holstein, *Z. Erzberg. Metallhut.* **16**, 209 (1963).
64. K. Ekler, *Can. J. Chem.* **42**, 1355 (1964).
65. C. G. Fink and C. H. Eldridge, *Trans. Electrochem. Soc.* **40**, 51 (1921).
66. M. Maja, *Electrochim. Metal.* **3**, 63 (1968).
67. U. B. Thomas and F. T. Forster, *Trans. Amer. Electrochem. Soc.* **92**, 313 (1947).
68. T. Rogatchev and S. Ruevski, *J. Appl. Electrochem.* 33 (1976).
69. T. E. Parker, *J. Electrochem. Soc.* **119**, 13C (1972).
70. S. Palanicharny and H. V. K. Udupa, Symp. Lead–acid Batteries, New Delhi, p. 77 (1971).
71. S. Matylla and J. D. Milewski, Chem. Power Sources Symp. Poznan, June 12th (1973).
72. G. W. Mao and J. G. Larson, *J. Electrochem. Soc.* **120**, 11 (1973).
73. J. J. Lander, *J. Electrochem. Soc.* **98**, 220 (1951).
74. S. Feliu and L. Galan, *Werkst. Korros.* **23**, 554 (1972).
75. A. A. Abdul Azim and K. M. El-Sobki, *Corrosion Sci.* **11**, 821 (1971).
76. G. W. Mao and T. L. Wilson, *J. Electrochem. Soc.* **117**, 1323 (1970).
77. J. J. Lander, *J. Electrochem. Soc.* **99**, 467 (1954).
78. J. Leibrandt, Abstr. 6th Int. Congr. Metallic Corros. Sydney, Aust., Dec. (1975).
79. T. N. Andersen and D. L. Adamson, *Met. Trans.* **5**, 1345 (1974).
80. G. Z. Kiryakov and V. V. Stender, *J. Appl. Chem. USSR* **24**, 12 (1951).
81. P. Ruetschi and B. D. Cahan, *J. Electrochem. Soc.* **105**, 369 (1958).
82. G. Kaden and B. Beyer, *Korrosion* 18 (1974).
83. D. Pavlov and T. Rogatschev, *Werkst. Korros.* **19**, 677 (1968).
84. G. Z. Kiryakov and N. A. Pilipchuk, *Tsvetn. Metal.* (1) 24–25 (1975).
85. V. N. Bryntseva and Yu. Dunaev, *Izv. Akad. Nauk. Kaz. SSSR Ser. Khim.* **4**, 27 (1967).
86. V. N. Bryntseva, G. Bundzhe *et al. Prot. Met.* **3**, 431 (1957).
87. D. Marshall and W. Tiedemann, *J. Electrochem. Soc.* **123**, 1849 (1976).
88. S. Feliu and M. Morcillo, *Corrosion Sci.* **15**, 593 (1975).
89. A. A. Abdul Azim and K. M. El-Sobki, *Corrosion Sci.* **12**, 371 (1972).
90. A. A. Abdul Azim and K. M. El-Sobki, *Corrosion Sci.* **16**, 209 (1976).
91. V. I. Bryntseva, Yu. Dunaev and G. Z. Kiryakov, *Izv. Akad. Nauk. SSSR Ser. Khim.* 3(43) (1968).

Additional Bibliography

Pb–Bi: K. W. Grosheim-Krysko, *Z. Metallkde.* 36, 85, 88 (1944).

Pb–Sb (with Ag, As, Fe, Ca): V. P. Mashovets *et al.*, *J. Appl. Chem. USSR* 21, 347, 441, 448 (1948); *J. Appl. Chem. USSR* 31, 1355 (1958).

Pb–Sb–Ag: C. G. Fink and D. J. Dornblatt, *Trans. Electrochem. Soc.* 79, 269 (1941).

Pb–Ag; Pb + Se/Te/Ba/Sr/Ca/Sn/Au/Hg/Tl (traces): G. V. Kiryakov and V. V. Stender, *J. Appl. Chem. USSR* 24, 1263 (1951); 25, 23, 30 (1952); *Izv. Akad. Nauk. Kaz. SSSR* (*Chem.*) 53 (1956).

Pb + traces of second metal: H. Ito and T. Shibano: *J. Mining and Metal Inst. Jap.* 73, 173 (1957); 75, 97 (1959).

Pb–As: G. W. Mao and J. G. Larson, *Metallurgia.* Dec. 236 (1968).

Pb–Co: J. J. Lander, *J. Electrochem. Soc.* 105, 289 (1958).

Pb–Ca: E. F. Schumacher and G. S. Phipps, *Trans. Electrochem. Soc.* 68 (1935).

    H. E. Haring and U. B. Thomas, *Trans. Electrochem. Soc.* 68, 293 (1935).

    U. B. Thomas and F. T. Foster, *Trans Electrochem. Soc.* 92, 313 (1947).

    J. J. Lander, *J. Electrochem. Soc.* 99, 467 (1952).

    Y. Z. Briggs, *Metals and Alloys* 9, 49–51.

Pb–Sb, Pb–Ca, Pb–Ca–Sn: J. L. Eeiniger and E. G. Siwek, *J. Electrochem. Soc.* 123, 602 (1976).

Pb–Sb, Pb–Sn, Pb–Tl: A. B. Sandybaeva, *Metal. Abstr.* 6, 35–0660.

Pb–Sb, Pb–Sn, Pb–Bi (also Ca, Co, Ag): K. S. Rajagopalan, *Metal. Abstr.* 4, 35–0336.

Pb–Sb + As: M. A. Dasoyan, *Metal. Abstr.* 5, 35–1415.

Sb-free alloys: S. Palanichamy, *Corrosion Abstr.* 13, (32) 25 (1974).

Pb–Sb–As: N. A. Novikova, *J. Appl. Chem. USSR* 44, 2447 (1971).

Pb–Pt, Pb–Pd: Yu. D. Dunaev, *Metal. Abstr.* 5, 34–0057.

Pb–Ag: O. Hyvarinen and M. H. Tikkanen: *Acta Polytech. Scand.* Ch. 89, 5 (1969).

Pb–Ag, Pb–Sn, Pb–Bi (also Sb, Tl): V. I. Bryntseva, *Zaschrift. Metal.* 3, 504 (1967).

Pb–Ca–Na: V. Aravamuthan, *Electrochim. Metal.* 3, 56 (1968).

Pb–K, Pb–Na: I. V. Sokolova, *Zaschrift. Metal.* 9, 332 (1973).

Pb–Ca–Sn: St. Matylla, Chem. Power Sources Symp., Poznan, June 12–13 (1973).

Pb–Ru: V. I. Bryntseva, *Metal. Abstr.* 10, 35–0429.

Pb–Tl: I. E. Titova, *Zaschrift. Metal.* 9, 702 (1973).

Pb–Ag: G. W. Rao and P. Rao, *Brit. Corrosion J.* 6, 122 (1971).

Pb–Ag, Pb–Tl: V. I. Bryntseva, *Izv. Akad. Nauk. Kaz. SSSR Ser. Khim.* 43 (1968).

Pb–Sb: P. Ruetschi and B. D. Cahan, *J. Electrochem. Soc.* 105, 7 (1958).

Pb–Ca, Pb–Sn–Sb: J. Burbank, *J. Electrochem. Soc.* 105, 693 (1957).

Pb–Sb, Pb–Ca: H. E. Haring and U. B. Thomas, *Trans. Electrochem. Soc.* 68, 293 (1935).

Pb–Ca, Pb–Li–Ca: S̈. I. Sklyarenko and O. S. Druzhinina, *J. Appl. Chem. USSR* **13**, 1794 (1970).

Pb–Sb, Pb–Te: F. R. Morral and J. Bray, *Trans. Electrochem. Soc.* **75**, 427 (1939).

Pb–1% Au: E. Mueller and K. Schwabe, *Z. Elektrochem.* **40**, 862 (1934).

14   Corrosion of Lead in Chloride Media

A. T. KUHN

Department of Dental Materials, Eastman Institute of Dental Surgery, London

In this chapter, we embrace a number of rather different situations—each relating directly to an important use or condition of lead.* Firstly, there is "open circuit" corrosion of lead or its alloys and this is a process generally taking place at fairly low potentials. Secondly, there is the now growing important use of lead or lead alloy electrodes as sacrificial anodes in seawater-activated batteries. The prevalent potentials here are somewhat more anodic than those in normal corrosion, in that the lead anode is "driven" to some extent by the chlorine cathode. Finally, there is the use of the lead or alloy anode in cathodic protection or process applications. This category distinguishes itself from the others because here alone does the electrode reach sufficiently anodic potentials to form PbO_2 in addition to the chlorides which are otherwise the main corrosion product. As Fink and Pan have stated,[28] there is little, if any, relationship between the behaviour of metals under free corrosion on the one hand, and driven corrosion on the other.

In general terms, we may summarize the chemistry of the system as follows. Lead corrodes to form the dichloride. Latimer[29] gives a standard Gibbs' free energy value of -75 kcal for the latter species in crystalline form, and von Fraunhofer has quoted[10] an E_0 value of -0.513 V (*vs* S.C.E.) for the chloride-forming reaction (equation 1). Latimer also quotes the solubility product $K_s = 1.6 \times 10^{-5}$ for the same reaction. Fromherz[30] has shown the existence of relatively high concentrations of $PbCl^+$ ions and has quoted:

$$PbCl^+ = Pb^{2+} + Cl^- \qquad\qquad K = 0.775$$

with data also given for the other halides. Concentrated halide solution

* The reader is advised to refer to Chapter 10 in relation to many aspects of this chapter.

dissolves lead halides forming complex ions, and Korenman[31] has given constants for $PbCl_3$ as $4·2 \times 10^{-2}$ with data for other halides in the same paper. Thus, even at low potentials where no oxides can form, we see the situation is complex. At higher temperatures and brine concentrations, the solubility of the $PbCl_2$ film, initially formed as a corrosion product, is further increased, and reference is made below to the works of Demassieux and Forbes in this context.

In the light of this information, it will be seen that corrosion rates of lead or its alloys at potentials cathodic to those where oxides might be formed, are mainly a function of the conditions prevailing and their effects on the dissolution of the $PbCl_2$ film, or formation of complex ions. Not only brine concentration and temperature, but also hydrodynamic conditions are important, and only the works of Barradas, cited below, make any attempt to define all of these parameters.

For this reason it can be seen that for each particular condition, for example, lead-sheathed cable buried in saline marshes, the kinetics and nature of corrosion will be particular to the circumstances. *Corrosion Abstracts*[1] and other publications[2] provide some sort of guide to such problems and should be consulted. Von Fraunhofer[3] has summarized the known behaviour of lead at open circuit in waters of varying salinity and other data is provided by Abdul Azim *et al.*[4] who studied NaCl solutions from 1 M down to 10^{-5} M. In the former, corrosion rates were $2·3$ mg dm^{-2} day^{-1}, in the latter they were far higher at 38 mg dm^{-2} day^{-1}. They report an empirical relationship $W = aC^{-b}$ where W is weight loss in the units used above, C is NaCl molarity and a and b are constants, b being $0·32$ in chloride media (similar relationships were obtained for sulphate solutions). However, it should be stressed that any change in their conditions would have provoked corresponding changes in the values of these constants.

It would be as well to discuss here the experimental problems confronting experimental determination of corrosion, which in the main is done by the method of weight loss. If samples of lead are corrosion tested, either at open circuit or under anodic load, they may either gain or lose weight depending on the conditions which may embrace the added weight of Pb as converted to its oxidation products, or reflect simple loss of material. Several papers in the field report simultaneous weight losses and gains over the course of one set of experiments. As Dawson[5] has pointed out, the only unequivocal method is that in which, at the measurement point, corrosion products are removed and what is then weighed is (as nearly as possible) the residual unoxidized metal. In such cases, only weight losses are reported. The data of Abdul Azim,[4] for example, are reliable in that he did strip corrosion products in a saturated solution of ammonium acetate, before washing and drying. The majority of cited papers did not, however, employ this procedure. In such cases, a second danger must be appreciated. The corrosion products, with larger specific volume than the metal from which they derive, swell and cause blisters which ultimately crack and fall off. In such circumstances, simple weight-change measurements are even less useful, since if a measurement is made just prior to blister disengagement at one point and just following it in another case, the result will be a nonsense. Regrettably, much of the data reported suffer from these pitfalls.

A renewed interest in the dissolution of lead or its alloys comes from the success of the primary seawater battery. Using a silver anode, this offers a highly effective underwater power source for naval and other marine applications. In an attempt to reduce the cost, silver has been successfully replaced either by lead or by lead alloys. This technology is described by Gray and Wojtowicz,[6] Coleman[7] and Vyselkov,[8] and Gray[6] and other authors show discharge curves for Pb and Pb–Cu alloys.

As has been stated above, the electrochemistry of the system centres around the formation of the poorly soluble film of $PbCl_2$ formed via the process:

$$Pb + 2\,Cl^- \rightarrow PbCl_2 + 2e \qquad (1)$$

We must, of course, make the analogy with the sulphate system as represented by the reaction:

$$Pb + SO_4^{2-} \rightarrow PbSO_4 + 2e \qquad (2)$$

which is treated at length in Chapters 12 and 13. The analogy ends when it is realized firstly that the chloride of lead is some 300 times more soluble in water or aqueous media than the sulphate,[9] and secondly, in practice, the importance of lead in sulphate media focuses on solutions high in sulphate concentration (strong sulphuric acid) which are relatively quiescent. In contrast, taking seawater situations, one is dealing with solutions dilute in Cl^- and frequently turbulent, both factors aiding removal, by dissolution of erosion, of the outer chloride film. The electrochemical behaviour of the system can therefore be described in terms of a competition between the rate of film formation (by equation 1) and its removal by a variety of mechanisms, both chemical and mechanical, with the properties of the film exerting the usual effects of passivation. High anion concentrations will repress the equilibrium between solid and dissolved $PbCl_2$, and also favour complex formation.

The nature of the $PbCl_2$ film and its thickness will influence the current–voltage data obtained in these systems, and if the film is thick, which is certainly the case for lead anodes which have been in service for some time, the $j–V$ characteristics will become predominantly linear, reflecting the ohmic character of the system, and not logarithmic as one would expect from a straight activation-controlled process. We should here correct an error in the literature where it is stated that the resistance of the $PbCl_2$ film is 40–50 MΩ.[10] This figure stems from a misreading of an early work of Shrier and refers to the very high conductivity $(40–50 \times 10^{-6}\ \Omega)$ of the dioxide.[11] The resistance of the film can undoubtedly be high, and its tendency to flake off shows itself as erratic long-term $j–V$ readings, notably in quiescent solutions, since in flow conditions there is a steady attrition of the outer surface of the film. Barradas and co-workers have deduced resistance values for the $PbCl_2$ film in its early stages[12,13] and von Fraunhofer[10] quotes the reversible potential of the equation (1) as:

$$PbCl_2 + 2e = Pb + 2\ Cl^- \qquad E_0 = -0.513\ V\ (vs\ S.C.E.)$$

Are any other chemical species involved in the $Pb–Cl^-–H_2O$ system? We must distinguish between speculations, such as those involving chloroplumbates or chloroplumbites raised by Sato,[14] and those which have been actually detected. For anodes of pure lead, only Shreir[37] has reliably reported the formation of PbO_2 and then patches rather than coherent coverings appear to be formed. X-ray analyses of surface-formed species by Barradas and co-workers[13] revealed Pb(OH)Cl but only at pH values less than 6.5. These workers also speculated the involvement of species such as $PbCl_3^-$ [12,13] and complexes of the type $PbCl_{n+2}^{n-}$. In the case of alloys of lead, there is much evidence for PbO_2 formation, while the ternary component gives rise to other chemical species. Shreir[37] has also reported the presence of orange–red compounds when pure lead anodes are used. He suggests these may be lower oxides of lead.

Scientific Studies of the Lead–Chloride System

The first significant paper is that of Briggs and Wynne-Jones[15] and covers the behaviour of the metal in aqueous solutions of all the halides. j–V data were obtained in 0·02–2 M solutions of NaCl, HCl (and corresponding Br⁻, F⁻ and I⁻ solutions) in the current density range 0·2–10 mA cm⁻². Photomicrography is used to record the growth of halide films on the surface. Modern instrumentation was not available to these workers and they found difficulty in obtaining stable current–voltage readings. They note that no plateau corresponding to the PbO_2 potential was seen. An approximately linear relationship was found between the logarithm of passivation time and the logarithm of the current density applied. Nearly ten years later, the work of Mindowicz[16] shows steady-state data in 0·1–1·5 M NaCl solutions. The corrosion current is highest in the most dilute solutions where the onset of oxygen evolution also takes place at the least anodic potentials. Thus in 0·1 M solution, the current density at the corrosion peak (j_p) is 4 mA cm⁻² and $j_{O_2} = 4$ mA cm⁻² at 1·1 V vs N.H.E. In 1·5 M NaCl, $j_p = 1$ mA cm⁻² (the peak potential is scarcely shifted) while $j_{O_2} = 1$ mA cm⁻² at 2·2 V, the data fitting in with a general picture in which high Cl⁻ concentrations repress film dissolution with the obvious result both in terms of lowering the corrosion current and raising the apparent overvoltage of oxygen because of the greater ohmic drop through the thicker film. Cyclic voltammetry showed the reverse sweep to be featureless, the corrosion peak being absent. A plot of log j_p vs $1/T$ is rectilinear and shows the effect of temperature on passivation. Data on O_2 evolution are treated in Chapter 9. The authors conclude with data obtained from cathodic galvanostatic transients. It was found that $Q_c > Q_a$ (170 mC cm⁻² and 240 mC cm⁻²), an observation which was taken further by Barradas and co-workers[4,10,13,17] as discussed below. Littauer and Shreir[37] show potential–time plots for pure lead anodes at 10 mA cm⁻² in brines of 0·2, 0·5 and 0·1 M concentration. In the last two cases, after an initial rise in potential, a steady plateau is attained. In the first case, the potential rises to 70 V, suggesting formation of a thick chloride layer. The authors have analysed their data in terms of the Peukert equation:

$$j^n \cdot t = k$$

where j is the applied current density; t is the time before the rise in potential is observed, and n and k are constants. A plot of log j vs log t is rectilinear in the case of 0.2 and 0.5 M NaCl, confirming the earlier work of Wynne-Jones. Steady state potentiostatic $j-V$ and $j-t$ data are reported by Abdul Aziz et al.[4] who studied $1, 0.1, 10^{-2}, 10^{-3}$ and 10^{-4} M NaCl solutions. In the range $1-10^{-2}$ M NaCl they report an empirical relationship $j = a - b$ log C for steady-state currents at 0.4 V vs S.C.E. In the range $10^{-2}-10^{-4}$ M NaCl, the concentration effect operates in the reverse sense. These workers conclude their paper with a model mechanism based on Langmuir adsorption and obtain straight line plots of $\theta/1 - \theta$ vs log C. However, θ values are somehow derived from weight-loss data and it is difficult to see how these authors justified a chemisorptive (i.e. monolayer) treatment in a system which is manifestly multilayer. Their other data are open to the same criticism as that of Mindowicz et al.[16] in that the hydrodynamic regime was ill-defined, and so leads to variable rates of chloride film dissolution. This shortcoming has been largely overcome by Barradas et al. who have published a series of exemplary papers on the subject.[12,13,17,18] Barradas and co-workers have largely succeeded in avoiding, either by use of fast transients or with a rotating disc electrode, the problems mentioned above. In essence, though, the overall understanding of the processes is not greatly altered. In the first paper with Ambrose, an HCl medium is used. Analysis of cyclic voltammetric data gives a square root dependency of sweep rate on j_p, suggesting a diffusion-controlled process. However, the value of D, the diffusion coefficient calculated from this data, is 10^{-3} times that of typical solution diffusion coefficients, suggesting that the reaction is controlled by diffusion processes in the solid $PbCl_2$. The disparity between Q_a and Q_c first found by Mindowicz[16] is attributed to reduction of dissolved plumbous species. $E_p \neq$ log (sweep rate) and $E_p =$ const. (sweep rate)$^{0.5}$, suggesting the importance of an ohmic component in the film. A value of $3\,\Omega$ is derived for the film resistance in this particular experiment. Further studies with a rotating disc electrode and potentiostatic transient techniques confirm the mechanism outlined here, suggesting $PbCl_3^-$ as a species involved in film dissolution, and the transient data confirms the importance of the solid state diffusional process. Two subsequent papers[13,17] extend the work to more concentrated HCl solutions and also NaCl. Using the same techniques as before, and also a ring-disc apparatus, the competitive situation between film formation and removal is more fully explored.

In the most recent publication, Barradas and co-workers[18] couple their electrochemical studies with scanning electron microphotographs (S.E.M.) of the electrode taken at various points along a potentiostatic time transient. They show clearly how the $PbCl_2$ deposit consists in fact of two distinct layers, an underlying fine-grained deposit with a much coarser grained $PbCl_2$ deposit above it. It is the latter which causes the electrode to passivate. The authors have placed on a quantitative footing the rate of the various processes taking place, which have been referred to earlier, and from this paper, the implication of raising a lead electrode to moderately high anodic potential on the one hand, or a high one on the other, can be seen.

Barradas and Fletcher[46] report temperature effects in HCl solutions using cycled Pb electrodes from $-6°C$ to $+30°C$. In a further paper from the same laboratories[47] the effect of dissolved gases is investigated. Oxygen in particular is shown to exert a further complicating effect on the equilibria of the sort mentioned earlier.

Behaviour of Lead Alloys in Chloride Solutions

The open-circuit behaviour of a series of Pb–Sb alloys with time in both aerated and de-aerated solutions of various pH, containing Cl^- ions (also other ionic species) has been reported by Khairy et al.[19] At lower pH values (1·5–3·5) the authors conclude that the electrodes behave as electrodes of the first kind (Pb:Pb^{2+}), whereas in more alkaline solutions their behaviour is governed by the M:MCl_x couple (for aerated solutions), with Pb determining the behaviour in Pb-rich alloys, and Sb in Sb-rich alloys. In de-aerated solutions, E vs $p\,Cl^-$ is shown for several alloys and above 0·05 M chloride ion concentration, the electrodes show potentials close to the theoretical Pb:$PbCl_2$ values though, as the concentration of Sb increases, in a series of alloys, the potential of the Sb:Sb_2O_3 alloy is approached. As von Fraunhofer has pointed out,[10] as little as 1% of a second metal can influence the corrosion behaviour of lead in halide solutions. The reasons for this are not fully understood. Mindowicz[16] shows steady state $j–V$ data for Pb, Pb+0·8% Ag, Pb+4·5% Ag and Pb+9·4% Ag in 0·5 M NaCl. From a value of $2\,mA\,cm^{-2}$ for Pb (the 0·8% Ag alloy appears to be identical) the corrosion current peak is halved for the 9·4% Ag, the transpassive corrosion current being halved in the same manner. The same author[45]

also reports results of galvanostatic pulses on Pb–Sb alloys in 3 and
0·1 M NaCl.

There is a body of work describing the corrosion behaviour of lead
and lead alloy anodes in chloride media. The original study appears to
be that of Fink and Pan[28] in which they recommend an alloy with at least
5% Tl for strong brines (although they themselves quote some
extremely early work). In many ways, these papers are the best treatment of the subject and far more thorough than the somewhat trivial
work which was published 30 years later. Other early work on strong
brine anodes is cited in refs 42–44. The work is notable in that it covers
the full compositional range from 0% Pb to 100% Pb. All subsequent
papers have restricted themselves to not more than 10% Ag. The weight
loss method is used (tests of 2–3 weeks duration) but without stripping
of corrosion products. In broad terms, the corrosion resistance at
1·6 A dm^{-2} and 22°C improves from 82 400 mg cm^{-2} at 1000 A h^{-1} for
pure Pb to approximately 1000 mg cm^{-2} for the 5% Ag alloy, halving
again at the 10% Ag and dropping to a minimum figure of 46 mg cm^{-2}
for the 55% Ag alloy, after which it rises again to 62 000 mg cm^{-2} for the
pure Ag. A plot of the data shows a kink at 2% Ag where corrosion rate is
below the expected. Fink points out that this composition corresponds
with a eutectic point in the melting curve. Other conclusions from this
paper are that doubling of the current density results in doubling of the
corrosion rate, and that corrosion increases with increasing temperature
(80°C). Reference 28 is probably the longest and most thorough
investigation in print on the corrosion of lead and alloys in chloride
media, and contains a wealth of observations which appear to have been
overlooked by later authors. In respect of corrosion rates, the findings of
the earlier paper are confirmed and extended, and additional results are
reported with respect to the Pb–Hg alloys and the Pb–Ag–Hg system.
Corrosion rates in the former case are little changed, while the latter
system shows a marked deleterious effect due to Hg addition. Finally,
the Ag–Pb–Mn system was studied, albeit only the 61% Ag–39% Pb, to
which 0–5% Mn was added. The measured corrosion rate increased
with increasing Mn concentration.

The main emphasis of the paper lies otherwise in its description of
the films formed upon the metal surface, their physical and chemical
nature and the effect of formation conditions (in different solutions) on
their subsequent behaviour in brine.

Fink explains the beneficial influence of silver as follows. The silver oxide is readily formed (more so than the chloride) and this compound, it is stated, attacks the metallic lead to form PbO_2 (he quotes Glasstone[35] in support of this theory, though that work is now largely discredited[41]). There is no simple explanation for the anomalously high corrosion rate of the 10% Ag alloy which is emphasized at higher $(3 \cdot 2 \text{ A dm}^{-2})$ current densities, and by shortening the test period.

Regarding the character of the films formed, below 2% Ag, there were none readily visible, there being (according to Fink) no silver-rich phases to promote formation of the protective PbO_2 layer. Up to 3% Ag, thick but loose and spongy films of PbO_2 are formed, and from 3–15%, while still spongy, it is more adherent than hitherto. The PbO_2 forms around the silver-rich crystals and finally bridges the gaps between them, though the film is loose at these bridge points. From 15–99% Ag, the film is so hard and adherent that it can only be removed by an emery wheel or similar means. The authors also suggest that at high Ag ratios the mechanical properties of AgCl are beneficial in that it acts as a binder, but that at the highest Ag ratios this effect becomes detrimental, the AgCl film obscuring PbO_2 and raising anode potentials. The high corrosion rate of pure Pb is thus explained in terms of absence of the PbO_2 film, while that of pure Ag results from the insulating nature of the AgCl film with a very high observed anode potential, sparking at points, and an effectively very high current density. This interesting explanation seems to leave an unanswered question—why does Ag_2O form in the case of alloys, whereas AgCl is postulated for the pure Ag anodes? The authors appear to argue in circular fashion when they subsequently suggest that oxygen (from decomposition of higher oxides of lead to PbO_2 and molecular oxygen) converts AgCl to its oxide. Though this appears somewhat improbable, the authors show a microphotograph with two layers (cf. the findings of Barradas and Belinko[18]), the outer one with PbO_2 and Ag_2O, and the inner one with PbO_2 and AgCl, the nature of these layers having been confirmed by chemical analysis.

Fink also suggested that the presence or absence of a diaphragm profoundly affected the corrosion rate. For lead-rich alloys, the presence of a diaphragm improved corrosion behaviour; for silver-rich alloys the reverse was true. This was explained in terms of the lower OCl^- and ClO_3^- concentrations in diaphragm cells and so the reduced amount of molecular oxygen formed. Equally well, however, one might invoke undoubted differences of pH as an explanation. Further explanations are offered in terms of the higher solubility of $Pb(ClO_3)_2$ than the corresponding chloride or $AgClO_3$.

Fink and Pan also showed the effect of time on corrosion rates. Various alloys were tested for weight loss over 360 h. In all cases, save that of the highest Ag content (for which a satisfactory explanation is offered) the corrosion rate is seen to drop with time until it reaches a plateau. The time taken to attain this plateau is inversely proportional to the Ag content, that is, the 2% Ag alloy required some 300 h, the 40% Ag alloy only 40 h.

In other experiments the authors formed films on the Pb alloys in solutions other than chlorides, and the results were, in part, encouraging. However, such experiments relate only to the life of the film formed under artificial conditions, and once this film has broken down, the corrosion rate reverts to that of the substrate. Insofar as such anodes are 5 cm thick, the 30% improvement in corrosion based on a film a millimetre or so in thickness is of minimal significance. Finally, Fink and Pan show that current density up to 10 A dm^{-2} has comparatively little effect, while increase in temperature begins to show very damaging effects above 50°C up to 85°C (highest temperature studied), the relationship between corrosion and temperature being exponential. This is partly explained in terms of the higher solubilities of the metal chlorides at high temperatures as shown by Demassieux[32] and by Forbes[33]. Their solubilities are also shown to increase with NaCl concentrations.

Finally, in terms of impurities, it is shown that nitrates and chlorates affect corrosion detrimentally. Lead or silver salts in solution repress the corrosion in dramatic fashion, although this is a statement based only on weight loss (indeed, weight gain in certain cases) and what structural changes are seen on the anodes are not mentioned, except to say that a fine-grain PbO_2 is formed with good properties. This ties in with statements elsewhere in the paper that the best PbO_2 films are formed at low current densities (to avoid gassing) and in more dilute NaCl (to prevent $PbCl_2$ dissolution) and at lower temperatures (to prevent a loose oxide deposit which arises when film formation is too rapid), although initial low-temperature film formation followed by conditioning at a higher temperature is the preferred preparation method.

The opinion seems to be that PbO_2 does not form on pure Pb anodes in halide media, and the work of Fink and Pan may be the basis for this. Mindowicz[16] has been misquoted in saying that this phase does exist on Pb anodes (although the evidence of Shreir[37] to the contrary has been discussed earlier). Nevertheless, it may well be that after long service in certain conditions, PbO_2 is formed on Pb anodes. Certainly, the tenor of several papers cited below is to the effect that certain alloys of lead, notably those containing silver, are more corrosion-resistant precisely because they facilitate formation of the PbO_2 phase which has a greatly reduced corrosion rate. In this, they behave as do the bielectrodes (q.v.) and Shreir[11] has suggested that silver (either as a primary crystal or as a eutectic) acts as a microelectrode to promote PbO_2 adhesion. He also points out that alloy constituents may be incorporated in the PbO_2 film and so prevent ionic transport through it or in various ways affect the mechanical or physico-chemical properties of the film.

Apart from the general consensus of opinion that silver-based alloys are superior to non-silver anodes, there appear to be very few reliable sources of information. Sato[21] has examined rates of corrosion of lead alloys in seawater at high temperatures (44–52°C). He used the weight loss method but failed to strip his anodes and thus exposes himself to the criticism of this technique. Sato used anodes with up to 4% Ag and 1–5% Sb or 1–8% Sn. He concludes that 3–4·5% Ag is the optimal silver content and that the antimonial ternary, such as 4% Ag, 3% Sb is best. At 1 A dm^{-2} he reported a 0·085 kg A-year^{-1} weight loss, pointing out that this is a far lower rate than that reported by Meoller.[22] However, Sato's results are based on runs of only 240-h duration and here, too, may well be of limited value. Sato mentions that high Sn contents lead to excessive blistering. A similar type of paper again is by Tudor and Ticker[23] who use weight loss but without stripping. They report results at current densities up to 25 A dm^{-2}, which is probably the upper limit of usable current density for such anodes. Their results relate to saline waters at various concentrations at 35°C. These authors characterize their salinities in terms of the measured specific conductances of the solutions. In strong solutions (16 Ω cm) a good PbO_2 build-up is reported and, if damaged, is self-healing. As the solutions get weaker (50 : 1000 : 5000 Ω cm) the PbO_2 film is less well developed and loses its self-healing faculty. Alloys with 2–10% Ag were satisfactory. Ternary additions of Sn or Hg were reported to be of no additional benefit. An interesting observation by these authors is that at very low current densities (less than 1 A dm^{-2}), the PbO_2 coating fails to develop (presumably the anode potential is below the formation potential of the oxide) and the anodes therefore rapidly fail. The operation of these anodes is thus seen to be constrained between a lower current density of 1 A dm^{-2} and an upper limit of $15-25 \text{ A dm}^{-2}$. Tudor and Ticker also discuss the possibility of preforming PbO_2 in a strong chloride solution or perhaps even a sulphate solution, before putting the anode into service in weak brines. There would appear to be an advantage to be gained by so doing, albeit a very limited one. Finally, reporting on their measured weight changes, figures of -0.039 g $\text{dm}^{-2} \text{ day}^{-1}$ (strong brines) to $+9.7 \text{ g dm}^{-2} \text{ day}^{-1}$ are reported, the latter figure pertaining to brackish waters of $5000 \ \Omega$ cm resistivity.

Blaise[34] shows current–voltage data for Pb–Sb alloys in NaCl but it is not obvious how he deduces all his conclusions, and differences in behaviour do not appear to be large. Schmeling[35] considers pitting corrosion in lead and reports galvanostatic and potentiostatic $j-V$ data in NaCl and mixtures of NaCl and $NaHCO_3$.

Shreir[20] has studied alloys of 1% Ag and 0.5% Bi, Te or Sb. These alloys were used as substrates for Pt-implanted bielectrodes and although he reports the Bi or Sb (at 0.5%) addition to 1% Ag the best, it is not certain whether this finding would apply to straightforward alloys without a bielectrode. In other work, Hiller and Lipps[24] report a wear rate of $0.25 \text{ kg A-year}^{-1}$ at 1.5 A dm^{-2} for a 1% Ag+6% Sb alloy. Morgan[27] states that Pb–Sb (6–8%) alloys only form the protective PbO_2 layer at current densities above 20 mA cm^{-2}, when a hard, well adherent layer is formed. Corrosion rates of some $1.3-1.6 \text{ kg A-year}^{-1}$ are reported. With the 1% Ag alloys, on the other hand, the peroxide is formed at current densities as low as 1 mA cm^{-2} and at 6 mA cm^{-2} the corrosion rate is $0.5 \text{ kg A-year}^{-1}$. The peroxide layer is stated to be softer and more easily removed, and above 10 mA cm^{-2} blistering becomes serious. On these grounds, Morgan states that the ternary (6% Sb) embodies the best features of both binaries, forming peroxide at $0.5-1 \text{ mA cm}^{-2}$ and not blistering below 30 mA cm^{-2}, having been tested up to 100 mA cm^{-2}. Corrosion appears to be least in the $10-15 \text{ mA cm}^{-2}$ range. In an earlier paper,[38] Morgan mentions that PbO_2 is not formed on Pb, but is formed on the 8% Sb alloy at 20 mA cm^{-2}. Results are also quoted for an alloy "CX3" of unspecified composition, which may well be the Pb/Ag/Sb/Co/Te range cited in Morgan's British Patent 880,519 (1961).

Finally, one should mention the findings, by several workers, that anodic dissolution rates obtained in given conditions are increased when the anodes operate under (apparently) identical conditions but at great depths, i.e. under pressure. Reding and Boyce,[25] using the same alloy and current density as Hiller and Lipps, report a doubling in corrosion rate at 200-m depths (simulated) and also a 0·4-V increase in potential after 160 days. The potentially higher chlorine solubility is not responsible for these changes, since the systems never operate at even approaching the limiting solubility of this gas. These workers suggest that the PbO_2 layer may somehow become detached from the metal, though this is a speculation. Support for the idea that voltage increases may be due to detachment of the PbO_2 layer from the metal comes from a reported discussion[39] in which the current flowing through a lead anode was dramatically increased by touching the outer oxide film with a small Pt wire, connected to the identical power source. The context of this discussion actually related to the mechanism of the Pb–Pt bielec-trode, but the point may well apply here. The discussion also cites[40] work in which a Pt "tack" was used to improve contact between the inner metal and outer PbO_2 layer, again with increase in current.

The work of Reding and Boyce has been confirmed by Holland-sworth and Littauer[26] using the same ternary alloy and the method of weight loss with product removal in a 2% KI, 10% acetic acid and 5% sodium acetate solution which scarcely dissolves metallic Pb. They summarize their results by stating that cracking and striation of the PbO_2 layer increases with depth, as does blistering. PbO_2 forms thicker layers at depths and oxide separation from the metal is reported, leading to voltage increases.

The authors, like their predecessors, are clearly at pains to explain these results. From oxide growth rates, they deduce that while at ambient pressures only $6 \times 10^{-5}\%$ of current goes to oxidation of lead (the 99·99994% going to oxygen or chlorine evolution) at depths of 180 m, the figure increases by 30-fold to $2 \times 10^{-3}\%$. They suggest that the increased conversion of Pb to PbO_2 is the result of the chlorine evolution reaction being hindered at depths ("not thermodynamically but because of Le Chatelier's Principle"), thus forcing more coulombs to the corrosion reaction. One must be reluctant to accept this explana-tion, because in seawater one can scarcely envisage dissolved Cl_2 concentrations ever building up at any depth. Furthermore, if the predominant reaction is hindered for any reason, a change in cell voltage should be seen immediately, whereas in fact one sees (their Fig. 4) only a slow series of "saw teeth" resulting from blister formation (resistance increasing) followed by the abrupt voltage drop after rupture. The question remains an open one, we feel. One must conclude, with the sentiment implied by Littauer and Hollandsworth,[26] that the future lies less with massive alloy anodes and more with lead-based bielectrodes, the latter being capable of supporting much higher current densities (up to 200 mA cm^{-2}). Also, the cost of the Pt would be less than that of Ag in a bulk alloy, with the further advantage that the bielectrodes can be physically much smaller than massive anodes which operate perhaps only at 25 mA cm^{-2} at the same total current rating.

Further Bibliography

While this chapter represents an attempt to summarize the anodic corrosion of lead in chloride media, the following additional published works, which have been collated by Miss Brenda Marsden, of the Salford University Library staff, are added to make an essentially complete bibliography of the subject.

Titova has studied Pb–Cd[48] and Pb–3% Tl alloys in HCl up to 6 M.[49] Alloying increases stability. The work of Sato has been mentioned previously; a series of papers in Japanese complement his other papers and possibly repeat the same material. Thus, potentiodynamic sweeps are reported,[50] along with the effect of temperature (up to 70°C), where it is stated that satisfactory operation is found at 5 A dm^{-2} and the surface film is the β-form of PbO$_2$ at pH = 8,[51,54] and the effect of Ag alloying.[52] Limiting current density is discussed in ref. 53. Three other papers[55–57] are more general in subject. Mindowicz is another author whose work has been mentioned. Refs 58 and 59 also deal with passivation–depassivation behaviour of lead in chloride media. Khitrov[60] reports effect of temperature on corrosion resistance and electrode potentials of Pb in HCl, and Smol'yaninov and Khitrov[61,64] extend this. Passivation is, unsurprisingly, made more difficult by both temperature increase and acidity. Miuri and Murai[62] investigate the Pb–Ag alloys. Current efficiency (in Cl$_2$) of 55–75% is reported up to 200 mA cm^{-2}, when the electrode then breaks down. Behaviour in twice diluted seawater is poor—in ten-fold diluted, catastrophic. Juchniewicz[63] finds that a.c. has little effect when superimposed on d.c. for a

Pb–Sb–Ag alloy in 3% NaCl. This is perhaps surprising in the light of published work on a.c. and lead corrosion generally (see p. 407). Abdul Azim[65] reports weight loss and potentiostatic data in NaCl. Zhuk,[66] in a recent paper, investigates lead and alloys in seawater and their stability.* Afanas'ev has studied the effect of Ag addition on Pb and Pb–Sb in HCl.[67] Another paper by Titova[68] investigates alloys in HCl. Gouda has studied the Pb–Fe couple in $1-10^{-6}$ M NaCl,[69] while Storcaj has examined the Al solder and Pb–Sn couple in a mixture of 0·01 M NaCl, 0·001 N Na_2SO_4 and 0·001 N Na_2SO_3.[70] James[71] has studied disintegration of lead alloys in salt solutions.* Ekilik discussed corrosion in 1–3·5 M HCl and primary alcohols. Corrosion was more severe in these, reaching a constant value for alcohols of greater than four carbon atoms.[72]

References

1. *Corrosion Abstracts.* Published by NACE, Houston, Texas.
2. L. L. Shreir (ed.), "Corrosion". Newnes (1963).
3. J. A. von Fraunhofer, *Anticorrosion,* p. 21 (May, 1969).
4. A. A. Abdul Azim, *Brit. Corrosion J.* **8**(2), 76–80 (1973).
5. J. Dawson, UMIST Corrosion Centre, private communication.
6. T. J. Gray and J. Wojtowicz, "Ocean '72", 582–585.
7. J. R. Coleman, *J. Appl. Electrochem.* **1**, 65 (1971).
8. A. A. Vyselkov and I. Y. Kop'yev, *Elektrotekhnica* **12**, 50 (1969).
9. "Handbook of Chemistry and Physics", Rubber Publishing Company (1976).
10. J. A. von Fraunhofer, *Anticorrosion,* p. 5 (December, 1968).
11. L. L. Shreir and A. Weinraub, *Chem. Ind.* 1326 (1958).
12. J. Ambrose, R. G. Barradas *et al., J. Colloid Interfacial Sci.* **47**, 441 (1974).
13. R. G. Barradas and K. Belinko, *Can. J. Chem.* **53**, 389 (1974).
14. E. Sato, *Bull. Chem. Soc. Jap.* **39**, 1592 (1966).
15. G. W. Briggs and W. F. K. Wynne-Jones, *J. Chem. Soc.* 2966 (1956).
16. J. Mindowicz and S. Biallozor, *Electrochim. Acta* **9**, 1129 (1964).
17. R. G. Barradas and K. Belinko, *Can. J. Chem.* **53**, 407 (1974).
18. R. G. Barradas and K. Belinko, *Electrochim. Acta* **21**, 357 (1976).
19. E. M. Khairy and A. A. Abdul Azim, *J. Electroanal. Chem.* **12**, 27 (1966).
20. L. L. Shreir, *Corrosion Tech.* April (1964) 16.
21. E. Sato, *Anticorrosion Methods and Materials,* 27 (April, 1966).
22. G. T. Meoller and J. T. Patrick, *Mater. Prot.* **1**, 46 (1962).
23. S. Tudor and A. Ticker, *Mater. Prot.* **3**, 52 (1964).
24. A. E. Hiller and D. A. Lipps, *Mater. Prot.* **4**, 36 (1965).

* Not seen by the author.

25. J. T. Reding and T. D. Boyce, *Mater. Perf.* (*U.S.A.*) **13**(9), 37–40 (September, 1974).
26. R. P. Hollandsworth and E. L. Littauer, *J. Electrochem. Soc.* **119**, 1521 (1972).
27. J. H. Morgan, *Corrosion Tech.* **5**, 347 (1958).
28. C. G. Fink and L. C. Pan, *Trans. Amer. Electrochem. Soc.* **46**, 349 (1924); **49**, 85 (1926).
29. W. Latimer, "Oxidation Potentials", Prentice-Hall, New York (1961).
30. H. Fromherz, *Z. Phys. Chem.* **153**, 382 (1931).
31. I. M. Korenman, *J. Gen. Chem. USSR* **16**, 157 (1946).
32. M. Demassieux, *C. R. Acad. Sci.* **158**, 702 (1914).
33. C. S. Forbes, *J. Amer. Chem. Soc.* **33**, 1973 (1911).
34. M. R. Blaise, *Corrosion et Anticorrosion* **13**, 3 (1965).
35. E. L. Schmeling, *Werkst. Korros.* **12**, 215 (1961).
36. S. Glasstone, *J. Chem. Soc.* **121**, 2091 (1922).
37. L. L. Shreir and E. L. Littauer *in* "Proceedings of the 1st International Congress on Metallic Corrosion". Butterworth (1964).
38. J. H. Morgan, *Corrosion* 128–130 (1957).
39. L. L. Shreir, *Corrosion* 118t (1961).
40. W. C. G. Wheeler, *Chem. Ind.* 75 (1959).
41. W. F. K. Wynne-Jones, *Trans. Far. Soc.* **52**, 1003 (1956).
42. C. G. Fink and R. E. Lowe, U.S. Patent 1,740,291 (1929).
43. M. A. Rabinovich, *Ukr. Khim. Zh.* **6**, 245 (1931).
44. P. B. Zhivotinski, *Ukr. Khim. Zh.* **6**, 252 (1931).
45. E. M. Khairy and A. A. Abdul Azim, *J. Electroanal. Chem.* **11**(4), 282 (1966).
46. R. G. Barradas and S. Fletcher, *Electrochim. Acta* **22**, 237 (1977).
47. R. G. Barradas, *J. Electrochem. Soc.* **122**, 103C (1975).
48. I. E. Titova, *Zaschrift. Metal.* **5**, 567 (1969).
49. I. E. Titova, *Zaschrift. Metal.* **9**, 702 (1973).
50. E. Sato, *Boshoku Gijutsu* (*Corrosion Eng.*) **17**, 449 (1968).
51. E. Sato and S. Yoshida, *Corrosion Eng.* **17**, 528 (1968).
52. E. Sato, *Corrosion Eng.* **9**, 436 (1960).
53. E. Sato, *Corrosion Eng.* **9**, 439 (1960).
54. E. Sato, *Corrosion Eng.* **11**, 436 (1962).
55. E. Sato, *Corrosion Eng.* **10**, 2 (1961).
56. E. Sato, *Corrosion Eng.* **9**, 152 (1960).
57. E. Sato, *Corrosion Eng.* **8**, 367 (1959).
58. J. Mindowicz and S. Biallozor, *Zh. Fiz. Khim.* **38**, 2828 (1964).
59. J. Mindowicz, *Khim. Ind.* (*Sofia*) **38**, 249 (1966).
60. V. A. Khitrov, *Izv. Voron. Gos. Ped. Inst.* **29**, 5 (1960).
61. I. S. Smol'yaninov and V. A. Khitrov, *Izv. Vyssh. Uchebn. Zaved. Khim. i Khim. Tekhnol.* **11**, 657 (1968).
62. C. Miura and T. Murai, *Corrosion Eng.* **18**, 91969.
63. J. Juchniewicz, *in* "Proceedings of the 1st International Congress on Metallic Corrosion". Butterworth (1962).

64. I. S. Smol'yaninov and V. A. Khitrov, *Izv. Vyssh. Zaved. Khim. i Khim. Tekhnol.* **5**, 413 (1962).
65. A. A. Abdul Azim, *Brit. Corros. J.* **8**, 76 (1973).
66. A. P. Zhuk, *Zaschrift. Metal.* **12**, 442 (1976).
67. V. N. Afanas'ev, *Chem. Abstr.* **54**, 18124.
68. I. E. Titova, *Zaschrift. Metal.* **8**, 334 (1972).
69. V. K. Gouda, *Brit. Corrosion J.* **8**, 81 (1973).
70. E. I. Storcaj, *Chimi Nefty Masinostr.* (10), 15 (1973).
71. W. J. James, *Corrosion* **23**, 15 (1967).
72. V. V. Eklik, *Chem. Abstr.* **79**, 51797d.

15   Corrosion of Lead in Miscellaneous Media and the Effects of a.c.

A. T. KUHN

Department of Dental Materials, Eastman Institute of Dental Surgery, London

Apart from the data relating to the anodic behaviour of lead or its alloys in predominantly chloride-based or sulphuric acid solutions (either with or without minor additions), all of which are discussed elsewhere in this book, there exists a limited amount of information relating to the anodic behaviour of lead or its alloys in other aqueous electrolytes. In certain cases the rationale for carrying out this work is explicitly stated. In other cases it is not, and one must guess what reason, if any, underlay the work. It is possible that a search for better electropolishing media underlay some of the papers referred to below. It will be recognized that some overlap exists between this chapter and Chapter 4, and the reader should make cross-references where appropriate.

Sulphuric–Perchloric Acid Mixtures

This mixture has been studied by a number of authors, presumably because of its use in the manufacture of Planté plates for standby accumulators. It is sufficient to say that pitting corrosion takes place in this medium. Following an early paper by Gann and Knabenbauer,[1] later work was reported by Azim[2] and most recently by the same author.[3] In the latter paper, antimonial alloys were also studied and these alloys enhanced pitting corrosion. The German work is based largely on weight-loss measurements; that of the Egyptian authors rests more on purely electrochemical data, and variation of the $HClO_4:H_2SO_4$ ratio. Data on the anodic behaviour of lead in pure perchloric acid, though mainly passivation times rather than true corrosion data, were obtained by Bushrod and Hampson,[4] but for behaviour of this acid, see Chapter 4. One figure in ref. 5 also shows data in HClO$_4$. Similar work has been reported by Ragheb.[76]

Alkaline Media

Lead does passivate in alkaline media, forming a protective oxide, but only at sufficiently high current densities where the PbO is formed and ultimately PbO$_2$, and corrosion is largely but not completely inhibited. Azim and El-Sobki[6] reviewed the earlier work, including a substantial paper by Tourky and Shallaby[7] who studied the behaviour of Pb–Ag (up to 60% Ag) alloys in alkali. Azim and El-Sobki themselves carried out potential-kinetic measurements.

No direct corrosion data were, however, adduced by them, and apart from their mention of Popova's report finding that corrosion rates both in the active and passive states decreased with increasing alkalinity,[8] no data seem to be available. Lead is not only unsuited as anode material in alkaline media, but is also unneeded in that other more suitable materials are available. Hampson[9] has made a very detailed study of the PbO$_2$/Pb^{2+} system in alkali.

In another paper,[32] overvoltage-time plots and charge vs current density plots are reported. X-ray diffraction patterns of surface films and the role of solubility of PbO in KOH are considered. Galvanostatic and potentiostatic measurements over a wide range (1–11 M) of alkali concentrations from 25°C to 80°C are reported by Popova.[33] In an earlier paper, the author[34] measures i_{crit} and PbO solubility as a function of hydroxide concentration. Kemula[35] uses polarography to follow passivation processes in alkali of a lead electrode with and without added complexing agent such as tartrate. The tartrate accelerates corrosion by attacking the passive film. The behaviour of Pb alloyed with Sb or Sn or Ag in strong alkalis is studied by Sandybaeva.[36] $i-V$ data show that silver-containing alloys inhibit oxygen evolution in alkaline solutions (in contrast to acidic media). Above 4 M NaOH, corrosion is very severe. The same authors describe the role of micro-couples in Pb–Pt or Pb–Ag alloys,[37] and the effect of temperature in a further paper.[38] Of the three alloys, 2·5% Sn, 4·5% Sb and 1% Ag at

150 mA cm^{-2}, the last two are most satisfactory in 4 M KOH. Further work is listed in refs 39 and 40. An early wide-ranging study over the pH range 0–14 giving $i-V$ data is due to Pourbaix.[41] Tourky *et al.* use alternate anodic and cathodic pulses to investigate the passivation mechanism,[42] while Abdul Azim[43] reports on de-aerated 1–6 M NaOH solutions, in particular the effect of concentration and stirring.

Other Media

Phosphate and pyrophosphate solutions

The behaviour of lead and lead–tin alloy anodes has been described by Rama Char.[10] Awad[11] has reported potentiometric measurements in a series of phosphate solutions, although no direct corrosion data are given here. Corrosion data are reported by Azim and Anwar.[12,13] Hampson has studied impedances in phosphates.[56]

Sulphamic acid solutions

Chapter 4 should be consulted. The anodic behaviour of lead in 1 M sulphamic acid at 25°C is reported by Menzies *et al.*[14] together with its behaviour in sulphamic acid–formamide solutions for electropolishing.

Borax solutions, sodium carbonate solutions and other salt solutions

The behaviour of Pb–Sn alloys in 0·1 N solutions of borax and sodium carbonate has been studied by Azim and Anwar.[15] In neutral and alkaline media, the highest corrosion resistance occurred in the eutectic alloys.

Other papers are by Khairy[44] (Pb–Sb in 0·05 M borax and 0·05 M carbonate) and Shallaby and Ashour[45] (0·1 N carbonate, Pb–Ag alloys).

Concentrated solutions of fluosilicic, perchloric and fluoboric acids were used by Nikiforova.[57] Log t (passivation) vs $A - B$ log i relationships were found in some cases, the anode passivating most readily in H_2SiF_6, and not at all in $HClO_4$ and HBF_4. A rotating disc electrode was used. Semchenko[58] studied Pb corrosion in $HClO_4$ from $-20°C$ to $+20°C$ (40% $HClO_4$). Anode gas analysis showed that O_3 and F^- ions depolarized the gas evolution process. Tourky[59] studied film formation and electrode potentials in a range of buffer solutions. Ivanov[60] reports capacitance measurements in KF, Na_2SO_4, $Ca(ClO_4)_2$ and $NaClO_4$.

Nitrate and acetate solutions, nitric acid

Johnson et al.[16] have examined the anodic dissolution of lead in 1 N solutions of KNO_3, and ammonium acetate and lead acetate at 25°C and 50°C. Oxide films formed were analysed using X-rays, and the valence of corrosion products was always less than 2. The effects of pH were also measured (see also Chapter 4).

Sokolova et al.[46] have studied 0–25% Na–Pb alloys in 2 M HNO_3 as well as 0–15% K–Pb alloys. The corrosion resistance vs composition plot shows several maxima and these are correlated with hardness and the electronic properties of the alloy. Marshall and Hampson[47] report on a number of electrochemical parameters including i–V data for 10–95% Sn alloys with Pb in nitrate electrolytes. The eutectic compositions are most stable. X-ray and S.E.M. data for films formed during anodic cyclic voltammograms on Pb are studied by Vaidyanathan et al.[48] Behaviour of Ag–Pb alloys high in silver content at 25 mA cm^{-2} is described by Klochko[49] as are Pd–Pb alloys both in 1 M HNO_3. The Ag system shows a discontinuity at 3·3% Pb, the Pd at a composition corresponding to Pb_2Pd (21·3% Pd).[50] Surface-active agents and their behaviour in 1 M HNO_3 are discussed by Loshkarev.[51] Smol'yaninov[52] studied corrosion resistance and electrode potentials of Pb in 0·1–12 M nitric acid from 20 to 100°C. The maximum corrosion rate is reported in 5 M solutions and this is explained in terms of a lessened solubility of the lead nitrate in the very strong acids. Anodic polarization curves of Bi–Pb alloys in 0·5 M nitric acid are shown by Vidal.[53] Two papers of related interest to this area of corrosion are by Lorenz,[54] describing anodic passivation of Pb in supersaturated $Pb(NO_3)_2$ solutions, and Khan and Solov'eva,[55] who describe the anodic behaviour of lead in a lead nitrate solution. Pb–Sb alloys in $NaNO_3$ have been studied by Blaise.[78]

Chromic acid solutions

These solutions are used for electroplating of chromium, almost invariably with lead or lead alloy anodes. There has therefore been a strong drive to find corrosion-resistant alloys for this medium.

Two scientific papers in the English language deal with this subject, both published more or less simultaneously, and largely supporting one another. Awad[17] investigates pure lead in a series of chromic acid solutions from 0·002 to 2 M at 30°C. It is not clear whether this work is galvanostatic or potentiostatic, or how the data were obtained, and were it not for the supporting evidence of Parker (*below*), one might express some concern in this respect. The mechanism is deduced to be one of lead chromate formation at the less anodic potentials, the chromate being converted to PbO and subsequently PbO_2 as the potential is raised. Even a PbO_2 filmed surface, however, corrodes at a substantial rate, which is proportional to the acidity of the solution, over the range 0·002–2 M. In the latter case, the apparent corrosion current density is approximately 10^{-3} A cm^{-2}, although in service, a thicker oxide film would build up and this figure would decrease. At one large installation in the U.K., for example, anodes of C.P. lead have a life of 7 years, or 1 mm year^{-1}, operated at a current density of 15 mA cm^{-2}, although in service such anodes are not continuously on load and the corrosion rate quoted is thus only a guide.

Parker[18] studied Pb, Pb–Sb and Pb–Sn alloys in 250 g l^{-1} chromic acid by steady-state potentiostatic techniques. His general findings agree, where they can be compared, with those of Awad.[17] Thus at *c.* 1 mA cm^{-2}, he observes a current density plateau which he refers to as the "brightening current" and he contrasts this plateau with the "active" → "passive" behaviour observed in sulphuric acid solutions alone. This "brightening current" was *c.* 1 mA cm^{-2} for the 10% Sb alloy, and *c.* 1·5 mA cm^{-2} for the 3% Sb alloy. In the case of tin, this trend was reversed, the similar values being 0·7 mA cm^{-2} for the 9% Sn alloy and only some 0·2 mA cm^{-2} for the 6% Sn alloy. Subsequently, the author cautions against inferring too much from these figures, and it is unfortunate that he does not appear to have studied pure lead itself in this medium. Parker also shows that no great differences are observed between anodes pre-polarized in sulphuric acid and then tested in chromic acid, and those polarized directly in the chromic acid itself. In this, his findings differ from observations made in chloride media and in the presence of cobalt ions, which are reported elsewhere in this book. Parker also investigated the difference between runs made with electrodes polished and used immediately, and those which had first been allowed to dry. He observed a difference in these two cases of a factor of 2 or less. In work described elsewhere, Andersen investigated long-term anode corrosion by weight loss, and found no differences between dried and undried anodes. Parker's observation, while doubtless valid, would seem to relate to a short-term increase in surface roughness, which is accelerated by the drying process, but which over the long term would be the same in all cases.

A number of other technologically orientated studies and informative patents have been published as far back as 1916; Ishida[19] discussed the advantages of Pb–Ag anodes, while Pelzel[24] discussed corrosion of lead and alloys at open circuit in chromic acid. A 5% Sn alloy is found to be as cost-effective as an Ag alloy and in any case better than pure lead or the Pb–Te system. Whether this recommendation can be applied to driven anodes is doubtful. British Patents 811,404 and 814,445 both specify 0·02–0·1% Ag, the latter having only 0·04–0·065% Te in addition, presumably because the 0·2–12% Sb mentioned in the former was found to have little effect. DDR Patent 101,426 advocates a Pb–Sn–Ag alloy with 1·5% Sn and 0·05% Ag. Comparing this with the traditional 10% Sb alloy which can last 6 months (or in the presence of F$^-$ ions, half that time) for a 5-mm thick anode, the authors show that while the antimonial alloy corrodes at the rate of some 30 mm year^{-1}, their ternary corrodes at only 8 mm year^{-1} (no Ag or low Ag), decreasing linearly to 4 mm year^{-1} with 0·05% Ag, after which added Ag (up to 0·5%) shows no further benefit. The same authors also mention that thallium or tellurium, in fluoride-containing solutions, aggravate waterline corrosion. Their statement that the same type of corrosion in the Pb–Sn–Ag ternary increases with increasing current load, is quite acceptable. They aimed to create a material in which the silver content would be high enough to improve anodic corrosion resistance over the Pb–Sn binary, and yet not so high that waterline corrosion became worse. Their alloy, they state, is five times longer-lived than the Pb–Sn alloy and twice as good as the Pb–Ag or Pb–Sn–Ag alloys in which the Ag exceeds 0·5%. Their two graphs show how average corrosion and waterline corrosion respond to Ag content, the former being essentially constant after 0·05% Ag, but the latter rising sharply as Ag content increases from 0·1 to 0·5%.

Four good Polish papers in this area by Krystofowicz[20–23] discuss alloys and passivation behaviour. Apart from the foregoing papers, one should also mention a fundamental study of Saber and Shams El Din[25] in which Pb anodes were subjected to open-circuit decay cycles. X-ray data taken at various points showed PbO, PbO_2 and $PbCrO_4$ phases. Cyclic voltammetry confirmed the picture postulated by other workers. Several papers have been published by Falicheva[26] showing $i–V$ data for Pb and Pb–Sn–Co alloys. Alloys containing 10% Sn and 0·5% Co seem to show good resistance.[27] The addition of 1–2 g l^{-1} of Co^{2+} ions still further reduces corrosion, presumably in the same way as discussed on pp. 369–370.[28] Elsewhere, Falicheva states that an alloy with 0·5–2% Co and up to 38% Sn is more difficult to passivate than lead, requiring 5 mA instead of 1 mA to attain a potential of 1·7 V. Oxygen evolution is stated to take place more easily. At a potential of −0·25 V (is this a mistake in translation?), 11 mA on Pb compares with 190 mA on the alloy.[29] Further work by Falicheva is unavailable in English.[30] A recent paper by Krystofowicz[31] advocates Pb–Tl alloys (up to 48% Tl, preferably 3–6%, however) and makes the interesting point that corrosion behaviour is independent of any history of plastic working of the alloy. A U.S. Patent (3,794,570) advocates alloys of Pb–Ag or Pb–Sn with more than 20 p.p.m. Ca. The tin is said to give a good adherent film while the calcium lessens distortion. A Czech paper describing novel chromium plating baths mentions 5% Sn alloys as being better than the Pb–Sb ones. German Patent 1,908,566 again prescribes Pb–Sb or Pb–Sn–Ag alloys. German Patent 2,027,575 claims compositions in the ranges 94·5–97·9% Pb, 2–4% Sn, approx. 0·05% Te and 1% Tl, Cd less than 2%. Such alloys show 0·19 g cm^{-2} weight loss compared with 0·7 g cm^{-2} for the 10% Sb alloy and 0·79 g cm^{-2} for pure lead in chromium plating baths.

The Effect of a.c. on Corrosion of Lead Anodes

Many large d.c. power supplies carry an element of a.c. "ripple" superimposed on the d.c. This can, of course, be suppressed but only at a certain cost. Other things being equal, single-phase rectifier will contain more "ripple" than a three-phase one. The effect of such a.c. on the charging of a lead–acid battery, as well as on its lifetime, has occasioned a number of studies in this area. Though nothing of substance has yet been published, it is probable that further studies must appear as the result of increasing use of thyristor "choppers" for the control of battery-powered electric vehicles. These have the effect of causing the battery to be discharged under a form of a.c. (square wave) and what effect this may have on the life of the battery is presently unknown. Insofar as lead anodes are used for cathodic protection installations, they too may be subject to the effects of a.c., not so much from the power source that drives them as from "pick-up" from neighbouring power lines, which are notorious for giving rise to stray currents. These are some of the impetuses underlying the a.c. studies reported here. The emphasis is seen to vary. In some cases, attention focuses on the quantity of PbO_2 formed on the surface of the lead, or the thickness of this layer, as well as its porosity. Such matters affect the charge capacity of a battery. Indirectly, however, because of factors such as true surface area (roughness or porosity effects), it will also affect corrosion rates. Almost all the studies relate to 50 Hz a.c. Because this can be considered to be a low frequency, the subsequent electrochemical behaviour depends largely on the extent to which the corrosion or reaction products can diffuse away before the polarity is reversed, and also on the "reversibility" of the reactions, in particular oxygen evolution/reduction or metal oxidation/reduction. Factors affecting the rate of escape (by diffusion) of products are extremely complex. One author, Mikhailovskii,[61-65] has published several papers, in which attempts were made to quantify this situation, by means of rotating disc electrodes and otherwise. In the first,[61] he reported the effects of a.c. in general at 50 Hz in acid and alkaline media. Later,[62,63] he considered the effect of stirring with sinusoidal and square wave a.c. Working in HCl[64] he found corrosion rates of up to 10 mg cm^{-2} day^{-1}. Finally,[65] he produced a complete analysis of the diffusional situation in 1 M KNO_3, comparing experimental with predicted results.

Several Indian authors have studied the effect of a.c. A good and detailed paper by Ghosh[66] used various percentages (up to 400%) of a.c. on d.c. and analysed the surface layers using thermogravimetric, chemical (total PbO_2), X-ray and electrochemical methods. Alternating current did increase the amount of PbO_2 formed, as shown by discharge curves, though the effect was only a few percent. Vijayavalli has published at least four papers.[67-69,77] In one paper,[67] he likened the effect of a.c. to that of the "forming agent" used in Planté battery production, and indeed the analogy held on a semi-quantitative basis. Weight loss was doubled at $2\,mA\,cm^{-2}$, and resistance of the surface film also decreased when a.c. was used, presumably because of increased porosity. He suggested[68] that 10–20% a.c. provided the greatest increase in film growth. An early though thorough paper is that by Costa and Hoar[70] who found that up to 4% a.c. increased PbO_2 growth, but that total metal corrosion was reduced very slightly under most conditions, both findings agreeing with those of Vijayavalli.

A number of other papers exist. Sapozhnikov[71] used a.c. without d.c. in $10\,N$ H_2SO_4 from 10–70°C at $0\cdot15\,A\,cm^{-2}$. Yamuna and Subramanyan[72] worked at 5 and 25 mA cm^{-2} in $1\,N$ sulphuric acid at 50 Hz. Amy and Mounjos[73] reported weight losses in $CaSO_4$ solutions at 1–2 A dm^{-2} (a study of anode corrosion in simulated earth conditions). Pereyaslov[74] studied the effect of an asymmetric voltage function of 6 and 12% Sb alloys at current densities of 10–50 mA cm^{-2} or greater. Again the corrosion was less than with d.c. The work of Juchniewicz in HCl has been referred to on p. 396. Sekido and Katoh[75] used a.c. methods to estimate corrosion rates on lead–antimony alloys (square wave technique).

References

1. W. Gann and W. Knabenbauer, *Werkst. Korros.* **18**, 597 (1967).
2. A. A. Azim, *Corrosion Sci.* **10**, 421–433 (1970).
3. A. A. Azim and S. E. Afifi, *Corrosion Sci.* **12**, 603 (1972).
4. C. J. Bushrod and N. A. Hampson, *Brit. Corrosion J.* **6**, 87 (1971).
5. A. Awad, *J. Electroanal. Chem.* **34**, 431 (1972).
6. A. A. Abdul-Azim and K. M. El-Sobki, *Corrosion Sci.* **12**, 207 (1972).
7. A. R. Tourky and L. A. Shallaby, *J. Chem. U.A.R.* **11**, 191 (1968).
8. S. S. Popova, cited in *Chem. Abstr.* **65**, 8338h (1966).
9. N. A. Hampson and J. P. Carr, *Ber. Bunsenges.* **74**, 557 (1970).
10. T. L. Rama Char, *Corrosion Prevent. Control* **5**, 37 (1958).
11. A. Awad and Z. A. Elhady, *J. Electroanal. Interfacial Chem.* **20**, 79 (1969).
12. A. A. Azim and M. M. Anwar, *Corrosion Sci.* **9**, 193 (1969).
13. A. A. Azim and M. M. Anwar, *Corrosion Sci.* **9**, 245 (1969).
14. I. A. Menzies and G. W. Marshall, *Corrosion Sci.* **9**, 287 (1969).
15. A. A. Azim and M. M. Anwar, *Corrosion Sci.* **9**, 193 (1969).
16. J. W. Johnson, C. K. Wu and W. J. James, *Corrosion Sci.* **8**, 309 (1968).
17. S. A. Awad and Kh. Kamel, *J. Electroanal. Interfacial Chem.* **34**, 431 (1972).
18. T. E. Parker, *J. Electrochem. Soc.* **119**, 13C (1972).
19. T. Ishida, *Jap. J. Soc. Chem. Eng.* **39**, 484 (1936).

20. K. Krzysztofowicz, *Prace Inst. Hutnic.* (*Poland*) **14**(5), 253–261 (1962).
21. K. Krzysztofowicz, *Ochrona Przed. Korozja* **16**(6), 165–172 (1973).
22. K. Krzysztofowicz, *Prace Inst. Hutnic.* (*Poland*) **22**(4), 215–221 (1970).
23. K. Krzysztofowicz, *Prace Inst. Hutnic* (*Poland*) **22**(3), 159–168 (1970).
24. E. Pelzel, *Metallwissenschaft* **21**, 1039 (1967).
25. T. M. Saber and A. M. Shams El Din, *Electrochim. Acta* **13**, 937 (1968).
26. A. I. Falicheva *et al.*, *Zaschrift. Metal.* **11**, 377 (1975).
27. A. I. Falicheva *et al.*, *Izv. V. U. Z. Khim. Tekhnol.* **19**, 433 (1976) (*Metal. Abstr.* **10**, 35–1124).
28. A. I. Falicheva *et al.*, *Zaschrift. Metal.* **11**, 507 (1975).
29. A. I. Falicheva *et al.*, *Zaschrift. Metal.* **11**, 377 (1975).
30. A. I. Falicheva *et al.*, *Chem. Abstr.* **83**, 67756h.
31. K. Krystofowicz, *Prace Inst. Hutnic.* **22**, 65 (1970).
32. J. P. G. Farr and N. Hampson, *Electrochem. Technol.* **6**, 10 (1968).
33. S. S. Popova and A. V. Fortunatov, *Sov. Electrochem.* **2**, 678 (1966).
34. S. S. Popova and A. V. Fortunatov, *Sov. Electrochem.* **2**, 446 (1966).
35. W. Kemula, *Roczniki Chem.* **31**, 205 (1957).
36. A. B. Sandybaeva *et al.*, *Chem. Abstr.* **81**, 130163p.
37. A. B. Sandybaeva *et al.*, *Metal. Abstr.* **6**, 35–0659.
38. A. B. Sandybaeva *et al.*, *Metal. Abstr.* **8**, 35–1719 (1975).
39. A. B. Sandybaeva *et al.*, *Metal. Abstr.* **6**, 35–0660.
40. A. B. Sandybaeva *et al.*, *Chem. Abstr.* **83**, 17620p.
41. M. Pourbaix, *Cebelcor Rapport Tech.* No. 13 (1953).
42. A. R. Tourky, *J. Chem. U.A.R.* **11**, 177 (1968).
43. A. A. Abdul Azim, *Corrosion Sci.* **12**, 207 (1972).
44. E. M. Khairy, *J. Electroanal. Chem.* **11**, 282 (1966).
45. L. A. Shallaby and S. Ashour, *J. Chem. U.A.R.* **12**, 335 (1969).
46. I. V. Sokolova, *Zaschrift. Metal.* **9**, 332 (1973).
47. A. Marshall and N. Hampson, *Corrosion Sci.* **15**, 23 (1975).
48. H. Vaidyanathan, *J. Electrochem. Soc.* **121**, 876 (1974).
49. M. A. Klochko and Z. S. Medvedeva, *Chem. Abstr.* **49**, 5998f.
50. M. A. Klochko and Z. S. Medvedeva, *Chem. Abstr.* **49**, 5998d.
51. M. A. Loshkarev, *Chem. Abstr.* **53**, 1068c.
52. I. S. Smolyaninov, *Chem. Abstr.* **74**, 150154.
53. C. Vidal, *Mem. Sci. Rev. Metal.* **67**, 809 (1970).
54. W. Lorenz, *Z. Phys. Chem.* **20**, 95 (1959).
55. O. A. T. Khan and V. I. Solov'eva, *Zh. Prikl. Khim.* **34**, 1793 (1961).
56. J. P. Carr and N. Hampson, *J. Electroanal. Chem.* **28**, 65 (1970).
57. M. M. Nikiforova and Z. A. Iofa, *Dokl. Akad. Nauk. SSSR* **115**, 1131 (1957).
58. D. P. Semchenko, *Chem. Abstr.* **85**, 132728.
59. A. R. Tourky, *J. Chem. U.A.R.* **10**, 253 (1967).
60. V. F. Ivanov and Z. N. Ushakova, *Electrokhimya* **9**, 787 (1973).
61. Yu. N. Mikhailovskii, *Zh. Fiz. Khim.* **37**, 340 (1963).
62. Yu. N. Mikhailovskii, *Zh. Fiz. Khim.* **38**, 995 (1964).
63. Yu. N. Mikhailovskii, *Zh. Prikl. Khim.* **37**, 789 (1964).
64. Yu. N. Mikhailovskii, *Chem. Abstr.* **71**, 97642n.

65. Yu. N. Mikhailovskii, *Chem. Abstr.* **81**, 71743c.
66. S. Ghosh, *Electrochim. Acta* **14**, 161 (1969).
67. R. Vijayavalli, *J. Electrochem. Soc.* **110**, 1 (1963).
68. R. Vijayavalli, *Bull. Acad. Polon. Sci. Ser. Sci. Chim.* **10**, 13 (1962).
69. R. Vijayavalli, *Batterien* **19**, 780 (1965).
70. J. M. Costa and T. P. Hoar, *Corrosion Sci.* **2**, 269 (1962).
71. A. V. Sapozhnikov, *Metal. Abstr.* **7**, 35–1156 (1974).
72. A. R. Yamuna and N. Subramanian, *Werkst. Korros.* **21**, 607 (1970).
73. L. Amy and C. Mounjos, *Rev. Gen. Elec.* **66**, 187 (1957).
74. P. P. Pereyaslov, *Chem. Abstr.* **86**, 62579e.
75. S. Sekido and S. Katoh, *Chem. Abstr.* **72**, 9607k.
76. A. Ragheb, *Werkst. Korros.* **16**, 755 (1965).
77. R. Vijayavalli and H. V. K. Udupa, Symp. Lead–acid Batteries, Calcutta, 1968 (*Metal. Abstr.* **4**, 35–0337).
78. M. R. Blaise, *Corrosion-Anticorrosion* **13**, 3 (1965).

General Electrochemical Bibliography of Lead and its Compounds

Numerous references relating to the electrochemical behaviour of lead, together with additional works covering certain physicochemical aspects likely to be of interest to electrochemists, as well as battery-related and industrial processes, have been collated by the editor. Invaluable assistance was given by Miss Brenda Marsden of Salford University Library, while Mrs Liz Williams, of Chloride Technical Ltd, kindly allowed (limited) access to their information retrieval system. Thanks are also due to the librarian of the Lead Development Association in London.

The classification is, to some extent, arbitrary and readers should peruse several sections to ensure that they be alerted to material of interest to them. These sections are: molten salts (pp. 414–418); related physical properties (pp. 418–419); batteries (pp. 419–434); anodes and corrosion (pp. 434–439); applied and industrial (pp. 439–444); lead electrodeposition (pp. 444–448); and various (pp. 448–460). The editor is well aware that a greater degree of classification could have been introduced, but the final decision was to go for the most up-to-date possible citations rather than "polishing" less up-to-date material.

Some of the titles are, frankly, uninformative. Others must arouse the curiosity of us all. At least one intention in assembling this mass of references is that the reader might thereby be allowed to scan the literature in a way not possible when undertaking systematic searches, whether these be manual or computer-based. Perhaps one might describe this ideal as an acceleration of serendipity!

References are followed (in the case of less easily accessible material) by a *Chemical Abstracts* or *Lead Abstracts* reference number. In the former, for example, **82** 1145 denotes volume number and abstract. In the latter case, the numeration system was changed and early lead abstracts are denoted as, say, 1966 3421 while later ones read 16-0443, denoting 1976 and so on. Finally, it should be noted that one of the four most important journals in the field changed its title from *Journal of Electroanalytical Chemistry* to *Journal of Electroanalytical and Interfacial Chemistry*. For consistency, the former title (abbreviated) is used.

Molten Salts

Potentials of metals in molten potassium.

A. J. Arvia and H. A. Videla, *Electrochim. Acta* 11, 537–544 (1966).

Formation of metal layers.

D. J. Astley, J. A. Harrison and H. R. Thirsk, *J. Electroanal. Chem.* 19 (4) 325–334 (1968).

Fused salt systems.

"Encyclopaedia of the Electrochemistry of the Elements" (A. J. Bard, ed.), Vol. 10. Marcel Dekker, New York (1976).

Linear sweep voltammetry of Ni(II), Co(II), Ca(II) and Pb(II), at glassy electrodes in molten lithium chloride–potassium chloride eutectic.

W. K. Behl, *J. Electrochem. Soc.* 118 (6), 889–894 (1971).

Significance of the solubility of metals in salt melts with respect to technical electrochemistry.

R. Bertram, D. Wiebe and P. Lambrecht, *Metalloberfläche* 24, 277–281 (1970).

Electrolyte for obtaining a lead–sodium–potassium alloy.

V. B. Busse-Machukas *et al.*, USSR Patent 529, 262 (1976).

Cathodic implantation of lead in platinum from molten salts studied by stripping polarography.

N. G. Chovnyk and A. M. Formichev, *Fiz. Khim. Elektrokhim. Rasplavl. Solei Tverd. Elektrolitov.* No. 2, 44–46 (1973); *Chem. Abstr.* 81, 144607 (1974).

Electrorefining of lead–zinc alloys by use of packed-bed electrodes.

J. H. Cleland and D. J. Fray, *in* "Advances in Extractive Metallurgy 1977" (M. J. Jones, ed), pp. 141–146. Institution of Mining and Metallurgy, London (1977).

Chronopotentiometric measurements in a fused aluminising electrolyte.

Yu. K. Delimarskii and V. F. Makogon, *Protect. Metals* (translation of *Zasch. Metal.*) 5, 107–109 (1969).

Anodic polarization of pure and impure lead in silicate melts.

Yu. K. Delimarskii, Y. E. Kosmaty and A. V. Gorodysky, *Ukr. Khim. Zh.* 34 (7), 673–677 (1968); *Met. Abstr.* 1, 1369 (1968).

Electrolytic recovery of lead from molten salts.

Yu. K. Delimarskii, O. G. Zarubitskii and V. G. Budnik, *Tsvetn. Metal.* No. 4, 27–30 (1968).

Polarographic and chronopotentiometric studies of lead oxide in fused sodium hydroxide.

Yu. K. Delimarskii, O. G. Zarubitskii and V. G. Budnik, *Zh. Prikl. Khim.* 42, 2493–2498 (1969).

Production of lead powder by the electrolysis of fused salts.

Yu. K. Delimarskii, V. F. Grishchenko and L. I. Zarubitskaya, *Poroshk. Metall.* 4, 26–27 (1977).

Electrochemical removal of bismuth from lead in sodium hydroxide melts.

W. T. Denholm, R. Dorin and H. J. Gardner, *in* "Advances in Extractive

Metallurgy 1977" (M. J. Jones, ed.), pp. 235–244. Institution of Mining and Metallurgy, London (1977).
Thermodynamics of cadmium halide–sodium halide and lead halide–sodium halide molten mixtures by e.m.f. measurements.
C. G. M. Dijkhuis and J. A. A. Ketelaar, *Electrochim. Acta* **11** (11), 1607–1627 (1966).
Thermodynamic analysis of the Pb–Cd system by measuring the electromotive forces.
B. Dobovisek and B. Pretnar, *Rudarsko. Met. Zbornik* **3**, 319–330 (1961); *Lead Abstr.* **3**; 1985 (1963).
Behaviour of lead(II)ions in molten NaOH + KOH eutectic.
A. Eluard and B. Tremillon, *J. Electroanal. Chem.* **30** (2), 323–326 (1971).
Ionic conductivity in some bivalent metal sulphides.
R. Galli and F. Garbassi, *Nature* **353** (5494), 720–722 (1975).
Lead chloride melt. Current yield in electrolysis.
A. A. Galn'bek *et al.*, *Zh. Prikl. Khim.* **4**, 787–795 (1962).
Regenerative battery.
General Motors Corporation, U.S. Patent 3,245,836 (1960); *Lead Abstr.* **6**, 4222 (1966).
A thermodynamic study of the $PbO–Na_2O–SiO_2$ system.
A. E. Grau and S. N. Flengas, *J. Electrochem. Soc.* **123**, 352–358 (1976).
Electrical double layer in molten salts. Part 1. The potential of zero charge.
A. D. Graves, *J. Electroanal. Chem.* **25**, 349–356 (1970).
Electrical double layer in molten salts. Part 2. The double layer capacitance.
A. D. Graves and D. Inman, *J. Electroanal. Chem.* **25**, 357–372 (1970).
Differential composition of alloys released electro-chemically at a liquid cathode.
V. Z. Grebenik and F. I. L'vovich, *Izv. Vyssh. Ucheb. Zaved. Tsvetn. Met. USSR*, No. 5, 40–43 (1971).
Thermopotential measurements for molten $CdCl_2$, $CdBr_2$, and $PbCl_2$.
J. Greenberg, D. E. Weber and L. H. Thaller, *J. Phys. Chem.* **67**, 88–91 (1963).
Polarization of liquid metallic electrodes in the separation of lead–antimony alloys.
I. T. Gul'din and A. A. Rozlovskii, *Izv. Vysshikh Uchebn. Zavedenii, Tsvet. Met.* **3**, 67–70 (1969).
Recovery of lead from lead chloride by fused-salt electrolysis.
F. P. Haver, C. H. Elges, D. L. Bixby and M. M. Wong, U.S. Bureau of Mines, Rep. Invest. R I 8166, 18 pp (1976).
Chronopotentiometry: a review of applications in molten salts.
R. K. Jain, H. C. Gaur, E. J. Frazer and B. J. Welch, *J. Electroanal. Chem.* **78**, 1–30 (1977).
Molten Salts: Vol. 4, Part 2. Chlorides and mixtures: electrical conductance, density, viscosity and surface tension data.
G. J. Janz, R. P. T. Tomkins, C. B. Allen, J. R. Downey, Jr., G. L. Gardner, U. Krebs and S. K. Singer, *J. Phys. Chem. Ref. Data* **4**, 871–1178 (1975).

Molten Salts: Vol. 4, Part 3. Bromides and mixtures, iodides and mixtures. Electrical conductance, density, viscosity and surface tension data.

    G. J. Janz, R. P. T. Tomkins, C. B. Allen, J. R. Downey, Jr. and S. K. Singer, *J. Phys. Chem. Ref. Data* **6**, 409–596 (1977).

Information sources for molten salts.

    G. J. Janz, C. B. Allen and R. P. T. Tomkins, *J. Electrochem. Soc.* **124**, 51C–54C (1977).

Kinetic study of fused lead electrode in molten salts.

    K. Kitazawa *et al.*, *J. Electrochem. Soc. Japan* **37** (1) 45–51 (1969); *Met. Abstr.* **2**, 1684 (1969).

Electrical conductivity of molten lead silicate and borate.

    K. A. Kostanyan and O. K. Geokchyan, *Arm. Khim. Zh.* **21**, 230–240 (1968); *Chem. Abstr.* **69**, 99014 (1968).

Effects of sulphates on the production of lead–sodium or lead–potassium alloys by electrolysis of chloride–carbonate melts.

    F. I. L'vovich, V. B. Busse-Machukas, A. G. Morachevskii and V. I. Markin, *Zh. Prikl. Khim.* **47**, 1654–1655 (1974); *Chem. Abstr.* **81**, 130168 (1974).

Standard electrode potentials of Cd/Cd(II), In/In(III), Pb/Pb(II), TI/Tl(I), Zn/Zn(II), in molten alkali acetates.

    R. Marassi *et al.*, *J. Electroanal. Chem.* **22** (2), 215–219 (1969).

Effect of the electrolyte composition on the rate of the electrolysis process in the production of a lead–sodium potassium alloy.

    V. I. Markin *et al.*, *Tr. Leningrad Politekh. Inst.* **348**, 74–76 (1976); *Chem. Abstr.* **85**, 00043 (1976).

Preparation of lead–alkali metal alloys from melts.

    S. P. A. Montecatini, French Patent 1,318,364, Belgian Patent 615,693 (1962).

Ways of improving the production of a lead–sodium–potassium alloy by electrolysis of a chloride–soda melt under industrial conditions.

    A. G. Moracherskii *et al.*, *V. Sb.*, *Ionnye Rasplavy* (*Kiev*) **3**, 194–196 (1975); *Chem. Abstr.* **85**, 111607 (1976).

Preparation of lead by fused-salt electrolysis.

    J. E. Murphy, F. P. Haver and M. M. Wong, Metal–Slag–Gas Reaction Processes, Proceedings of the International Symposium (1975); *Chem. Abstr.* **83**, 67685 (1975).

Lead potentials in a mixture of molten sodium and potassium chlorides containing lead ions.

    I. F. Nichkov, S. P. Rasponin and S. V. Karzhavin, *Izr. Vysshickh Uchebn. Zavedenii, Tsvetn. Met.* **6**, 83–86 (1963).

Behaviour of lead and lead (II) oxide electrodes in molten alkali metal nitrates.

    J. J. Podesta, R. C. V. Piatti and A. J. Arvia, *An. Assoc. Quim. Argent.* **62**, 267–277 (1974); *Chem. Abstr.* **83**, 87183 (1975).

Process for producing fluorine and sodium lead alloy.

    E. I. du Pont de Nemours and Co. U.S. Patents 3,196,090 and 3,196,091 (1964).

Electrodeposition of molten low-melting metals and alloys from fused-salt systems.

G. L. Schnable and J. G. Javes, *Electrochem. Technol.* **2**, 201–206 (1964).
Anionic conductivity in halogen-containing lead silicate glasses.

P. C. Schultz and M. S. Mizzoni, *J. Amer. Ceram. Soc.* **56** (2) 65–68 (1973).
The e.m.f. method for determining thermodynamic properties of molten alloys whose components have similar electrochemical properties. Study of the lead–tin system.

N. I. Shurov, N. G. Ilyushchenko, A-I. Arsfinogenov and T. L. Mitrofanova, Deposited Doc. VINITI 1039–1074, 10 pp. (1974); *Chem. Abstr.* **86**, 162056 (1977).
Cathodic polarisation of a lead cathode during lead, copper, and zinc extraction from slag.

Y. M. Sizov, O. A. Esin, A. N. Kvyatovskii and M. A. Abdeev, *Izv. Akad. Nauk Kaz. SSSR, Ser. Khim.* **17**, 14–18 (1967); *Lead Abstr.* **8**, 5154 (1968); *Chem. Abstr.* **67**, 110757 (1967).
Preparation of lead–potassium–sodium alloys by electrolysis.

Soc. Minière et Metallurgique de Penarroya, French Patent 1,311,228 (1961).
Concentration cells with quartz membranes in molten salts: $PbCl_2$–$NaCl$, $PbBr_2$–$NaBr$, PbI_2–NaI systems.

S. Sternberg and C. Hrdlička, *Electrochim. Acta* **13**, 863–871 (1968).
Kinetics of parasitic processes in the electrolysis of molten sodium chloride with a liquid cathode.

L. Suski and K. Kubisz, *Electrochim. Acta* **12**, 1161–1170 (1967).
Some laws of the formation of deposits of molten metal on a solid cathode.

E. A. Ukshe and V. N. Devyotkin, *Zh. Prikl. Khim.* **38**, 431–439 (1965).
Influence of the nature of an electrolyte in electrocapillary curves of lead in salt melts.

E. A. Ukshe and I. V. Tomskikh, *Dokl. Akad. Nauk SSSR* **150**, 347–348 (1963).
Investigation of the electric double layer in salt melts.

E. A. Ukshe *et al.*, *Electrochim. Acta* **9**, 431–439 (1964).
The electronic commutator determination of E^0 of formation and related thermodynamic quantities for molten lead chloride.

T. B. Warner and R. L. Seifert, *J. Phys. Chem.* **69**, 1034–1039 (1965).
Manufacture of lead–sodium alloys by melt electrolysis.

G. Wehner and J. Liebig, *Freiberg. Forschungsh. B.* **165**, 5–53 (1973).
Electrode potential measurements in molten salts.

D. G. Winter and A. M. Strachan, *in* "Advances in Extractive Metallurgy 1977" (M. J. Jones, ed.). Institution of Mining and Metallurgy, London (1977).
Fused-salt electrolysis for production of lead and zinc metals.

M. M. Wong and F. P. Haver, *in* "Molten Salt Electrolysis in Metal Production 1977", pp. 21–29. Institution of Mining and Metallurgy, London (1977).
Cathodic solution and anodic deposition of lead in electrolysis of fused potassium hydroxide.

N. F. Zakharchenko, *Ukr. Khim. Zh.* **40**, 764 (1974).

Electrochemical investigations of liquid lead–arsenic solutions.
> E. Zaleska, *Roczniki Chemii* **48**, 195–200 (1974).

Electrolytic production of lead power in melted sodium hydroxide.
> O. G. Zarubitskii, N. F. Zakharchenko and V. G. Budnik, *Poroshk. Metall.* **4**, 5–7 (1975).

Cathodic–anodic separation of lead bismuthide in ionic melts.
> O. G. Zarubitskii, V. G. Budnik and A. A. Omel'chuk, *Zh. Prikl. Khim.* **49**, 1894 (1976).

Electrical conductivity of melts of the $NaCl–CaCl_2–PbCl_2$ system.
> V. L. Zolotarev and O. I. Egerev, *Zh. Prikl. Khim.* **42**, 1184–1185 (1969).

Related Physical Properties

Physicochemical properties of oxide films on Pb and alloys.
> I. A. Aguf, *Sb. Rab. Khim. Istochnikam. Toka* (*USSR*) (10) 34; *Lead Abstr.* 16–2719 (1975).

Electrochemical properties of grey lead oxide.
> K. Appelt, *Electrochim. Acta* **13**, 1521 (1968).

Surface conductivity of grey lead oxide.
> K. Appelt, *Electrochim. Acta* **13**, 1727 (1968).

Novel lead titanate, $PbTi_3O_7$.
> K. Aykan, *J. Amer. Ceram. Soc.* **51**, 577 (1968).

Electrical conductivity of $PbCl_2–PbS$ mixtures.
> M. C. Bell, *J. Electrochem. Soc.* **113**, 31 (1966).

Ferroelectric properties of lead titanate.
> V. G. Bhide, *J. Amer. Ceram. Soc.* **51**, 565 (1968).

Conductivity of PbO.
> S. B. Brody, *J. Chem. Phys.* **52**, 1000 (1970).

Thermodynamics properties of $PbCl_2$, $SnCl_2$, and CnCl from e.m.f. measurements as galvanic cells with a solid electrolyte.
> W. G. Bugden and R. A. J. Shelton, *Inst. Mining Met. Trans.* **79**, C215–C220 (1970).

Oxides of lead-solubility in alkalis.
> P. Chartier, *Bull. Soc. Chim. Fr.* (7) 2253 (1969).

Thermodynamics of lead–oxygen at 25°C—pseudocubic oxides.
> P. Chartier, *C.R. Acad. Sci.* **258**, 4495 (1964).

Electrical properties of lead titanate.
> W. R. Cook, *J. Appl. Phys.* **34**, 1392 (1963).

Semiconductor properties of PbTe, PbSe, PbS, PbO.
> R. Dalven, *Infrared Phys.* **9**, 141 (1969).

Electrical properties of system $PbZrO_3–PbTiO_3–PbNb_2O_6$.
> R. H. Dungan, *J. Amer. Ceram. Soc.* **45**, 382 (1962).

Thermo E.M.F. of molten couples.
> Ya. I. Dutchak, O. P. Stets'kev and V. Ya. Prokhorenko, *Phys. Metals Metallog.* **14**, 135–137 (1963).

Anomalous electrical conductivity of PbO_2 powder under pressure.
> J. Hempelmann, *Naturwissenchaften* **62**, 343 (1975).

Pressure and electrical conductivity of PbO_2.
P. Herger, *J. Appl. Electrochem.* **7**, 417 (1977).
Conductivity of doped PbF_2.
A. V. Joshi, *J. Electrochem. Soc.* **124**, 1253 (1977).
Ionic conductivity of beta–lead fluoride.
J. H. Kennedy, *J. Electrochem. Soc.* **123**, 47 (1976); also R. W. Bonne, *J. Electrochem. Soc.* **124**, 28 (1977).
Solid electrolyte properties and crystal forms of lead fluoride.
J. H. Kennedy, R. Miles and J. Hunter, *J. Electrochem. Soc.* **120**, 1441–1446 (1973).
Physical properties of sputtered PbO_2 films.
F. Lappe, *Phys. Chem. Solids* **23**, 1563 (1962).
Properties of lead titanate–lead zirconate ceramics.
P. D. Levett, *Amer. Ceram. Soc. Bull.* **42**, 348 (1963).
Conduction of polycrystalline PbF_2.
G. C. Liang, *J. Electrochem. Soc.* **122**, 466 (1975).
Metastable form of $PbTiO_3$.
F. W. Martin, *Phys. Chem. Glasses* **6**, 43 (1965).
$PbO-TiO_2$ system.
Y. Matsuo, *J. Amer. Ceram. Soc.* **46**, 409 (1963).
Mn-doped $PbTiO_3$ ceramics.
Y. Matsuo, *J. Amer. Ceram. Soc.* **48**, 111 (1965).
Microcracking of $PbTiO_3$.
Y. Matsuo, *J. Amer. Ceram. Soc.* **49**, 229 (1966).
Free energies of formation of compounds in system $PbO-TiO_2$.
G. M. Mehrotra, *Scripta Metal* **7**, 1047 (1973).
E.M.F. of PbO_2–Ti at high temperatures.
G. M. Mehrotra, *Metall.* **30**, 839 (1976).
Electrical conductivity of PbO_2.
U. B. Thomas, *J. Electrochem. Soc.* **94**, 42 (1948).
Effect of additives on properties of $PbTiO_3$.
T. Y. Tien, *J. Amer. Ceram. Soc.* **45**, 567 and 572 (1962).
Piezo ceramic–$PbTiO_3$.
British Patent 888,740 (1962); French Patent 1,343,275 (1963).
$PbTiO_3$ as an electromechanical transducer.
German Patent 1,125,341 (1963).

Batteries

Evaluation of battery negative expanders.
A. A. Abdul, *J. Appl. Electrochem.* **4**, 351 (1974).
Inorganic additives and positive plate cycle life.
A. A. Abdul Azim, *J. Appl. Electrochem.* **7**, 119 (1977).
The formation of lead dioxide electrodes by the Planté process.
S. E. Afifi, *Surf. Technol.* **4**, 173 (1976).

The formation of lead dioxide electrodes by the Planté process.
  S. E. Afifi, W. H. Edwards and N. A. Hampson, *Surf. Technol.* **4**, 173–185 (1976).
Comparative study of depassivators for battery negative plates.
  I. A. Aguf, *Sb. Rab. Khim. Istochik. Toka. Uses* No. 7, 35 (1972); *Chem. Abstr.* **80**, 9724.
Some problems of the thermodynamics of the lead dioxide electrode.
  I. A. Aguf, *Russ. J. Phys. Chem.* **39**, 598 (1965).
Effect of Na_2SO_4 on battery positives.
  I. A. Aguf, *Sov. Electrochem.* **13**, 1613 (1977).
Liquid Pb electrode for oxidation of carbonaceous fuels.
  M. Anbar, *Rec. 10th Intersoc. Energy conversion* 48 (1975); *Chem. Abstr.* **84**, 108405.
The intermediate oxides of lead.
  J. S. Anderson and M. Sterns, *J. Inorg. Nucl. Chem.* **11**, 272 (1959).
Electrode potentials and thermal decomposition of α and β-PbO_2.
  R. T. Angstadt, *J. Electrochem. Soc.* **109**, 177 (1962).
Investigation of the crystal structure and electrochemical properties of grey oxide: consideration of the thermodynamics of the PbO_2/H_2O system in the presence of Cl^- ions.
  K. Appelt, ZDA/LDA Translation No. 69/47 (1969).
The anodic dissolution of Pb in H_2SO_4.
  G. Archdale, *J. Electroanal. Chem.* **39**, 357 (1972).
The electrochemical dissolution of Pb to form $PbSO_4$ by a solution precipitation mechanism.
  G. Archdale and J. A. Harrison, *J. Electroanal. Chem.* **34**, 21 (1972).
Phase changes during the manufacture of lead–acid battery plates.
  J. Armstrong *et al.*, *in* "Power Sources 1966" (D. H. Collins, ed.), p. 163. Pergamon, Oxford (1967).
Anodic corrosion of lead in sulphuric acid.
  I. I. Astakhov, *Amer. Soc. Metals Rev. Metal Lit.* **21**, 45 (1964).
Anodic passivation of Pb in various media.
  A. A. A. Azim, *Corrosion Sci.* **9**, 245 (1969).
Optical microscopy of battery plates.
  N. E. Bagshaw and K. P. Wilson, *Electrochim. Acta* **10**, 867 (1965).
Model of battery with porous electrode.
  B. S. Baker, *J. Electrochem. Soc.* **120**, 1005 (1973).
Conditions for the formation of α or β lead dioxide during the anodic oxidation of lead.
  R. A. Baker, *J. Electrochem. Soc.* **109**, 337 (1962).
Solar activity and effect on battery capacity.
  D. R. Barber, *Nature* **195**, 684 (1962).
"Encyclopedia of Electrochemistry of the Elements" Vol. 1 (Lead).
  A. J. Bard (ed.). Dekker, New York (1973).
Potential-pH diagram of lead in the presence of sulphate ions.
  S. C. Barnes, *in* "Batteries 2" (D. H. Collins, ed.), p. 41. Pergamon, Oxford (1964).

Pourbaix diagram of Pb in presence of sulphate.
S. C. Barnes, *4th Int. Battery Symp.*, *Brighton* Sept. (1964); *Lead Abstr.* (1964) 2884.

The behaviour of the lead dioxide electrode.
W. H. Beck, *Trans. Far. Soc.* **50**, 136 (1954).

The behaviour of the lead dioxide electrode. Part 2. The irreversible discharge of the lead accumulator.
W. H. Beck *et al.*, *Trans. Far. Soc.* **50**, 147–152 (1954).

Voltage characteristics of lead-acid battery during discharge.
D. Berndt, *4th Int. Battery Symp. Brighton*, Sept. (1964); *Lead Abstr.* 1964 2883.

The effect of temperature and current density on the utilisation of lead and lead oxide electrodes.
D. Berndt, *in* "Power Sources 1968" (D. H. Collins, ed.).

Lead–acid batteries with low antimony alloys.
D. Berndt, *J. Power Sources* **1**, 3 (1976).

New data on basic Pb sulphates.
H. W. Billhardt, *J. Electrochem. Soc.* **117**, 690 (1970).

New data on basic lead sulphates.
H. W. Billhardt, *J. Electrochem. Soc.* **117**, 690 (1970).

Radiographic study of current distribution in battery plates.
H. Bode, *Electrochim. Acta* **11**, 1211 (1966).

Some potential and exchange studies of α-lead dioxide.
S. J. Bone, *Electrochim. Acta* **4**, 292 (1961).

Use of carbon replicas in electron microscopy.
D. E. Bradley, *J. Appl. Phys.* **27**, 1399 (1956).

Expander action in lead–acid battery.
M. P. J. Brennan, *J. Electroanal. Chem.* **48**, 465 (1973).

The role of antimony in the lead–acid battery.
M. P. J. Brennan, *J. Appl. Electrochem.* **4**, 49 (1974).

Expander action in lead–acid battery.
M. P. J. Brennan, *J. Electroanal. Chem.* **52**, 1 (1974).

Effect of phosphoric acid on battery positives.
K. P. Bullock, *J. Electrochem. Soc.* **124**, 1478 (1977).

Anodization of lead in sulfuric acid.
J. Burbank, *J. Electrochem. Soc.* **103** (2), 87–91 (1956).

Anodization of lead and lead alloys in sulphuric acid.
J. Burbank, *J. Electrochem. Soc.* **104**, 693 (1957).

The anodic oxides of lead.
J. Burbank, *J. Electrochem. Soc.* **106** (5), 369–376 (1959).

Morphology of PbO_2 crystals in the positive plates of some lead–calcium cells.
J. Burbank, *Electrochem. Soc. Ext. Abstr.* (1962).

Morphology of PbO_2 in the positive plate of lead–acid cells.
J. Burbank, *J. Electrochem. Soc.* **111** No. 7, 765 (1964).

Role of Sb in battery positive plates.
J. Burbank, *J. Electrochem. Soc.* **111**, 1112 (1964).

Anodic oxidation of $PbSO_4$, Pb_3O_4, $2PbCO_3 \cdot Pb(OH)_2$, PbO.
  J. Burbank, *in* "Power Sources 1966" (D. H. Collins, ed.), p. 147. Pergamon, Oxford.
The plate materials of the lead acid cell. Parts 1 and 2. Anodic oxidation of $PbSO_4$, Pb_3SO_4, $2PbCO_3 \cdot Pb(OH)_2$ and orthorhombic PbO.
  J. Burbank, NRL Report 6345 (1965) (Part 1); 6450 (1966) (Part 2).
Anodic oxidation of the basic sulphates of lead.
  J. Burbank, *J. Electrochem. Soc.* **113**, 10 (1966).
The plate materials of the lead–acid cell.
  J. Burbank, NRL Report 6613 (1967).
Anodic oxidation of tetragonal PbO.
  J. Burbank, *US Gov. Res. Dev. Rep.* **68** (7) 664, 827 (1968); *Lead Abstr.* 1968 5488.
Cycling and overcharge of positive plates.
  J. Burbank, *J. Electrochem. Soc.* **117**, 299 (1970).
Crystallisation of $PbSO_4$ on Pb–Sb alloy.
  J. Burbank, *J. Electrochem. Soc.* **118**, 525 (1971).
The lead–acid battery—a review.
  J. Burbank, *in* "Advances in Electrochemistry and Electrochemical Engineering", Vol. 8 (P. Delahay, ed.). Wiley, New York (1971).
Cycling anodic coatings on pure and antimonial lead in H_2SO_4.
  J. Burbank, *in* "Power Sources 3" (D. H. Collins, ed.), p. 13. Oriel Press, Newcastle upon Tyne (1971).
PbO_2 in the lead–acid cell. Formation from three typical oxides.
  J. Burbank and E. J. Ritchie, *J. Electrochem. Soc.* **116**, 125 (1969).
The relation of the anodic corrosion of lead and lead antimony alloys to microstructure.
  J. Burbank and A. C. Simon, *J. Electrochem. Soc.* **100** No. 1, 11 (1953).
Stress in anodically formed lead dioxide.
  C. J. Bushrod, *Brit. Corrosion J.* **6**, 129 (1971).
Electron microscope study of positive lead–acid electrodes during formation.
  W. O. Butler, *J. Electrochem. Soc.* **117**, 1339 (1970).
A new experimental technique for the study of films produced at electrochemical interfaces.
  B. D. Cahan, *J. Electrochem. Soc.* **106**, 543 (1959).
Fast linear sweep voltammetry studies on polycrystalline lead and electrodeposited lead dioxide (α and β) in aqueous sulphuric acid.
  J. P. Carr, *J. Electroanal. Chem.* **33**, 109 (1971).
Differential capacitance of polycrystalline lead in some aqueous electrolytes.
  J. P. Carr, *J. Electroanal. Chem.* **32**, 345 (1971).
Differential capacitance and linear sweep voltammetry studies on polycrystalline lead and electrodeposited lead dioxide.
  J. P. Carr, *J. Electrochem. Soc.* **118**, 1262 (1971).
The lead dioxide electrode.
  J. P. Carr and N. A. Hampson, *Chem. Rev.* **72**, 679 (1972).
The differential capacitance of binary alloys in aqueous solutions.
  J. P. Carr *et al.*, *J. Electroanal. Chem.* **34**, 425 (1972).

Lead–acid cells (Part 5). ($PbSO_4$ oxidation to PbO_2.)
P. Casson and N. A. Hampson, *J. Electroanal. Chem.* **92**, 191 (1978).
Hydrogen-loss concept of PbO_2 plate failure.
S. M. Caulder, *J. Electrochem. Soc.* **120**, 1515 (1973).
Thermal decomposition of formed and cycled lead dioxide electrodes.
S. M. Caulder, *J. Electrochem. Soc.* **121**, 1546 (1974).
The migration of antimony and tin and their relationship to the structure and performance of the $PbO_2/PbSO_4$ electrode.
S. M. Caulder, *J. Electrochem. Soc.* ext. abs. (1974).
The cyclic voltammetry of lead and a lead–antimony battery grid alloy in aqueous sulphuric acid at 25° to −40°C.
T. G. Chang, M. M. Wright and E. M. L. Valeriote, *in* "Power Sources 6" (D. H. Collins, ed.), p. 69. Academic Press, London and New York (1977).
X-ray study of the active material in the lead–acid cell.
T. Chiku, *J. Electrochem. Soc.* **115**, 10, 982 (1968).
Electron microprobe study of the active materials in the lead–acid storage battery.
T. Chiku, *J. Electrochem. Soc.* **116**, 1407 (1969).
Formation of $PbSO_4$ crystallites on Pb in the lead–acid cell.
T. Chiku, *J. Electrochem. Soc.* **118**, 1395 (1971).
Thermal analysis of lead–acid batteries.
K. W. Choi, *J. Electrochem. Soc.* **125**, 1011 (1978).
Mg–$PbCl_2$ batteries.
J. H. Coleman, *J. Appl. Electrochem.* **1**, 65 (1971).
Pb–Ag_2O cell.
D. H. Collins, *7th Int. Power Sources Symp., Brighton* 15 Sept. (1970).
The influence of superimposed a.c. on the anodic corrosion of lead in aqueous sulphuric acid.
J. M. Costa, *Corrosion Sci.* **2**, 269 (1962).
Activity coefficient and standard potential pf $PbO_2/PbSO_4$.
A. K. Covington, *Trans. Far. Soc.* **61**, 2050 (1965).
Variable composition of basic lead sulphates.
R. F. Dapo, *J. Electrochem. Soc.* **121**, 253 (1974).
Fundamental studies of the behaviour of antimony in the lead–acid battery.
J. L. Dawson, PhD thesis, Salford (1967).
Role of Sb in lead battery.
J. L. Dawson, *7th Int. Power Sources Symp., Brighton* 15 Sept. (1970); *Lead Abstr.* 1971 921.
Cohesion of positive plate material.
J. L. Dawson and M. E. Rana, Typed Paper in CTL Library.
Potential-pH diagram of lead and its applications to the study of lead corrosion and to the lead storage battery.
P. Delahay, *J. Electrochem. Soc.* **98**, 57 (1951).
A polarographic study of the influence of temperature on the rate of oxygen consumption by iron, lead and zinc.
P. Delahay, *J. Electrochem. Soc.* **99**, 414 (1952).

Analysis of battery pastes (Pb).
M. Denby, Booklet, publ. LDA Dec. 1964; *Lead Abstr.* 1965 3095.
Some important factors that influence the composition of the positive plate material in the lead–acid battery.
V. H. Dobson, *J. Electrochem. Soc.* **108**, 401 (1961).
Phase changes in battery plate manufacture.
I. Dugdale, Preprint 5th Int. Battery Symp. Brighton 1966; *Lead Abstr.* 1966-4443.
Thermodynamics of the lead storage cell. The heat capacity and entropy of lead dioxide from 15 to 318 K.
J. A. Duisman and W. F. Giauque, *J. Phys. Chem.* **72**, 562–573 (1968).
The use of Planté cells by the British Post Office.
P. J. Edwards, "Lead 71", Proc. 4th Conf., Hamburg (1971).
The behaviour of lead electrodes in sulphuric acid solutions.
K. Ekler, *Can. J. Chem.* **42**, 1355 (1964).
Current distribution in porous Pb or PbO_2 electrodes.
K. J. Euler, *Metalloberflaeche* **27**, 87 (1973); *Lead Abstr.* 13-2482.
Current distribution in porous positive plates.
K. J. Euler, *Electrochim. Acta* **13**, 1533 (1968).
Radiography of battery plates.
K. J. Euler, *Electrochim. Acta* **13**, 2245 (1968); also W. H. Lange, *Electrochem. Technol.* **6**, 405 (1968).
The anodic behaviour of cadmium and lead in alkali.
J. P. G. Farr and N. A. Hampson, *Electrochem. Technol.* **6**, 10 (1968).
Effect of grid cast structure on the anodic corrosion of lead.
S. Feliu, *Werkst. Korros.* **23**, 554 (1972).
The effect of ageing of grids on anodic corrosion.
S. Feliu, J. J. Regidor, M. Torralba and E. Otero, ILZRO Report no. LE-169 No. 3 (1971).
The polarisation of lead in sulphuric acid and the reactions occurring at the positive plate of the lead–acid battery.
W. Feitknecht and A. Gaumann, *J. Chim. Phys.* **49**, C135–144 (1952).
Anodic passivation and depassivation of lead in H_2SO_4.
W. Feitknecht, *Z. Elektrochem.* **62**, 95–803 (1958).
Discharge characteristics of the lead/lead–sulphate electrode.
M. I. Gillibrand and G. R. Lomax, *Electrochim. Acta* **11**, 281 (1966).
Pore volume and density of battery plates.
C. W. Fleischmann, *J. Electrochem. Soc.* **123**, 969 (1976); **124**, 1487 (1977).
The behaviour of the lead dioxide electrode.
M. Fleischmann, *Trans. Far. Soc.* **51**, 71 (1955).
Anodic oxidation of solutions of plumbous salts.
M. Fleischmann, *Trans. Far. Soc.* **54**, 1370 (1958).
The potentiostatic study of the growth of deposits on electrodes.
M. Fleischmann, *Electrochim. Acta* **1**, 146 (1959).
The kinetics of anodic lead dissolution in H_2SO_4.
A. N. Fleming, *in* "Power Sources 5" (D. H. Collins, ed.), p. 1. Academic Press, London and New York (1975).

The electrochemical oxidation of Pb to form $PbSO_4$.

A. N. Fleming and J. A. Harrison, *Electrochim. Acta* **21** (11), 905–912 (1976).

The examination of corrosion products and processes.

J. A. von Fraunhofer, *Anti Corrosion*, 17 (1967).

NMR measurements of the Knight shift in conducting PbO_2.

D. A. Frey, *J. Electrochem. Soc.* **107**, 930 (1960).

Sealed $Cu-PbO_2$ secondary battery.

S. Furumi, *J. Electrochem. Soc. Japan* **32**, 207 (1964); *Lead Abstr.* 1966 3795.

Superimposition of a.c. on d.c. in the preparation of α and β lead dioxide by anodic deposition.

S. Ghosh, *Electrochim. Acta* **14**, 161 (1969).

Additions to Pb–Zn alkaline battery.

P. N. Ghosh, *Appl. Phys. Quart. (India)* **11**, 25 (1971); *Lead Abstr.* 12-1833.

Discharge characteristics of $Pb/PbSO_4$ electrode.

M. I. Gillibrand, *Electrochim. Acta* **11**, 281 (1966).

Potentiostatic plate formation (positive).

H. W. Glaser, *Electrochim. Acta* **13**, 2013 (1968).

Stress measurements in electrodeposited lead dioxide.

K. S. A. Gnanasekaran, *Electrochim. Acta* **15**, 1615 (1970).

Behaviour of Pb-based electrodes with and without Pb(IV) layer.

J. A. Gonzalez, *Rev. Met. (Madrid)* **12**, 190 (1976); *Chem. Abstr.* **86**, 142836.

Some further experiments on atmospheric action in fatigue.

H. J. Gough, *J. Inot. Metals* **56**, 55 (1935).

SEM for battery plate examination.

R. H. Hammar, *7th Int. Power Sources Symp., Brighton* 15 Sept. (1970); *Lead Abstr.* 1971-923.

Expander action-impedance and passivation studies.

N. A. Hampson, *J. Electroanal. Chem.* **54**, 263 (1974).

Role of Sb on anodic behaviour of battery.

N. A. Hampson, *J. Appl. Electrochem.* **4**, 49 (1974).

Formation of PbO_2 by Planté process.

N. A. Hampson, *Surf. Technol.* **4**, 173 (1976).

Pb oxidation in presence of expanders.

J. A. Harrison, *J. Electroanal. Chem.* **47**, 93 (1973).

Anodic dissolution of Pb in H_2SO_4.

J. A. Harrison, *in* "Power Sources 5" (D. H. Collins, ed.). Academic Press, London and New York (1975).

SEM study of indexed location on battery plates.

S. Hattori, *in* "Power Sources 5" (D. H. Collins, ed.) p. 139. Academic Press, London and New York (1975).

PbO solid electrolyte cell.

J. R. Houston, *ILZRO Res. Final Rept Proj. Le-98* Oct. (1967); *Lead Abstr.* 1968 5434.

Direct electrical pulses from PbO_2–Pt–HCHO cell.

H. F. Hunger, AD 632,103 (1966); *Lead Abstr.* 1967 4549.

Investigation of the behaviour of anodic material in lead–acid cells.

S. Ikari, *J. Electrochem. Soc. Japan* **27**, E247 (1959).

Studies on the combined water in paste for lead storage batteries.
S. Ikari, *J. Electrochem. Soc. Japan* **27**, E150 (1959).
Anodic corrosion of lead and the potential of the positive electrode in sulphuric acid.
S. I. Ikari, *J. Electrochem. Soc. Japan* **28**, E192 (1960).
Anodic behaviour of α-PbO$_2$ substrates containing different percentages of lower-lead oxides.
I. M. Issa, *J. Appl. Electrochem.* **5**, 271 (1975).
Anodic disintegration of metals undergoing electrolysis in aqueous salt solutions.
W. J. James, *Corrosion* **23**, 15 (1967).
The behaviour of lead–acid batteries under pulsed discharge conditions.
M. G. Jayne, *in* "Power Sources 6" (D. H. Collins, ed.), p. 35. Academic Press, London and New York (1977).
The anodic dissolution of Pb in aqueous solutions.
J. W. Johnson, *Corrosion Sci.* **8**, 309 (1968).
Behaviour of the lead dioxide electrode.
P. Jones, *Trans. Far. Soc.* **50**, 972 (1954).
Oxide formation and overvoltage of oxygen on lead and silver anodes in alkaline solution.
P. Jones, *Trans. Far. Soc.* **52**, 1003 (1956).
Role of BaSO$_4$ in battery negatives.
B. N. Kabanov, *Zh. Prikl. Khim.* **37**, 1936 (1964); *Lead Abstr.* 1965 3421.
Anodic corrosion of Pb in H$_2$SO$_4$.
B. N. Kabanov, *Dokl. Akad. Nauk. SSR Fiz. Khim.* **154**, 1414 (1964); *Lead Abstr.* 1964 2888.
Anodic diffusion of oxygen through lead dioxide.
B. N. Kabanov, *Electrochim. Acta* **9**, 1197 (1964).
Thermal decomposition of lead peroxide.
N. Kameyama, *J. Soc. Chem. Ind. Japan* **49**, 154 (1946).
Improvement of lead–acid battery maintenance.
H. Kawamoto, *Japan Telecommun. Rev.* **17**, 140 (1975).
Impedance measurement on Pb/H$_2$SO$_4$ battery.
M. Keddam, *J. Appl. Electrochem.* **7**, 539 (1977).
Galvanic thin layer Pb/PbF$_2$ cell.
J. H. Kennedy, *J. Electrochem. Soc.* **123**, 10 (1976).
The effect of antimony and different anions on the potential of lead.
E. M. Khairy, *J. Electroanal. Chem.* **12**, 27 (1966).
Polarisation of Pb–Sb alloys at low current densities.
E. M. Khairy, *J. Electroanal. Chem.* **11**, 282 (1966).
Role of lead dioxide films in process of lead anode corrosion.
G. Kiryakov, *J. Appl. Chem. USSR* **27**, 847 (1953).
Modelling lead–acid batteries.
K. R. Kleckner, SAE Paper 730251 (International Automotive Congress Detroit U.S.A. 8–12 Jan, 1973); *Lead Abstr.* 16-1819.

A study of the decrease of hydrogen overvoltage on a spongy lead electrode in sulphuric acid.

A. M. Kolotyrkin, *J. Phys. Chem. USSR* **20**, 667–677 (1946).

Oxidation kinetics of freshly pasted plates.

P. M. Kornienko, *Protsessy Perenosa Tepla Massy* 151 (1974); *Chem. Abstr.* 83-134931.

Forming battery plates with asymmetrical AC.

D. A. Kozlov, *Electrotekhnika* 49 (1965); *Lead Abstr.* 1966 4058.

An investigation of the corrosion of lead in oxidizing media.

E. V. Krivolapova, *Chem. Abstr.* **49**, 12161 (1955).

Gas-recombination in lead–acid batteries.

E. C. Laird, *J. Electrochem. Soc.* **121**, 13 (1974).

Constant potential anodization of lead in sulphuric acid solutions.

J. J. Lander, NRL report 3549 (1949).

The basic sulphates of lead.

J. J. Lander, *J. Electrochem. Soc.* **95**, 174 (1949).

Some preliminary studies of positive grid corrosion in the lead–acid cell.

J. J. Lander, *J. Electrochem. Soc.* **98**, 220 (1951).

Anodic corrosion of lead in H_2SO_4 solutions.

J. J. Lander, *J. Electrochem. Soc.* **98** (6), 213 (1951).

Anodic corrosion of lead in H_2SO_4 solutions and some preliminary studies of positive grid corrosion in the lead–acid cell.

J. J. Lander, *J. Electrochem. Soc.* **98**, 522–523 (1951).

Further studies on the anodic corrosion of lead in H_2SO_4 solutions.

J. J. Lander, *J. Electrochem. Soc.* **103**, 1–8 (1956).

Impedance measurements on $Pb/PbSO_4$ system.

D. I. Leikis, *Sov. Electrochem.* **11**, 1439 (1975).

Phase composition of the products of anodic corrosion of lead and its alloys.

L. M. Levinzon, *J. Appl. Chem. USSR* **39**, 525 (1966).

Cathodic reduction of PbO_2.

L. I. Lyamina, *Sov. Electrochem.* **10**, 463 (1974).

Lead–fluoride acid battery.

G. D. MacDonald, *J. Electrochem. Soc.* **119**, 660 (1972).

Theory of lead acid batteries.

W. Machu, *Werkst. Korros.* **16**, 676 (1965); *Lead Abstr.* 1965 3637.

Studies on the passive film of lead.

M. Maeda, *J. Electrochem. Soc. Japan* **26**, E183 (1958).

Role of lignin additions in battery plates.

B. K. Mahato, *J. Electrochem. Soc.* **124**, 1663 (1977).

Study of lithium–lead alloys as grid material for lead–acid batteries.

G. W. Mao, *J. Electrochem. Soc.* **117**, 1323 (1970).

Recent results on lead calcium alloys as grid materials for lead–acid batteries.

G. W. Mao, *J. Electrochem. Soc.* **120**, 1,11 (1973).

Influence of paste preparation on charge/discharge.

G. Manoim, *Sb. Rabot. Khim. Istokhim. Toka* 41 (1972); *Chem. Abstr.* **80**, 33116.

The discharge mechanism of certain oxide electrodes.
   H. B. Mark, *J. Electrochem. Soc.* **108**, 615 (1961).
Discharge properties of α PbO_2 in dilute H_2SO_4 electrolyte.
   H. B. Mark, *J. Electrochem. Soc.* **109**, 634 (1962).
The recovery from polarisation of α and β-PbO_2 in H_2SO_4 electrolyte.
   H. B. Mark, *J. Electrochem. Soc.* **110**, 945 (1963).
Microstructural aspects of PbO_2 grid corrosion.
   D. Marshall, *J. Electrochem. Soc.* **123**, 1849 (1976).
Influence of impurities in lead antimony alloys on the work of the grid in the
   lead acid accumulator. III. Over-tension of hydrogen on alloys and over-
   tension of oxygen on the oxidised alloys.
W. P. Maschowez and W. N. Fateew, *J. Appl. Chem. USSR* **21** (5)
   (1948).
Anodic corrosion of the Pb–Ca–Sn alloys.
   St. Matylla, *Chem. Power Sources Symp.*, *Poznan* (1973).
Oberflacheveranderung von bleielektroden durch electrolyse.
   H. Metzler, *Electrochim. Acta* **11**, 111 (1966).
Theory of porous electrodes.
   K. Micka, *Chem. Listy* **65**, 673 (1971); *Lead Abstr.* 1971 1401.
Mechanism of PbO_2 reduction.
   K. Micka, *Chem. Listy* **65**, 449 (1971); *Lead Abstr.* 1971 1400.
Theory of porous Pb electrodes.
   K. Micka, *Electrochim. Acta* **19**, 499 (1974).
Theory of porous electrodes and Pb battery plates.
   K. Micka, *Electrochim. Acta* **21**, 599 (1976), also **18**, 629 (1973).
Theory of positive plate.
   K. Micka and I. Rousar, *Coll. Czech. Commun.* **40**, 921 (1975).
Battery with PbO_2 on Ti grid (with RuO_2 interlayer).
   Yu. Micron, *Chem. Lett.* p. 1407. Dec (1977).
Electrical properties of electrodeposited PbO_2 films.
   W. Mindt, *J. Electrochem. Soc.* **116**, 1076 (1969).
The positive electrode in the lead cell.
   P. Ness, *Electrochim. Acta* **12**, 161 (1967).
Analysis of porous Pb electrode.
   J. Newman, *J. Electrochem. Soc.* **110**, 1251 (1971).
Double-layer capacity of porous Pb electrodes.
   J. Newman, *J. Electrochem. Soc.* **122**, 70 (1975).
Voltammetric investigations of negative lead battery electrodes: the effect of
   organic expanders.
   H. Niklas and H. Jacobljevich, *J. Appl. Electrochem.* **2**, 165–167 (1972).
Discharge characteristics of lead chloride-magnesium sea water batteries.
   H. Nomura, M. Takeuchi and H. Abe, *GS News* (in Japanese) **26** (1),
   47–53 + Translation (1967).
Nucleation and growth of electrodeposited PbO_2.
   W. A. Nystrom, *Plating* **56**, 285 (1969); *Lead Abstr.* 1969 6084.
Interferometric study of low temperature battery charging.
   R. N. O'Brien, *J. Electrochem. Soc.* **124**, 96 (1977).
Activity coefficients in system $PbSO_4$-H_2SO_4-H_2O.

K. B. Oldham, *Trans. Far. Soc.* **60**, 1646 (1964).
Electrolyte flow and mass-utilisation of porous PbO_2.
H. Panesar, *Electrochim. Acta* **15**, 1421 (1970).
$PbSO_4$ distribution in discharged battery plates.
H. Panesar, *Metalloberflaeche* **26**, 252 (1972); *Lead Abstr.* 13-2143.
Preliminary investigation into the behaviour of lead alloy anodes in sulphuric and chromic acid solutions.
T. E. Parker, *J. Electrochem. Soc.* **119**, 13c (1972).
Pb battery electrodes at low temperatures.
E. V. Parshikova, *Elektrotekhnika* (10), 54 (1975); *Lead Abstr.* 17-4540.
Mechanism of the anodic oxidation of lead in sulphuric acid solutions.
D. Pavlov, *Ber. Bunsenges. Phys. Chem.* **71**, No. 4, 398 (1967).
Processes of formation of divalent lead oxide compounds on anodic oxidation of lead in sulphuric acid.
D. Pavlov, *Electrochim. Acta* **13**, 2051–2061 (1968).
Mechanism of passivation processes of the lead sulphate electrode.
D. Pavlov, *Electrochim. Acta* **15**, 1483 (1970).
Mechanism of positive plate formation.
D. Pavlov, *J. Electrochem. Soc.* **119**, 8 (1972).
Formation of battery negative plates.
D. Pávlov, *J. Electrochem. Soc.* **121**, 854 (1974).
Study of positive plates.
D. Pavlov, *J. Appl. Electrochem.* **6**, 339 (1976).
Mechanism of initial negative plate formation.
D. Pavlov, *J. Electroanal. Chem.* **72**, 319 (1976).
Growth processes of the anodic crystalline layer on potentiostatic oxidation of lead in H_2SO_4.
D. Pavlov and N. Iordanov, *J. Electrochem. Soc.* **117**, 1103 (1970).
Mechanism of passivation processes of the lead sulphate electrode.
D. Pavlov and R. Popova, *Electrochim. Acta* **15** (9), 1483–1491 (1970).
Dependence of the composition of the anodic layer on the oxidation potential of lead in sulphuric acid.
D. Pavlov, C. N. Poulieff, E. Klaja and N. Iordanov, *J. Electrochem. Soc.* **116**, 316 (1969).
Microstructural control in lead alloys for storage battery application.
J. Perkins, *J. Mater. Sci.* **10**, 136 (1975).
Substructural network in battery positive plates.
J. Perkins, *J. Electrochem. Soc.* **124**, 524 (1977).
Charge acceptance rate as function of charge current and temperature.
K. Peters, *6th Int. Power Sources Symp. Brighton*, 24 Sept. (1968); *Lead Abstr.* 1969 6140.
Temperature, current density and plate capacity.
K. Peters, *Electrochim. Acta* **17**, 839 (1972).
Crystallographic changes in battery plates.
J. R. Pierson, *Electrochem. Technol.* **5**, 323 (1967).
Crystallography and microscopy of battery plates.
J. R. Pierson, *6th Int. Power Sources Symp. Brighton*, 24 Sept. (1968); *Lead Abstr.* 1969 6143.

Morphology of negative electrode with lignins.

J. R. Pierson, *J. Electrochem. Soc.* **117**, 1463 (1970).

Acceptable contamination levels in lead–acid battery.

J. R. Pierson, C. E. Weinlein and C. E. Wright, *in* "Power Sources 5" (D. H. Collins, ed.). Academic Press, London and New York (1975).

Kinetics of Pb(IV) electrodes—exchange currents.

J. P. Pohl and H. Rickert, *Z. Phys. Chem.* (*Frankfurt*) **95**, 47 and 59 (1975).

On the kinetics of the lead dioxide electrode.

J. P. Pohl and H. Rickert, *in* "Power Sources 5" (D. H. Collins, ed.), pp. 15–22. Academic Press, London and New York (1975).

Potential distribution in lead–acid battery.

J. E. Puzey, *6th Int. Power Sources Symp., Brighton* 24 Sept. (1968); *Lead Abstr.* 1969 6144.

Study of Pb–$HClO_4$–H_2SO_4 system.

A. Ragheb, *Werkst. Korros.* **16**, 755 (1965); *Lead Abstr.* 1965 3638.

Corrosion and corrosion products on lead and lead alloys.

M. E. Rana, PhD thesis, UMIST (1975).

Alkaline electrolyte for Zn–PbO_2 cells.

M. L. Rao, *J. Electrochem. Soc.* **120**, 855 (1973).

Electrochemical discharge behaviour of porous PbO_2 electrodes.

P. Reinhardt, *Z. Phys. Chem.* (*Leipzig*) **257**, 412 (1976); *Chem. Abstr.* **85**, 266 71.

Structure and electrochemistry of positive plates.

P. Reinhardt and K. Wiesener, *Chem. Tech.* **27**, 616 (1975); *Lead Abstr.* 16-3346.

Zum verlauf der entladereaktion positiver activmassen der bleiakkumulators.

P. Reinhardt, M. Vogt and K. Weisener, *J. Power Sources* **1** (2), 127–139 (1976).

PbO_2 in the lead–acid cell.

E. J. Ritchie, *J. Electrochem. Soc.* ext. abs. (1969).

PbO_2 in the lead–acid cell. II. Cycling and overcharge on pure and antimonial lead grids.

E. J. Ritchie and J. Burbank, *J. Electrochem. Soc.* **117** (3), 299–305 (1970).

Mechanical and corrosion properties of low-antimony lead.

T. Rogatchev, *J. Appl. Electrochem.* **6**, 33 (1976).

Resistance study of porous Pb plates.

I. L. Romanova, *Sov. Electrochem.* **6**, 1776 (1970).

Anodic oxides of lead (discussion).

P. Rüetschi, *J. Electrochem. Soc.* **106**, 1079 (1959).

Oxygen overvoltage and electrode potentials of α and β PbO_2.

P. Rüetschi, *J. Electrochem. Soc.* **106**, 547 (1959).

Stability and reactivity of lead oxides.

P. Rüetschi, *in* "Batteries" (D. H. Collins, ed.), p. 89. Pergamon, Oxford (1962).

Stability and reactivity of lead oxides (self-discharge).

P. Rüetschi, *Electrochim. Acta* **8**, 333 (1963).

Ion selectivity and diffusion potentials in corrosion layers.

P. Rüetschi, *J. Electrochem. Soc.* **120**, 331 (1973).
Anodic oxidation of lead at constant potential.

P. Rüetschi and R. T. Angstadt, *J. Electrochem. Soc.* **111** (12), 1323–1330 (1964).
Anodic corrosion and hydrogen and oxygen overvoltage on lead, and lead antimony alloys.

P. Rüetschi and B. D. Cahan, *J. Electrochem. Soc.* **104**, 406 (1957).
Electrochemical properties of PbO_2 and the anodic corrosion of lead and lead alloys.

P. Rüetschi and B. D. Cahan, *J. Electrochem. Soc.* **105**, No. 7, 369 (1958).
Phase composition of electrode pastes of lead storage cells.

A. I. Rusin and M. A. Dayosan, *Zh. Prikl. Khim.* **46** (12), 2643–2647 (1973).
Modelling of lead–acid batteries.

W. Runge, *Archiv. Elektrotechnik.* **57**, 235 (1975); *Lead Abstr.* 16-2720.
Polarisation curves for Pb and alloys in H_2SO_4.

E. Sato, *J. Electrochem. Soc. Japan* **32**, 148 (1964); *Lead Abstr.* 1966 3816.
The polarisation curve of lead and lead alloys in sulphuric acid solution.

E. Sato, *J. Electrochem. Soc. Japan* **32**, 148 (1964).
Differential capacitance and cyclic voltametric studies on smooth lead in H_2SO_4 solutions.

T. F. Sharpe, *J. Electrochem. Soc.* **116**, 1639 (1969).
Low rate cathodic linear sweep voltammetry studies on anodised lead.

T. F. Sharpe, *J. Electrochem. Soc.* **122**, 845 (1975).
Textures of electrodeposited lead dioxide.

Y. Shibasaki, *J. Electrochem. Soc.* **105**, 624 (1958).
Microsegregation in the lead–antimony alloys.

A. C. Simon, *J. Electrochem. Soc.* **100**, 1 and 11 (1953).
Investigation of storage battery failure by a method of plastic impregnation.

A. C. Simon, *J. Electrochem. Soc.* **102**, 279 (1955).
Crystallogenesis in the forming of plates for the lead–acid storage battery.

A. C. Simon, *J. Electrochem. Soc.* **109**, 760 (1962).
Initial voltage drop ("Spannungsack") in Pb battery discharge.

A. C. Simon, *Electrochem. Technol.* **3**, 307 (1965).
Stress corrosion on lead battery grids.

A. C. Simon, *J. Electrochem. Soc.* **114**, 1 (1967).
Microscopy of negative plate.

A. C. Simon, *6th Int. Power Sources Symp.*, Brighton 24th Sept. (1968); *Lead Abstr.* 1969 6145.
Morphological changes in the lead dioxide electrode during its reduction and reoxidation.

A. C. Simon, *J. Electrochem. Soc.* **117**, 987 (1970).
Effect of Sb_2O_3 on the microstructure of the PbO_2 electrode.

A. C. Simon, *J. Electrochem. Soc.* **117**, 1264 (1970).
The crystallography of formed and cycled PbO_2 electrodes.

A. C. Simon, *J. Electrochem. Soc.* **118**, 659 (1971).
Structure of $Pb/PbSO_4$ system and changes due to additives.

A. C. Simon, *Electrochim. Acta* **19**, 739 (1974). Also *J. Electrochem. Soc.* **121**, 463 (1974).

Influence of active material thickness on positive grid corrosion in the lead–acid battery.
  A. C. Simon, *J. Electrochem. Soc.* **121**, 531 (1974).
Structural transformations of the PbO_2 active material during cycling.
  A. C. Simon, *J. Electrochem. Soc.* **122**, 461 (1975).
The lead–acid battery.
  A. C. Simon, *Proc. Symp. Adv. Batt. Res., Argonne Lab.* pA-34 (1976).
Mathematical model of porous PbO_2 electrode.
  D. Simonsson, *J. Appl. Electrochem.* **3**, 261 (1973).
Current distribution in the porous lead dioxide electrode.
  D. Simonsson, *J. Electrochem. Soc.* **120**, 151 (1973).
Process of accumulator paste formation.
  M. F. Skalozubov, *J. Appl. Chem. USSR* **26**, 679 (1953).
Factors affecting porosity of positive plate.
  E. Skoluda, *Electrochim. Acta* **17**, 1353 (1972).
The influence of different factors on the kinetics of potentiostatic forming of pasted positive lead electrodes.
  E. Skoluda, *Int. Symp. Power Sources, Poznan* (1973).
Passivation of lead in sulfuric acid. Nucleation and growth of the lead sulfate film.
  J. Stange, *Electrochim. Acta* **19** (3), 111–116 (1974).
Morphology of Pb and PbO_2 with SEM.
  G. Sterr, *Electrochim. Acta* **15**, 1221 (1970).
Thermodynamic study of bivalent metal halides in aqueous solution.
  R. Stokes, *Trans. Far. Soc.* **44**, 295 (1948).
Antimony in the lead–acid battery.
  D. E. Swets, *J. Electrochem. Soc.* **120**, 925 (1973).
Behaviour of lead electrodes in sulphuric acid solutions.
  E. Tarter and K. Ekler, *Can. J. Chem.* **47**, 2191 (1969).
Electron transfer at lead electrodes.
  H. R. Thirsk, *J. Chim. Phys.* **49**, C131 (1952).
Corrosion and growth of lead–calcium alloy storage battery grids as a function of calcium content.
  U. B. Thomas, *Trans. Electrochem. Soc.* **92**, 313 (1947).
The electrical conductivity of lead dioxide.
  U. B. Thomas, *J. Electrochem. Soc.* **94**, 42 (1948).
Self-diffusion of oxygen in lead oxide.
  B. A. Thompson, *J. Phys. Chem.* **67**, 594 (1963).
Positive plate sulphation in Pb–Ca plates.
  S. Tudor, *Electrochem. Technol.* **5**, 21 (1967).
Cyclovoltammetric study of lead battery.
  J. L. Urrutia, *Rev. Met.* (*Madrid*) **10**, 241 (1972); *Chem. Abstr.* **82**, 142628.
Film formation on anodically polarised lead.
  H. Vaidyanathan, *J. Electrochem. Soc.* **121**, 876 (1974).
Study of the influence of antimony on the character of passivation of lead in H_2SO_4 solutions.
  E. S. Vaisberg, E. V. Kribolapova and B. N. Kabanov, *Zh. Prikl. Khim.* No. 10, 2354 (1959).

Potentiostatic oxidation of $PbSO_4$ films.
　E. M. L. Valeriote, *J. Electrochem. Soc.* **124**, 370 (1977).
Low temperature oxidation kinetics of anodic films on Pb.
　E. M. L. Valeriote, *J. Electrochem. Soc.* **124**, 380 (1977).
The kinetics of the $PbSO_4/PbO_2$ electrode at low temperatures.
　E. M. L. Valeriote and L. D. Gallop, *in* "Power Sources 5" (D. H. Collins, ed.), p. 55. Academic Press, London and New York (1975).
Function of a.c. superimposed on d.c. in the anodic oxidation of lead in sulphuric acid.
　R. Vijayavalli, *J. Electrochem. Soc.* **110**, 1 (1963).
Effect of frequency of a.c. superimposed on d.c. in the anodic oxidation of lead in sulphuric acid.
　R. Vijayavalli, *Batterien* **19**, 780 (1965).
Cyclic voltammetry on lead electrodes in sulphuric acid solution.
　W. Visscher, *J. Power Sources* **1** (3), 257–266 (1977).
Discharge capacities of alpha and beta PbO_2.
　E. Voss, Preprint 3rd Int. Battery Symp. Brighton 1962; *Lead Abstr.* 1963-1612.
Surface areas of α and β PbO_2.
　E. Voss and H. Freundlich, *in* "Batteries" (D. H. Collins, ed.). Pergamon, New York (1964).
Solution-precipitation mechanism in lead–acid cell electrode reactions.
　J. L. Weininger, *J. Electrochem. Soc.* **121**, 1454 (1974).
Corrosion of lead alloys at high anodic potentials.
　J. L. Weininger, *J. Electrochem. Soc.* **123**, 602 (1976).
Carbon fibre–lead wire positive grids.
　J. L. Weininger and C. R. Morelock, *J. Electrochem. Soc.* **122**, 1161 (1975).
High pressure high temperature polymorphism of the oxides of lead.
　W. B. White, *J. Amer. Ceram. Soc.* **44**, 170 (1961).
Radiographic and SEM study of positive plates.
　K. Wiesener, *Electrochim. Acta* **18**, 913 (1973).
Fundamental studies of the lead–acid battery.
　B. R. Wilkinson, PhD thesis, Salford University (1970).
Battery charging at low temperatures.
　E. Willihnganz, *in* "Power Sources 5" (D. H. Collins, ed.), p. 43. Academic Press, London and New York (1975).
Effect of additives on cathodic processes in battery negative plates.
　E. G. Yampolskaya, *Zh. Prikl. Khim.* **49**, 2421 (1976); *Chem. Abstr.* **86**, 48478.
Potentiostatic charging of Pb battery.
　V. S. Yanchenk, *Sov. Electrochem.* **7**, 879 (1971).
Blister formation in batteries.
　C. F. Yarnell, *J. Electrochem. Soc.* **119**, 19 (1972).
Cast lead–calcium–tin grids for maintenance-free batteries.
　J. A. Young, Paper, Battery Council International (1973).
Reactive additive for sulphate in batteries.
　Belgian Patent 648,498 (1964); *Lead Abstr.* 1965 2980.

Pb–PbO$_2$ ammoniacal electrolyte cell.
British Patent 899,538 (1959); *Lead Abstr.* 1963 1369.
Pb batteries with anhydrous organic electrolytes.
British Patent 910,930 (1962); *Lead Abstr.* 1963 1754.
Battery with Pb fibres.
British Patent 1,377,009 (1974); *Lead Abstr.* 15-1181.
Lead–chromate battery.
British Patent 1,357,397 (1974); *Lead Abstr.* 15-0033.
Lead–perchlorate battery.
British Patent 1,393,147 (1975); *Lead Abstr.* 15-1478.
Dished accumulator plates.
British Patent 26,680 (1897).
PbO$_2$–Sb battery.
French Patent 1,300,174 (1962); *Lead Abstr.* 1963 1491.
Effect of thallium on battery plates.
Ger. Offen Patent 2,356,212 (1975); *Lead Abstr.* 16-2375.
Additive to lead battery (disodium nitrilotriacetate etc.).
Japanese Patent (Kokai) 50,119,938 (1975); *Lead Abstr.* 17-3695.
Battery with sulphuric–phosphoric electrolyte.
U.S. Patent 3,765,942 (1970); *Lead Abstr.* 14-2759.
Lead fibres for cell electrodes.
U.S. Patent 3,690,866 (1970); *Lead Abstr.* 13-2220.
Battery with fluoboric electrolyte.
U.S. Patent 3,770,507 (1972); *Lead Abstr.* 14-2761.
Brine battery.
U.S. Patent 3,785,871 (1972); *Lead Abstr.* 14-2851.
Seawater (PbCl$_2$) battery.
U.S. Patent 3,468,710 (1968).
Sponge Pb electrodes for plumbic acid battery.
U.S.S.R. Patent 281584 (1973); *Lead Abstr.* 14-2655.
Prevention of decay of PbO$_2$ positive plates in storage.
U.S. Patent 3,887,398 (1975).
Battery with PbF$_2$ cathode.
U.S. Patent 3,989,543 (1976).
Lead battery with dissolving electrolyte.
U.S. Patent 4,001,037 (1977).

Anodes and Corrosion

Anodic corrosion of Cd–Sb–Pb alloys in H$_2$SO$_4$.
A. A. Abdul-Azim, *Corrosion Sci.* **16**, 209 (1976).
Anodic corrosion of Pb–Sb–Ag alloy.
A. A. Abdul Azim, *Corrosion Sci.* **17**, 415 (1977).
Rapid potentiostatic method for corrosion testing of Pb alloys.
A. Aleksandrov, *Metallurgia* **29**, 19 (1974); *Chem. Abstr.* **82**, 23429d.
Kinetics of anodic Pb dissolution in chlorides and thiocyanate media.
V. K. Altukhov, *Mater. Nauch. Techn. Konf. Voronezh Politekh.* 93 (1972); *Chem. Abstr.* **82**, 49195.

New anodes for electrochemical processes (PbO_2-graphite).
  Anon., *Chem. Eng.* **72** (1965) (15) 82; *Lead Abstr.* 1965 3591.
Composite anodes of PbO_2–Ni. Detection of Pb ions in solution.
  N. G. Bakchitsaraitsyan, *Tr. Mosk. Khim. Tekhnol.* **54**, 149 (1967); *Chem. Abstr.* **68**, 26300.
Anodes of PbS for Pb electrowinning.
  I. J. Bear, *Trans. Inst. Min. Metall. Section C.* **85**, C49 (1976); *Lead Abstr.* 16-3426.
Corrosion of Pb in HBF_4.
  F. Beck, *Werkst. Korros.* **28**, 688 (1977).
Relationships on stability of Pb alloys to sulphate corrosion.
  V. I. Bryntseva, *Zasch. Metal.* **3**, 504 (1967).
Effect of added metal ions on Pb corrosion rate.
  V. I. Bryntseva, *Katal. Reakts. Zh. Faze* **2**, 298 (1974); *Chem. Abstr.* **86**, 62515.
Anodic properties of Pb–Tl and Pb–Ag alloys.
  V. I. Bryntseva, ZDA/LDA Transl. 70/232.
Relation of anodic corrosion of Pb, Pb–Sb alloys to microstructure.
  J. B. Burbank and A. C. Simon, *J. Electrochem. Soc.* **100**, 11 (1953).
Pb–Ag anodes in zinc electrowinning (added Mn ions).
  D. Buttinelli, *Mineraria* **25**, 118 (1974); *Lead Abstr.* 15-1591.
Effect of Pb structure on corrosion in H_2SO_4.
  M. A. Dasoyan, *Dokl. Akad. Sci. USSR Chem. Tech. Sect.* **107**, 863 (1956).
Pb anodes for Cu electrowinning.
  G. Eggett, *Chem. Ind.* 1019 (1976).
Anodes for solvent-extraction copper electrowinning.
  G. Eggett and D. Naden, *Hydrometal.* **1**, 123 (1975).
Anodic dissolution of Pb in chloride and thiocyanate solutions.
  D. E. Emel'yanov, *Mater. Nauch. Techn. Konf. Voronezh Politekh. Inst.* 93 (1972); *Chem. Abstr.* **82**, 49195.
Anodic behaviour of Pb and Pb–Sn–Co alloys in chromic acid.
  A. I. Falicheva, *Zasch. Metal.* **11**, 377 (1975).
Pb–Sn–Co anodes for chromium plating.
  A. I. Falicheva, *Izv. Vyssh. Ucheb. Zaved. Khim.* **19**, 433 (1976); *Chem. Abstr.* **85**,11559; also A. I. Falicheva, *Zasch. Metal.* **11**, 507 (1975); *Chem. Abstr.* **83**, 170045.
Pb–Sn–Co alloys for chromium plating.
  A. I. Falicheva, *Zasch. Metal.* **11**, 377 (1975); *Chem. Abstr.* **83**, 123038.
Effect of grid cast structure on anodic Pb corrosion.
  S. Feliu, *Werkst. Korros.* **23**, 554 (1972); *Lead Abstr.* 13-2062.
Calculation of corrosion rate of a binary Pb alloy.
  S. Feliu, *Revista Metal. (Spain)* **12**, 245 (1976); *Lead Abstr.* 17-4592.
Anodic corrosion of 2-phase (one Pb) metals.
  S. Feliu and M. Morcillo, *Corrosion Sci.* **15**, 593 (1975).
PbO_2 composite anodes.
  S. O. Fujii, *Jap. Kokai* 74-134575; *Chem. Abstr.* **84**, 10474.
Lead anodes in chromic acid.
  D. W. Hardesty, *Plating* **56**, 705 (1969); *Lead Abstr.* 9-6383.

Pb–Sb anodes for SX Cu electrowinning.
  T. H. Jeffers and R. D. Groves, Paper A, 76–80, available at AIME, New York; *Lead Abstr.* 17-5021.
Anodic corrosion of low-Sb alloys.
  G. Kaden, *Korrosion* **7**, 29 (1976); also **5**, 18–30 (1974).
Anodic behaviour of Pb–Ag and Pb–Pt systems in seawater.
  Y. Kikkuchi, *Boshuku Gijutsu* **24**, 459 (1975); *Lead Abstr.* 17-3737.
Effect of Cu^{++} on anodic corrosion of Pb–Sb alloys in H_2SO_4.
  A. Kirow, *Metalloberfläche* **26**, 234 (1972); *Lead Abstr.* 12 1986.
Prospects for improved Pb anodes in Zn electrowinning (4 component).
  G. Z. Kir'yakov, *Tsvetn. Metal.* (1), 21 (1975); *Lead Abstr.* 15-1592.
Migration of Pb from anode to cathode.
  G. Z. Kir'yakov, *Tsvetn. Metal.* (8), 18 (1975); *Chem. Abstr.* **84**, 23695.
Effect of Cl^- ions on anodic dissolution of Pb.
  T. N. Klepinina, *Sb. Tr. Kafedr. Fiz. i Obsch. Khim. Voronezh. Politekn.* **1**, 182 (1973); *Chem. Abstr.* **83**, 87065.
Ternary Pb (including Tl, Ag, Co, Sn, As) alloys for Zn electrowinning. Effect of AC.
  D. Krupkowa, *Rudy Metale Niezelazne* **12**, 296 (1967); *Lead Abstr.* 7-4989; also I. Enchev, *Rudodobir Met.* **23**, 44 (1968); *Lead Abstr.* 10-328.
PbO_2–Ti composite anode with interlayers.
  J. Kubicki, *C.R. Congr. 4th Int. Corros. Mar., Salissures* 313 (1976); *Chem. Abstr.* **87**, 143236,7.
Pb–Ca anodes in Cu electrowinning
  H. Kudelka *CIM Bull. (Can.)* **70**, 186 (1977); *Lead Abstr.* 18-0095.
Lead dioxide anodes from fluosilicate electrolyte.
  V. F. Lazarev, *Izv. Vyssh. Uchebn, Zaved. Khim. Khim. Tekhnol.* **18**, 1336 (1975); *Chem. Abstr.* **84**, 92635; **85**, 200041.
Composition of Pb and alloy corrosion products.
  L. M. Levinzon, *J. Appl. Chem.* **39**, 525 (1966).
Effect of As on Sb–Pb alloy characteristics.
  G. W. Mao and J. G. Larson, *Metallurgia* **28**, 236 (1968).
Inhibition effect of Ag on anodic Pb corrosion.
  G. W. Mao and P. Rao, *Brit. Corrosion J.* **6**, 122 (1971).
Effect of AC on Pb corrosion.
  Y. N. Mikhailovskii, *Zh. Fiz. Khim.* **37**, 340 (1963); *Lead Abstr.* 1963 1955; also C. E. Galimberti, *Corrosion* **20**, 150t (1964); *Lead Abstr.* 1964 2684.
Pb and Pb–Ag alloys in artificial seawater and sulphates.
  T. Mitamura, *Denki Kagaku* **45**, 176 (1977); *Chem. Abstr.* **87**, 124488.
Theory of passivity (XIII) of Pb.
  W. J. Muller and W. Machu, *Sitzungsber. Akad. Wiss–Wien Math. Naturwiss. Kl. Abt.* IIb **142**, 557 (1933).
Sn and Bi in Pb–Ca alloys.
  M. Myers, *Proc. 87th Battery Council Int., Hollywood Fla, USA*, p. 135 (1975); *Lead Abstr.* 16-2359.
Pb–Co–S alloy for Zn electrowinning.
  K. Naganathan, *East. Metal. Rev. (India)*, 145 (1972); *Lead Abstr.* 12-2043.

Pb–Ag bielectrode in seawater.
R. Nakamura, *Kurume Kogyo Koto* **25**, 21 (1976); *Chem. Abstr.* **87**, 174749.
Pb–PbO$_2$ bielectrodes.
K. C. Narasinham, *Electrochim. Acta* **22**, 197 (1977).
Computer-assisted corrosion predictions for Pb–Sb–Sr alloy system.
Z. U. Niyazova, *Zavod. Lab.* **42**, 584 (1976); *Lead Abstr.* 17-4544.
Effect of geometrical surface irregularities on anodic Pb corrosion.
A. de la Orden, *Rev. Metal. (Spain)* **7**, 39 (1971).
Pb anode in chlorate–carbonate solutions.
K. I. Popov, *Surface Technol.* **6**, 51 (1977).
Pb–Ag anodes in HF–H$_2$SO$_4$ mixtures.
N. F. Razina, *Izv. Akad. Nauk. Kaz. SSR Ser. Khim.* **24**, 17 (1974); *Chem. Abstr.* **80**, 127468.
Alloys of Pb with Sn, Sb, Ag as anodes in acid, neutral, alkali solutions.
A. B. Sandybaeva, *Izv. Akad. Nauk. Kaz. SSR* **24**, 45 (1974); *Chem. Abstr.* **81**, 130163.
Pt-plated PbO$_2$ anode for seawater electrolysis.
E. Sato, *Boshoku Gijutsu* **24**, 459 (1975); *Chem. Abstr.* **85**, 11597.
Microporosity in Pb–Sb alloys.
A. C. Simon, *US Naval Res. Lab. Rept* 6723 (1968); *Lead Abstr.* 9-6206.
Pb–Sn–Co anodes for chromium plating.
B. S. Spirodonov and A. I. Falicheva, *Sb. Tr. Kafedr. Fiz. Obsch. Khim. Voronezh. Politekhn.* **1**, 169 (1973); *Chem. Abstr.* **83**, 67756.
Pb–Sn anodes by electrodeposition on basis metal.
R. Subramanian, *Metal Finish.* **64**, 53 (1966); *Lead Abstr.* 1966 4174.
PbO$_2$ anodes for large-scale KClO$_3$ manufacture.
H. V. K. Udupa, *J. Appl. Chem. Biotechnol.* **24**, 43 (1974).
PbO$_2$-graphite anode.
H. V. K. Udupa, *J. Electrochem. Soc.* **123**, 1294 (1976).
PbO$_2$ anodes for electrosynthesis—pretreatment of Ti.
D. Wabner, *Z. Naturforsch. B. Anorg. Chem. Org. Chem.* **31**B, 45 (1976).
PbO$_2$ composite anodes for electrosynthesis.
D. Wabner, *Z. Naturforsch. B. Anorg. Chem. Org. Chem.* **31**B, 39 (1976).
Anodic dissolution of Pb–Ag in phosphite–fluoride–hypophosphite.
N. White and S. Das, *Gupta Abstr.* 76. ECS Meeting, Seattle, May 21 (1978).
PbO$_2$–graphite anode.
British Patent 893,823 (1959).
Improved Pb electrodes (with Ag net for seawater electrolysis).
British Patent 1,376,820 (1971).
Process anode (Pb–Sn–Cu–Pd).
British Patent 1,369,707 (1974).
Sacrificial Pb anode for replenishment of chemical reducing baths.
British Patent 1,348,118 (1975); *Lead Abstr.* 15-1508.
Pb–PbO$_2$ bielectrode.
British Patent 1,423,093 (1976).

PbO$_2$–Ni composite anode.
British Patent 1,452,276 (1976).
Preconditioning of Pb alloy anodes.
British Patent 1,452,907 (1976).
PbO$_2$–Ti composite anode.
British Patent 1,476,487 (1977).
Descaling Pb anodes (in gluconate solutions).
British Patent 958,410; *Chem. Abstr.* **61**, 3930.
Preconditioning of Pb anodes.
French Patent 1,419,356.
Washing lead anodes.
French Patent 1,304,775 (1976); *Lead Abstr.* 1963 2054.
Lead alloy anodes for sulphuric acid.
German Patent 1,161,037 (1958); *Lead Abstr.* 1964 2422.
PbO$_2$ anodes.
Ger. Offen. Patent 2,537,100 (1976); *Chem. Abstr.* **85**, 101296.
PbO$_2$–Ti anode with interfacial treatment.
Ger. Offen. Patent 2,537,100 (1976); *Chem. Abstr.* **85**, 101296.
Anode for chromic acid.
U.S. Patent 3,794,570 (1971).
Lead alloy anodes for chromium plating.
U.S. Patent 3,755,094 (1971) *Lead Abstr.* 14-2729.
Pb anode for chromium plating.
U.S. Patent 3,794,580 (1972); *Chem. Abstr.* **81**, 9149.
Ca–Pb anodes for electrowinning.
U.S. Patent 3,859,185 (1974).
PbO$_2$–Ti composite anode with boride interlayer.
U.S. Patent 3,880,728 (1975).
PbO$_2$–Ti composite anode.
U.S. Patent 3,935,082 (1975).
Porous ceramic–PbO$_2$ electrode.
U.S. Patent 4,008,144 (1977).
PbO$_2$–Ti composite anode.
U.S. Patent 4,019,970 (1977).
PbO$_2$-graphite anode with Bi addition.
U.S. Patent 4,038,170; *Chem. Abstr.* **87**, 92677.
PbO$_2$-graphite anode.
U.S. Patent 4,038,170 (1977) to Pepcon.
Pb bielectrodes.
U.S. Patent 3,284,333 (1966); *Chem. Abstr.* **66**, 34259.
Cathode–Pd–Ag/Pb.
U.S. Patent 3,947,333 (1974).
Anode pretreatment.
U.S.S.R. Patent 328,198 (1973); *Chem. Abstr.* **76**, 157487.
Pb–Co–Tl anode.
U.S.S.R. Patent 185,074 (1966); *Chem. Abstr.* **66**, 79056.

Quaternary Pb alloy (Ag, Pb, Be, Tl).
 U.S.S.R. Patent 309,966 (1969); *Lead Abstr.* 12-2024.

Applied and Industrial

Joint extraction of Pb and MnO_2 from Mn anode slimes.
 R. I. Agladze, *Soobshch. Akad. Nauk. Gruz. SSR* **77**, 617 (1975); *Chem. Abstr.* **83**, 67682.
Electrowinning of Pb sulphides.
 E. S. Allen, Paper TIVb4 *MMIJ–AIME Joint Meeting Tokyo*, 24 May (1972); *Lead Abstr.* 15-0939.
Lead electrorefining.
 Anon. *Cominco Magazine* **23**, 17 (1962); *Lead Abstr.* 1963 1481.
Peruvian Pb refinery.
 C. A. Aranda, AIME Symposium Mining/Metallurgy of Pb, Vol. 1. "Mining and Concentration of Pb". AIME, New York (1970); *Lead Abstr.* 1971 1500.
Electrowinning of Pb with MnO_2 anodes.
 V. Aravamuthan, *Chem. Proc. Eng. (India)* **1**, 29 (1968); *Lead Abstr.* 1970 646.
Cathodic corrosion of Pb in soil.
 W. G. von Baeckmann, *Werkst. Korros.* **20**, 578 (1969); *Lead Abstr.* 1970 122.
Electrowinning of Pb.
 C. C. Banks, Rept LR 192(ME). Warren Springs Lab. UK (1973); *Lead Abstr.* 14-2985.
Electrorefining for 99·9999% Pb production.
 G. Baralis, *Metal Ital.* **59**, 494 (1967); *Lead Abstr.* 1968 5280.
Electrorefining of Pb.
 K. R. Barrett, *9th Commonwlth Mining Congr. Lond.* May (1969); *Lead Abstr.* 1971 1138.
Pb production by ammoniacal leaching and electrolysis.
 G. C. Bratt, *98th Ann. Meeting AIME Washington DC*, 17 Feb (1969); *Lead Abstr.* 1970 408.
Electrometallurgical review of Pb.
 A. R. Burkin, *Metal. Mater. Metal Rev.* **116**, 47 (1971); *Lead Abstr.* 1971-1250.
Direct electrowinning of Pb from PbS concentrates.
 K. J. Cathro, *Proc. Aust. Inst. Min. Met.* (260) 9 (1976); *Lead Abstr.* 17-4633.
Alkaline electrolytic extraction of As from Pb converter dust.
 E. V. Davydov, *Tsvetn. Metal.*, 68 July (1967); *Lead Abstr.* 1967 4975.
Electroformed Pb.
 J. W. Dini, *Metal Finish.* **67**, 53 (1969); *Lead Abstr.* 1969 6393.
Electrolysis of Pb from PbS and galena concentrate.
 S. Dixit, *Trans. Ind. Inst. Met.* **28**, 293 (1975); *Lead Abstr.* 16-3425.
Electrolytic Pb recovery from slimes.
 J. G. Donaldson, *RI 6263 Us Bur. Mines* (1963); Lead Abstr. 1964 2298.
Lead electrometallurgy progress report.
 M. R. Edwards, *Int. Met. Rev.* **21**, 123 (1976).

Electrorefining plant.

K. Emicke, *Erzmetall.* **24**, 205 (1971); *Lead Abstr.* 11-1376.

Pb powders by electrolysis.

S. Fazekas, *Kohasz. Lapok* (Hungarian) **96**, 371 (1963); *Lead Abstr.* 1964 2565.

Electrorefining of Pb at Monreale smelter.

K. Frenai, ZDA/LDA Transl. No. 70/175.

Electrorefining of Pb.

German Patent 1,180,140 (1964); *Lead Abstr.* 1965 3148; also French Patent 1,378,715 (1964); E. R. Freni, *J. Metal.* **17**, 1206L (1965); *Lead Abstr.* 1966 3915; *Lead Abstr.* 1965 3150.

Possible electrolysis of Pb–Cu powders.

V. P. Galushko, *Sov. Electrochem.* **7**, 844 (1971).

Electrowinning of Pb with aqueous amine solution.

L. S. Getskin, *Tsvetn. Metal.*, 20 (1965); *Lead Abstr.* 1965 3520.

Pb by electrolysis of PbO.

K. S. A. Gnarasekaren, *Indian J. Technol.* **8**, 297 (1970); *Lead Abstr.* 1971 1210.

Pb–Sn alloy for electrotyping.

N. A. Grekova, *Zasch. Metal.* **13**, 236 (1977); *Chem. Abstr.* **86**, 179417.

Protection of lead artefacts.

V. F. Griz, *Zasch. Metal.* **12**, 199 (1976); *Chem. Abstr.* **85**, 48349.

Indirect electrowinning of Pb from galena with ferric leach.

F. P. Haver, *RI 7360 U.S. Bur. Mines*; *Lead Abstr.* 1970 578.

Electrochemistry of galena/flotation reagent system.

B. C. Haydon, *Trans. Inst. Mining Met.* **78**, C 181 (1969); *Lead Abstr.* 1970 507.

PbS as ion-selective (Pb) electrode.

H. Hirata, *Anal. Chem.* **43**, 279 (1971).

Electrolysis of Pb dusts with Pb-xylitol complexes.

S. Ishankhodzhaev, *Uzb. Khim. Zh.* (1) 3 (1976); *Lead Abstr.* 17-4103.

Electrorefining of Pb with suspension electrodes.

H. Kametani, *Nippon Kogyo Kaishi* **91** (1051), 601 (1975); *Chem. Abstr.* **87**, 143243.

Electrochemical production of Pb pigments.

N. A. Karnaev, *Khim. Tekhnol. Primenenie Soedin. Khroma* 96 (1975); *Chem. Abstr.* **84**, 61264.

Pb recovery by chlorination and electrolysis of galena.

H. Kenworthy, *RI 6554 US Bur. Mines*: (1964); *Lead Abstr.* 1965 3397.

Organic levelling agents for Pb electrorefining.

R. C. Kerby, 107th AIME Meeting, Denver, 26th Feb (1978).

Electrochemical study of galena–xanthate–oxygen flotation system.

J. A. Kitchener, *Trans. Inst. Mining Met.* **73**, 313 (1964); *Lead Abstr.* 1964 2940.

Electrorefining of Pb.

R. Kleinert, *Erzmetall.* **22**, 327 (1969); *Lead Abstr.* 1970 73.

Present state of Pb refining technology.

R. Kleinert, ZDA/LDA Transl. No. 70/152.

Amalgam electrorefining of Pb.
L. F. Kozin, *Tr. Inst. Khim. Nauk. Akad. Nauk. Kaz. SSR* **12**, 37 (1964); *Lead Abstr.* 1965 3466.
Electroleaching of dusts from Pb production.
Yu. A. Kozmin, *Fiz. Khim. Issled. Biol. Khim.* 96 (1972); *Chem. Abstr.* **81**, 57605.
Analysis of the Betts (Pb refining) process.
C. J. Krauss, *J. Metals* **28** (11), 4 (1976).
Electrolysis of Pb-containing mattes and speisses.
J. Kruger, *Erzmetall.* **26**, 268 (1973); *Lead Abstr.* 14-2631.
Electroforming of Pb foil.
L. Krushev, *Bulg. Metallurg.* **2**, 16 (1972); *Chem. Abstr.* **80**, 22012.
Electrochemistry of impurities in Pb electrorefining.
G. A. Kuznetsova, *Tsvetn. Metal.* 64 (1967); *Lead Abstr.* 1967 4916.
Amalgam electrorefining of lead.
W. H. Lee and Y. Z. You, *Kumsok Hakhoe Chi* **12**, 399 (1974); *Lead Abstr.* 16-3081.
Amines for Pb extraction.
H. Mimura, *Nippon Kogyo Kaishi* **83**, 36 (1967); *Lead Abstr.* 1970 145.
Effect of impurities in Pb electrorefining.
F. Miyashita, *Techn. Rept Kansai Univ.* (14) 61 (1973); *Lead Abstr.* 14-2820.
Electrowinning of Pb from galena.
J. E. Murphy, *RI 7913 US Bur. Mines* (1974); *Lead Abstr.* 15-0630.
High purity Ag production at Trepca Pb refinery.
B. Nikolic, *Serbo-Croat Khem. Ind.* **10**, 1542 (1975); *Chem. Abstr.* **84**, 182888.
Treatment of battery scrap with organic amines.
T. Nikolov, *Neue Huette* **19**, 591 (1974); *Lead Abstr.* 15-1177.
Pb electrorefining in Japan.
E. Nomura, paper A74-21 (1974) AIME; *Lead Abstr.* 15-0968.
$Pb-AgO_2$ analyser.
D. A. Okun, *J. Electroanal. Chem.* **4**, 65 (1962).
Electropolymerisation of Pb paints.
J. Oliver, *Prof. Finish* **34**, 66 (1970); *Lead Abstr.* 1970 829.
Lead from electrolysis of scrap batteries.
F. Ovari, *Zh. Prikl. Khim.* **49** 1078 (1976).
Lead electrorefining.
A. J. Phillips, *AIME Met. Soc. Trans.* **224**, 657 (1962).
Compatibility of paints (Pb-based) with cathodic protection.
A. F. Routley, *Paint Technol.* **31**, 28 (1967); *Lead Abstr.* 1967 4829.
Electrolysis of Cu–Pb speiss.
E. N. Sazhin, *V. Sb Metallurg. i Obogashchenie Alma–Ata* **10**, 78 (1975); *Chem. Abstr.* **85**, 164054.
Lead–zinc extraction by electro-oxidation.
B. J. Scheiner, *RI 8092 US Bur. Mines* (1975); *Lead Abstr.* 17-3762.
Electrorefining of Pb.
A. Schmidt, *in* "Angewandte Elektrochemie" (p. 215 *et seq.*) Verlag Chemie, Weinheim, Germany (1976).

Electrolytic Pb extraction.
　W. Schulz, *Chem. Ing. Tech.* **40**, 570 (1968); *Lead Abstr.* 1968 5673.
Electrorefining of Pb.
　M. P. Smirnov, *Tsvetn. Metal.* 24 (1973); *Lead Abstr.* 14-2988.
Electrorefining of Pb with nitrate baths.
　V. I. Solov'eva, *Tr. Altaisk Gorno-met. Nauchn–Issled. Inst. Akad. Nauk. Kaz. SSR* **14**, 75 (1963); *Lead Abstr.* 1964 2413.
Recovery of Pt and Pb from electrolytic H_2O_2 waste sludges.
　J. Steindor, *Proc. Nauk. Inst. Chem. Nieorg. Metal.* **29**, 287 (1976); *Chem. Abstr.* **85**, 127646.
Electrodeposition of Pb paints.
　L. Tasker, *Paint Oil Colour J.* **145**, 457 (1964); *Lead Abstr.* 1964 2500.
Electrochemistry of galena/O_2/xanthate flotation system.
　D. Toperi, *Trans. Inst. Mining Met.* **78**, C 191 (1969); *Lead Abstr.* 1970 508.
Lead refining with sulphamate bath.
　G. Tremolada, *Electrochim. Metallorum* **1**, 457 (1966); *Lead Abstr.* 1967 4832.
Pb powder by electroreduction of PbO and PbO_2.
　H. V. K. Udupa, *Trans. Ind. Inst. Metal.* **19**, 99 (1966); *Lead Abstr.* 1967 4785.
Electroreduction of Pb oxides from wastes.
　H. V. K. Udupa, *Electrochim. Metallorum* **4**, 255 (1969); *Lead Abstr.* 1970 798.
Electrolytic production of Pb powder.
　H. V. K. Udupa, Indian Patents 127,956 and 131,959; *Chem. Abstr.* **82**, 9372 and 23758.
Electrochemical preparation of lead telluride.
　Ya. A. Ugai, *Sov. Electrochem.* **12**, 1626 (1976).
Review of Pb electrorefining processes.
　F. Vogel, *Z. Erzbergbau Metallhuettenwes.* **15**, 656 (1962); *Lead Abstr.* 1963 2047.
Electrolysis of Pb from sulphide ores.
　B. J. Welch, *Scand. J. Metals* **1**, 49 (1972); *Lead Abstr.* 12-2042.
Effect of anode sludge on purity of refined Pb.
　F. Wenzel, *Neue Huette* **19**, 263 (1974); *Lead Abstr.* 15-0635.
Leach electrolysis method for Pb production.
　M. M. Wong, *World Mining Metal Techn. Proc. Joint MMIJ/AIME* **2**, 603 (1976); *Chem. Abstr.* **87**, 42276.
Electrowinning and refining of Pb.
　S. E. Woods, *Met. Mater.* **8**, 187 (1974); *Lead Abstr.* 14-2986.
Industrial effluent treatment with PbO_2 anode.
　A. Yoshiyama, *Denki Kagaku Oyobi Butsuri Kagaku* **43**, 521 (1975); *Chem. Abstr.* **85**, 9940.
Electrolytic colouring of metals by PbO deposition.
　British Patent 1,010,065 (1961); *Lead Abstr.* 1966 3814.
Electrotype production process.
　British Patent 931,309 (1962); *Lead Abstr.* 1964 2199.

Electrolytic mfc of powder Pb.
  British Patent 989,475 (1965).
Electrorefining of Pb.
  British Patent 1,251,290 (1969); *Lead Abstr.* 12-1883.
Recovery of Pb from battery scrap (electrolytic method).
  British Patent 1,368,423 (1970); *Lead Abstr.* 0298.
Electrowinning and refining of Pb.
  British Patent 1,345,411 (1971); *Lead Abstr.* 14-289ς.
Pb recovery by cathodic reduction of ores.
  British Patent 1,349,672 (1971); *Lead Abstr.* 14-3086.
Electrolytic recovery of Pb from sulphides.
  British Patent 1,345,102 (1971).
Electrolytic process for PbS treatment.
  British Patent 1,361,767 (1974); 1,362,943 (1974); *Lead Abstr.* 15-0345/6.
Electrolytic recovery of Pb.
  British Patent 1,427,228 (1976).
Electrolytic recovery of lead.
  British Patent 1,428,957 (1976).
Electrode assembly for O_2 analysis.
  British Patent 913,412.
Electrochemical manufacture of white lead.
  British Patent 144,819 (1919); *Chem. Abstr.* **14**, 3022; U.S. Patent 1,452,620 (1922); *Chem. Abstr.* **17**, 2197.
  See also J. J. Mattielo, "Protective and Decorative Coatings", Harper, New York (1962); and H. F. Payne, "Organic Coating Technology", Wiley, New York (1961).
Electrolytic cell for Pb refining.
  British Patent 1,462,572.
Electrorefining of Pb.
  French Patent 1,338,526 (1962); *Lead Abstr.* 1964 2136.
Pb production by ammoniacal leaching and electrolysis.
  German Patent 1,533,071 (1966); *Lead Abstr.* 1970-515.
Bath for electrolytic removal of Pb.
  German Offen Patent 2,254,398 (1972); *Lead Abstr.* 15-0319.
Electrorefining of Pb.
  German Offen Patent 2,355,144 (1974); *Lead Abstr.* 15-0638.
Anodic oxidation of aqueous solutions with PbO_2 anodes.
  German Patent 1,070,600 (1959) BASF.
Production of Pb electrodes.
  Japanese Patent 75 01812; *Chem. Abstr.* **84**, 167302.
SO_2 removal from waste gas with PbO_2 electrode.
  Japanese Patent 76 23,468; *Chem. Abstr.* **87**, 89967.
Electrolytic precipitation of Pb sulphides from leach liquors.
  U.S. Patent 3,162,587 (1964); *Lead Abstr.* 1965 3152.
Electrolytic Pb recovery from ores.
  U.S. Patent 3,772,003 (1972); *Lead Abstr.* 14-2828.
Electrolytic production of Pb salts.
  U.S. Patent 3,795,595 (1974); *Chem. Abstr.* **80**, 127579.

Production of Pb from sulphide ores.
U.S. Patent 3,929,597 (1975).
Sulphur removal from scrap battery.
U.S. Patent 3,883,348 (1975); 3,892,563 (1975).
Electrochemical treatment of scrap batteries.
U.S. Patent 3,985,630 (1976).
Electrowinning of Pb from fluorinated PbS in H_2SiF_6.
U.S. Patent 3,972,790 (1976); *Chem. Abstr.* **85**, 125933.
Sewage treatment with PbO_2 electrodes.
U.S. Patent 3,764,500 (1973) Pepcon.
Improved Betts process for high purity Pb.
U.S. Patent 3,960,681 (1976).
Electrolytic production of lead containing powder.
U.S.S.R. Patent 406,956 (1974); *Chem. Abstr.* **81**, 57660.
Electrolytic production of yellow Pb chromate.
U.S.S.R. Patent 558,067 (1975); *Chem. Abstr.* **87**, 45983.

Lead Electrodeposition

Electrocrystallization of lead on copper single crystal planes: some effects of superimposed 'a.c.' on 'd.c.'.
M. F. Ahmed, B. S. Sheshadri and F. Pushpanaden, *Electrodeposition Surface Treatment* **3** (1), 65–73 (1975).
Fabrication of fibre-reinforced metal by filament winding and electrodeposition. An evaluation of some electro-plating solutions.
A. A. Baker, M. P. B. Allery and S. J. Harris, *J. Mater. Sci.* **4** (3), 242–251 (1969).
Electrodeposition of Pb from fluoborate solutions.
H. Benninghof, *Metall.* **30**, 344 (1976).
Electrodeposition of Pb and alloys from fluoborate.
H. Benninghof, *Metall.* **30**, 813 (1976).
Electrodeposition of thin Pb films on semiconductors.
P. Bindra, *J. Electrochem. Soc.* **124**, 1012 (1977).
Deposition of Pb–Sb alloys from fluoborate.
V. V. Bogoslovskii, *Zasch. Metal.* **11**, 499 (1975).
Tin–lead electrodeposits. The effect of impurities in the solution.
G. W. Cavanaugh and J. P. Langan, *Plating* **57** (4), 369–371 (1970).
Use of electrochemical intermetallic processes for electroplating from molten electrolytes.
Y. K. Delimarskii, I. G. Pavlenko and O. G. Zarubitsky, *Ukr. Khim. Zh.* **33** (8), 863–864 (1967); *Met. Abstr.* **1**, 109 (1968).
Electrodeposition of Pb–Sn alloys from phenol sulphonate etc.
N. Dohi, *Kinzoku Hyomen Gijutsu* **26**, 309 (1975); *Lead Abstr.* 17-3723. Pb–Sn.
N. Dohi, *J. Met. Fin. Soc. Jap.* **26**, 309 (1975).
Cathodic deposition of zinc and lead with high current densities.

G. A. Emel'yanenko, E. Ya. Baibarova and G. G. Simulin, *Ukr. Khim. Zh.* 29 (5), 515–518 (1963); *Chem. Abstr.* 59 (8), 8357a (1963).
Inhibited electrocrystallisation of Pb in acid media.

F. Elsner, *Metalloberfläche* 31, 171 (1977).
Zinc–lead electrodeposit.

J. P. G. Farr and S. V. Kulkarni, *Znn. Conf. Inst. Metal Finish.*, Torquay 4th May (1966).
Phase structure of binary alloys produced by electrodeposition.

N. P. Fedotev and P. M. Vyacheslavov, *Plating* 57 (7), 700–706 (1970).
Electrodeposition of metals in porous glass.

F. P. Fehlner, *J. Electrochem. Soc.* 117 (3), 411–413 (1970).
Effect of temperature on the rate of discharge of lead from trilonate electrolytes in the presence of various anions.

V. S. Galinker, O. K. Kudra and L. W. Novikova, *Ukr. Khim. Zh.* 36 (2), 150–154 (1970); *Met. Abstr.* 3, 923 (1970).
Some problems in electrodeposition.

J. A. Harrison, S. K. Rangarajan and H. R. Thirsk, *J. Electrochem. Soc.* 113 (11), 1120–1133 (1966).
Electrocrystallisation of Pb onto Cu single Xtals.

S. Itoh, *Surf. Technol.* 5, 27 (1977).
Growth of intermetallics on Sn–Pb electrodeposits.

P. J. Kuy and C. A. MacKay, *Trans. Inst. Metal. Fin.* 54, 68 (1976).
Growth of electrodeposits.

J. M. Keen and J. P. Farr, *J. Electrochem. Soc.* 109, 668 (1962).
Zn–Pb alloys.

E. D. Kochman, *Rhef. Zhurn. Korr. Zasch. Korr.* 4K427 (1975).
Electrodeposition of Pb–Sn from fluoborate.

N. T. Kudryaev, *Technik* 31, 321 (1976).
Electrodeposition of Pb from organic aprotic media.

V. V. Kuznetsov, *Zasch. Metal.* 11, 631 (1975).
Lead electroplating in organic aprotic solvents.

V. V. Kuznetsov, *Zasch Metal.* 11, 631 (1975).
Electrodeposition study.

V. I. Lainer, *et al.*, *Zh. Prikl. Khim.* 36 (2), 350–356 (1963); *Corrosion Technol.* 10 (9), 235 (1963).
Laws governing the electrodeposition of alloys.

V. I. Layner and Y. Tszu-Zhan, *Zh. Prikl. Khim.* 36 (1), 121–129 (1963); *Metal Finish. Abstr.* 5 (2), 54 (1963).
Cathodic polarization in the electrolytic deposition of lead–tin alloys.

V. I. Layner and Y. Tszu-Zhan, *Zh. Prikl. Khim.* 36 (2), 250–256 (1963); *Metal Finish. Abstr.* 5 (2), 54 (1963).
Electrocrystallization of lead from a pyrophosphate solution.

M. A. Loshkarev and T. F. D'Yachenko, *Zh. Prikl. Khim.* 37 (1), 70–76 (1964); *Metal Finish. Abstr.* 6 (2), 48.
Electrolytes.

K. G. Mahle, German Patent 1,255,431 (1966); *Met. Pat. J.* 8 (1) (1968); *Met. Abstr.* p. 2.

Pb.

B. F. Muller, *Galvanotechnik* (*Saulgau*) **65**, 180 (1974).

Structure of electrodeposited lead dendrites.

F. Ogburn *et al.*, *J. Electrochem. Soc.* **112** (6), 574–577 (1965).

Electrodeposition of Cu–Pb alloys (review).

L. Oniciu, *Rev. Chim.* **26**, 842, 852 (1975); *Lead Abstr.* 18-0217.

Some problems in the theory of electrodeposition of alloys. X. Phase studies of electrodeposited copper–lead alloys produced from complex electrolytes.

Y. M. Polukarov and V. V. Grinina, *Elektrokhimiya* **1** (1), 31–35 (1965); *Metal Finish. Abstr.* **7** (6), 289 (1965).

Effect of surface-active compounds on electrodeposition of Pb from aqueous chloride solutions.

A. V. Pomosov and A. I. Levin, *Tvsetn. Metal.* (6), 121–125 (1959) (Transl. L.L.U. RTS 1814).

Electrodeposition of Ag–Pb alloys.

M. Pushpavanam, *Electroplating Metal Fin.* **29**, 3–5, 8–10, 12–13 (1976).

Electrolytic deposition of a lead–thallium alloy.

V. T. Pustouit, USSR Patent 153,820 (1961); *Chem. Abstr.* **59** (13), 14, 904b (1963).

Alloy deposits.

E. Raub, Paper presented at the Proceedings of the International Conference on the Protection against Corrosion by Metal Finishing, pp. 73–80. Basle 22–25 Nov. 1966.

Electrocrystallisation of Pb.

E. Raub, *Bänder Bleche Rohre* (12) 520 (1976); *Lead Abstr.* 17-4911.

Structure and properties of electrodeposited tin–lead alloys.

W. Riedel, *Metalloberfläche* **23** (2), 42–44 (1969).

Electrodeposition of lead from salt melts.

A. Riesenkamf, *Rudy Metale Niezelazne* **10** (5), 271–274 (1965).

New dimension in tin–lead plating.

G. B. Rynne, *Plating* **58** (8), 867, 869, 872, 874, 876 (1971).

Electrodeposition of molten metals from glycerine.

C. Smelt, *Ind. Chem.* **38**, 628–630, 633 (1962); *J. Appl. Chem.* **13** (5), i–521.

Electrodeposition of binary alloys—recent developments.

R. D. Srivastava, *J. Appl. Electrochem.* **6**, 321 (1976).

Electrodeposition of Pb–Sn alloys.

H. J. Steeg, *Metall.* **30**, 834 (1976); also British Patent 1,438, 701 (1976).

Pb–Sb.

K. M. Tyutina, *Korroz. Zasch. neftegaz. prom. stil.* **5**, 19–20 (1973).

Kinetics of the cathodic deposition of lead from a perchloric acid electrolyte.

Yu. E. Udovenko and N. G. Krapivnyi, *Dnepropftr. Khim. Tekhnol. Inst. Dnepropetrovsk USSR*. Russian edition: *Ukr. Khim. Zh.* **42** (12), 1261–1264; *Chem. Abstr.* **86** (12), 080733.

Codeposition of Pb–Cu from nitrate bath.

H. V. K. Udupa and K. C. Narasinham, *Plating Surf. Finish.* **62**, 1150 (1975).

Effect of cathode rotation on deposition of lead from sodium plumbite solution.
K. S. Udupa, G. S. Subramanian and H. V. K. Udupa, *Plating (USA)* **49** (12), 1274–1278 (1962).
Electrodeposition of metals in the presence of polymers. Part 2. Formation of highly dispersed lead on a cathode in the presence of epoxy resin.
A. R. Ulberg *et al.*, *Ukr. Khim. Zh.* **35** (2), 148–151 (1969).
Pb electrodeposition in dispersions of polymers.
Z. P. Ulberg, *Ukr. Khim. Zh.* **40**, 1274 (1974).
Residual stresses in electrodeposits.
A. T. Vagramyan and Y. S. Tsareva, *Zh. Fiz. Khim.* **29** (1), 185–193 (1955).
Electropolishing of Pb.
M. Verhaege and R. Wild, *Prakt. Metall.* **10**, 220 (1973).
Rate of Pb electrodeposition and electrodissolution.
A. K. Vijh, *Surf. Technol.* **5**, 257 (1977).
The influence of colloidal lead sulphide on the properties of lead electrodeposits.
A. S. Voevodsky, V. A. Lyando and V. V. Ostroumov, *Zh. Fiz. Khim.* **36** (7), 1536–1539 (1962); *Met. Abstr.* **30** (10), 748 (1963).
Investigation of a process for the deposition of indium–lead alloys from tartrate electrolytes.
Yu. V. Zaitsev and P. S. Titov, *Tsvetn. Metal.* (6), 136–139 (1962).
Pb.
British Patent 1,351,875.
Electrodeposition of Pb from DMF-water solutions.
British Patent 1,410,764.
Sn–Pb barrel plating.
British Patent 1,361,269.
Electrodeposition of Pb–Sn.
British Patent 1,453,886 (1976); U.S. Patent 3,984,291 (1976).
Electrodeposition of Sn–Pb alloys.
British Patents 1,469,547 and 1,476,176 (1977).
Electrodeposition of Pb–Sb alloys.
Chem. Abstr. **81**, 71903e, **82**, 9202t.
Bright Pb.
D.O.S. Patent 2,356,678.
Pb.
D.O.S. Patent 2,424,070.
Pb.
D.O.S. Patent 2,449,189.
Deposition of Pb + In for bearings.
French Patent 2,266,750 (Vandervell).
Brightener for electrodeposition of Pb–Sn alloys.
Ger. Offen Patent 2,510,870; *Chem. Abstr.* **84**, 23757.
Sn–Pb.
Japanese Patent 74,034,569.

Process for depositing lead. Etablissements.
Eugene Arbez, U.S. Patent 3,616,305 (1967).
Process for surface treatment of lead and its alloys.
Bell Telephone Laboratories Inc., U.S. Patent 3,539,427 (1968).
Plating tin–lead alloy on printed circuits and electrolyte therefor.
North American Rockwell Corp., U.S. Patent 3,554,878 (1968); *Off. Gaz. U.S. Pat. Off.* **882** (2), p. 766 (1971).
Bath and method for electrodepositing tin and/or lead.
Conversion Chemical Corp., U.S. Patent 3,769,182 (1971).
Addition agent for the electrolyte used in the electrodeposition of lead.
ITT Rayonier Inc., U.S. Patent 3,554,884 (1971); *Off. Gaz. U.S. Pat. Off.* **882** (2), 767.
Electroplating bath for depositing tin–lead alloy plates.
Dr Ing. Max Schloeter, Fabrik für Galvanotechnik, U.S. Patent 3,730,853 (1971); *Off. Gaz. U.S. Pat. Off.* **910** (1), 283 (1973).
Coating printed circuit boards with tin or tin–lead alloy and tin–lead fluoborate plating baths.
Allied Chemical Corp., U.S. Patent 3,859,182, 4th Jan. 1973 (1975).
Electrodeposition of Co–Pb, Ni–Pb alloys.
U.S. Patent 4,028,200 (1977).
Sn–Pb.
U.S. Patent 3,749,649; also J. W. Tharp, Product Finishing (Cincinatti) **37**, 54–59 (1973).
U.S. Standards on Pb electroplate.
Galvano–Organo (Paris) **44**, No. 452, 89–91 (1975).
Additives for electrodeposition of Sn–Pb.
U.S. Patent 3,956,123 (1976).
Sn–Pb bath.
U.S. Patent 3,785,939.
Sn–Pb stripping bath.
U.S.S.R. Patent 398,702.
Pb–Cd.
U.S.S.R. Patent 399,577.
Zn–Pb (as fine Xtals).
U.S.S.R. Patent 411,158.

Various

Double layer and adsorption of MeOH on PbO_2.
M. M. Abou-Romia, *Indian J. Chem.* **12**, 1180 (1974).
Study of PbO_2 deposits with transmitted light microscopy.
R. G. Acton, *5th Int. Power Sources Symp., Brighton.* Sept. (1966).
AC polarography of Pb ions in different media.
H. P. Agarwal, *Proc. 14th Semin. Electrochem.* (1973); *Chem. Abstr.* **82**, 147081.
Potentiodynamic study of surface species on Pb.
I. A. Aguf, *Sov. Electrochem.* **6**, 239 (1970).

Effect of organics on anodic Pb passivation.
 I. A. Aguf, *Sov. Electrochem.* **6**, 1667 (1970).
Electrochemical properties of PbO_2 in H_2O_2.
 M. M. Andrusev, *Zh. Fiz. Khim.* **48**, 1198 (1974).
Pure beta PbO_2 anodes.
 Anon., Leaflet from Bart Manufacturing Corp. (1964); *Lead Abstr.* 1964 2778.
Etchants for Pb.
 Anon., *Metal Prog.* **104**, 73 (1973); *Lead Abstr.* **14**, 2639.
Potential measurements on powder electrodes and Pb dust.
 K. Appelt, *Electrochim. Acta* **1**, 337 (1959).
Cathodic penetration of Na into Pb.
 I. I. Astakhov, *Sov. Electrochem.* **6**, 379 (1970).
Preparation of PbO_2 for X-ray diffraction studies.
 N. E. Bagshaw, *J. Appl. Chem.* **16**, 180 (1966).
Electro-organic oxidations with alpha PbO_2.
 N. Bakhchisaraitsyan, *Zh. Prikl. Khim.* **35**, 1643 (1962); *Lead Abstr.* 1963 1805.
Rate of ozone formation from peroxodisulphate electrolysis.
 J. Balej, *Coll. Czech Chem. Commun.* **39**, 3409 (1974); *Chem. Abstr.* 82 117802.
Cinematographic study of Pb electrodes.
 M. Barak, *3rd Int. Battery Symp.*, Bournemouth (1962).
Adsorption of Pb^{++} on Hg induced by anions.
 D. J. Barclay, *J. Electroanal. Chem.* **28**, 71 (1970).
Rotating ring-disc study of $Pb-PbCl_2-HCl$.
 R. G. Barradas, *Can. J. Chem.* **53**, 389 and 407 (1975).
Dissolved gases and Pb–Pb(II) electrode in Cl^-.
 R. G. Barradas, *Chem. Phys. Aqu. Gas Sol.* (*Proc. Symp.*) 357 (1975); *Chem. Abstr.* 83 87286.
S.E.M. and electrochemical study of surface effects in $PbCl_2$ formation.
 R. G. Barradas, *Electrochim. Acta* **21**, 357 (1976).
Expander action at $Pb(Hg)/PbCl_2$ interface.
 R. G. Barradas, *J. Power Sources* **2**, 137 (1977).
Temperature effects of cycled Pb electrodes in HCl.
 R. G. Barradas, *Electrochim. Acta* **22**, 237 (1977).
Anodic behaviour of Pb–Hg electrodes in HCl.
 R. G. Barradas, *J. Electroanal. Chem.* **80**, 295 (1977).
Standard potential of $Pb/PbCl_2/HCl$ cell.
 F. L. Bates, *J. Electrochem. Soc.* **121**, 79 (1974).
Potential of Pb^+ (unstable)/Pb system.
 J. H. Baxendale, *Chem. Commun.* 19th Oct., 715 (1966).
Effect of proteins on Pb(II)–Hg electrode system.
 B. Behr, *J. Electroanal. Chem.* **46**, 223 (1973).
Oxidation of SO_2 on PbO_2 anode.
 G. Belanger, *Anal. Chem.* **46**, 1576 (1974).
Physico-chemical studies of Pb/Pb^{++} in Cl^-.
 K. Belinko, *Diss. Abstr. Int. B.* **36**, 6185 (1976); *Chem. Abstr.* 85 70010.

Redox potential of Pb(IV)/Pb(II) tetra-acetate.
  A. Berka, *J. Electroanal. Chem.* **4**, 150 (1962).
Electroreflectance of Pb/NaF.
  A. Bewick, *Surface Sci.* **55**, 349 (1976).
Electron-optical study of water adsorption on Pb.
  A. Bewick, *J. Electroanal. Chem.* **60**, 163 (1975).
Adsorption of *n*-decylamine on Pb.
  J. O'M. Bockris, *J. Electrochem. Soc.* **111**, 736 (1964).
Preparation of lead alloy specimens for metallography.
  M. A. Bol'shanina, *Zavod. Lab.* **30**, 315 (1964).
Ionic conductivity of beta-PbF_2.
  R. W. Bonne, *J. Electrochem. Soc.* **124**, 28 (1977).
Electrode processes at lead halides.
  R. W. Bonne, *J. Electroanal. Chem.* **89**, 87 (1978).
Pb/$PbCl_2$/AgCl/Ag solid state cell.
  R. S. Bradley, *Trans. Far. Soc.* **62**, 2242 (1966).
Pb(II) adsorption on Hg electrode in presence of organics.
  M. Branica, *J. Electroanal. Chem.* **44**, 401 (1973).
Potential-pH (Pourbaix) diagram for Pb–H_2O at high temperature.
  P. A. Brook, *Corrosion Sci.* **12**, 297 (1972).
Electrophoretic deposition of Pb.
  D. R. Brown, *J. Appl. Chem.* **15**, 40 (1965).
Rotating disk study of Pb^{++} on Au.
  S. Bruckenstein, *Anal. Chem.* **45**, 2036 (1973).
Anodic oxidation of basic Pb sulphates.
  J. Burbank, *J. Electrochem. Soc.* **113**, 10 (1966).
Electrochemistry of Pb in DMSO.
  J. N. Butler, *J. Electroanal. Chem.* **14**, 89 (1967).
Dissolution of Pb in HNO_3 with AC.
  L. M. Butorina, *Tr. Uses Nauchno. Issled.* **35**, 147 (1973); *Chem. Abstr.* **82**, 49174.
Effect of structure on H_2 evolution on Pb alloys.
  T. W. Caldwell, Paper 6, *9th Power Sources Int. Symp. Brighton* 16 Sept. (1974); *Lead Abstr.* 15-1172.
Exchange reaction at polycrystalline Pb electrode.
  G. Ceccaroni and F. Rallo, *Gazz. Chem. Ital.* **102**, 825 (1973).
Cyclic voltammetry of Pb and Pb–Sb alloys to −40°C.
  T. G. Chang, *10th Int. Power Sources Symp., Brighton* (1976).
Electrochemical equilibrium and diffusion study of Pb–O_2 system.
  H. Charle, *Z. Phys. Chem.* **99**, 199 (1976); *Chem. Abstr.* **85**, 10880.
Cathodic reduction of alpha and beta PbO_2 in alkali and their solubility.
  P. Chartier, *Bull. Soc. Chim. Fr.* 2253 (1969); *Lead Abstr.* 12-1539, 1540.
Electrochemistry of PbO_2 decomposition.
  P. Chartier, *Bull. Soc. Chim. Fr.* 2250 (1969); *Lead Abstr.* 12-1538.
Electrochemistry of Pb in chloro-acid electrolytes.
  V. I. Chernenko, *Ukr. Khim. Zh.* **42**, 1261 (1976).
Effect of surfactants on polarography of Pb.
  G. D. Christian, *Monatsh Chem.* **104**, 1214 (1973); *Lead Abstr.* 14-2976.

Kinetic study of Pb(II)/Pb system in Na pyrophosphate.

M. Collier, *J. Chim. Phys. (Phys-Chim. Biol.)* **74**, 174 (1977).

Anomalous h.e.r. Tafel slope on Pb.

B. E. Conway, *J. Electrochem. Soc.* **116**, 1665 (1969).

Pb selenide electrodes for potentiometry.

D. C. Cormos, *Rev. Roum. Chim.* **19**, 1949 (1974); *Chem. Abstr.* **83**, 67899.

Single Xtal PbS electrode for potentiometry.

D. C. Cormos, *Rev. Roum. Chim.* **20**, 259 (1975); *Chem. Abstr.* **83**, 90357.

Single Xtal PbS electrode.

D. C. Cormos, *Rom. Stud. Univ. babes-Bolyai. Ser. Chem.* **20**, 36 (1975); *Chem. Abstr.* **85**, 13325.

Pb anodes for preparation of pure electrolytic Cr.

J. J. Dale, *Inst. Met. Finish Trans.* **39**, 135 (1962); *Lead Abstr.* 1963 1433.

Adsorption of organic substances on Pb electrode.

M. A. Dasoyan, *Sb. Rabot. po Khim. Istochnik.* **11**, 3 (1976).

Electroreflectance and double layer effects at Pb electrodes.

H. P. Dhar, *Surf. Sci.* **66**, 449 (1977); *Chem. Abstr.* **87**, 173306.

Electrode processes on and in a porous Pb or PbO_2 electrode.

Yu. D. Dunaev, *Tr. Inst. Khim. Nauk. Akad. Nauk. Kaz. SSR* **9**, 18 (1962); *Lead Abstr.* 1964 2231.

Pb alloy with S and Co additions.

Yu. D. Dunaev, *USSR Patent 179,473* (1964); *Lead Abstr.* 1966 4361.

Lead oxide–lead sulphate system.

J. D. Esdaile, *J. Electrochem. Soc.* **113**, 71 (1966).

Amine inhibition of Pb^{++} reduction.

I. Filipovic, *Croat. Chem. Acta* **41**, 151 (1969); *Lead Abstr.* 1970 549.

E.M.F. determination of O_2 in molten Pb.

W. A. Fischer, *Archiv. Eisenhuettenwesen* **37**, 697 (1966); *Lead Abstr.* 1967 4602.

Thermodynamic studies of the PbO_2 electrode (German).

W. Fischer and H. Rickert, *Ber. Bunsenges.* **77**, 975 (1973).

Thermodynamic study of PbO_2 electrode.

W. Fischer and H. Rickert, *Ber. Bunsenges. Phys. Chem.* **77**, 975 (1973); *Chem. Abstr.* **80**, 90175.

Kinetics of PbO_2 deposition on Pt.

M. Fleischmann, *Electrochim. Acta* **12**, 967 (1967).

E.M.F. measurements of solid electrolyte PbO cells.

S. N. Flengas, *J. Electrochem. Soc.* **115**, 796 (1968).

Effect of AC on Pb electrode in alkali.

V. N. Flerov, *Zh. Fiz. Khim.* **37**, 862 (1963); *Lead Abstr.* 1964 2291.

Solid electrolyte $PbCl_2$ cells.

R. T. Foley, *J. Electrochem. Soc.* **116**, 13C (1969).

Potential of zero charge of Pb electrode in presence of organics.

A. N. Frumkin, *Electrochim. Acta* **19**, 69 and 75 (1974).

Anodic dissolution of woods metal in alkali.

V. V. Gerasimov, *Zasch. Metal.* **5**, 53 (1969); *Lead Abstr.* 1970 203.

Oscillopolarography of Pb.

I. A. Gershkovich, *Teoriya I Praktika Polyarogr.* 94 (1973); *Chem. Abstr.* **82**, 147090.

Etching and chemical polishing of Pb.
  R. C. Gifkins and J. A. Corbett, *Metallurgia* **74**, 239 (1966).
Pb/Pb^{++} exchange reaction in HClO$_4$.
  A. S. Gioda, *J. Electrochem. Soc.* **124**, 1324 (1977).
Polarisation resistance method applied to Pb corrosion.
  J. A. Gonzalez and S. Feliu, *Werkst. Korros.* **26**, 758 (1975).
Pb electrocrystallisation during reduction of Pb oxides.
  K. M. Gorbunova, *Electrochim. Acta* **15**, 1597 (1970).
Adsorption of alcohols on Pb by differential capacity.
  N. B. Grigorev, *Sov. Electrochem.* **5**, 87 (1969).
Adsorption of thiourea on Pb electrodes.
  N. B. Grigorev, *Sov. Electrochem.* **6**, 89 (1970).
Inhibition of electrode reactions by adsorbed Pb.
  D. J. Gross, *Anal. Chem.* **38**, 405 (1966).
Behaviour of pulsed Pb electrode.
  G. M. Gusel'nikov, *Antikorroz Zasch. Stroit. Konstr. Tekhnol* 13 (1975); *Chem. Abstr.* **86**, 196819.
Single Xtal Pb selenide electrode for potentiometry.
  I. Haiduc, *Stud. Babes-Bolyai Ser. Chim.* **21**, 56 (1976); *Chem. Abstr.* **85**, 171221.
Membrane (lead chromate) electrodes.
  E. J. Hakoila, *Suomen Kemistilehti B* **46** (1973).
Double layer capacitance of Pb vs Hg electrode.
  N. A. Hampson *Electrochim. Acta* **15**, 581 (1970).
Double layer of PbO$_2$ in KNO$_3$.
  N. A. Hampson, *J. Electroanal. Chem.* **27**, 109 (1970).
Impedance of PbO$_2$ in sulphate.
  N. A. Hampson, *J. Electroanal. Chem.* **27**, 201 (1970).
Differential capacitance of Pb in various electrolytes.
  N. A. Hampson, *J. Electroanal. Chem.* **32**, 345 (1971).
Fast LSV studies on Pb and PbO$_2$ in sulphuric acid.
  N. A. Hampson, *J. Electroanal. Chem.* **33**, 109 (1971).
Differential capacitance and LSV on Pb and PbO$_2$.
  N. A. Hampson, *J. Electrochem. Soc.* **118**, 1262 (1971).
Differential capacitance of Pb alloys in aqueous nitrates.
  N. A. Hampson, *J. Electroanal. Chem.* **34**, 425 (1972).
Capacity of Pb–PbO$_2$ couple in fluoboric, fluosilicic acids.
  N. A. Hampson, *J. Appl. Electrochem.* **4**, 1 (1974).
Pb–Sn alloys in nitrate media.
  N. A. Hampson, *Corrosion Sci.* **15**, 23 (1975).
Electrochemistry of beta-PbO$_2$.
  N. A. Hampson, *J. Electroanal. Chem.* **79**, 273 and 281 (1977).
Oxidation of PbSO$_4$ and PbO$_2$.
  N. A. Hampson, *J. Electroanal. Chem.* **79**, 281 (1977).
Impedance of PbO$_2$ films on Pt.
  N. A. Hampson, *Surf. Technol.* **5**, 163 (1977).

Study of system Pb:PbO$_2$:PbSO$_4$.

    N. A. Hampson, *J. Electrochem. Soc.* **124**, 1655 (1977).

Surface structure of PbSO$_4$/PbO$_2$.

    N. A. Hampson, *J. Appl. Electrochem.* **7**, 257 (1977).

Anodic oxidation of PbSO$_4$.

    N. A. Hampson, *J. Electroanal. Chem.* **83**, 87 (1977).

Dissolution of Pb amalgam in H$_2$SO$_4$.

    J. A. Harrison, *J. Electroanal. Chem.* **43**, 321 (1973).

Kinetics of anodic Pb dissolution in sulphuric acid.

    J. A. Harrison, *in* "Power Sources 5" (D. H. Collins, ed.). Academic Press, London and New York (1975).

Pb oxidation to PbSO$_4$.

    J. A. Harrison, *Electrochim. Acta* **21**, 905 (1976).

Raman spectra of passive films on Pb electrodes.

    R. H. Heidersbach, Paper 23 "Corrosion '77" NACE San Francisco, 14 March (1977); *Lead Abstr.* 18-0331. Also "Corrosion '78" NACE, Houston, 6 March (1978).

Raman study of corrosion products on Pb anodes in chloride media.

    P. J. Hendra, *J. Electroanal. Chem.* **80**, 405 (1977).

Electrosorption of Pb^{++} on a single Xtal.

    F. Hilbert, *Surf. Technol.* **5**, 135 (1977).

Electropolishing of Pb.

    D. I. G. Ives and F. R. Smith, *Trans. Far. Soc.* **63**, 317 (1967).

Effect of light on anodic oxidation of Pb.

    S. O. Izidinov, *Sov. Electrochem.* **5**, 593 (1969).

Electro-reduction of Pb–Mn alloy.

    A. V. Izmailov, *Izv. Vyssh. Uchebn. Zaved. Khim. Khim. Tekhnol.* **7**, 456 (1964); *Lead Abstr.* 1965 3407.

Tracer study of PbSO$_4$ dissolution kinetics.

    A. L. Jones, *Trans. Far. Soc.* **66**, 2088 (1970).

Chiselled (fresh surface) Pb electrode.

    B. N. Kabanov, *Zh. Fiz. Khim.* 341 (1939); also *Acta Physicochem. URSS* **10**, 617 (1939).

Anodic diffusion of O$_2$ through PbO$_2$.

    B. N. Kabanov, *Electrochim. Acta* **9**, 1197 (1964).

Incorporation of alkali metals into Pb cathodes.

    B. N. Kabanov, *Electrochim. Acta* **13**, 19 (1968).

Electrodissolution of Pb and hydrogen evolution in alkali.

    B. N. Kabanov, *Sov. Electrochem.* **8**, 140 (1972).

Low temperature efficiency of lead–acid battery.

    A. F. Kalish, *Elektrotekhnika* **10**, 54 (1975); *Chem. Abstr.* **85**, 163228.

Pb-graphite electrodes.

    A. I. Kamenev, *Vestn. Mosc. Univ. Khim.* **17**, 569 (1976); *Chem. Abstr.* **86**, 48536.

Electrochemical reduction of nitric acid on Pb–Sn alloys.

    D. E. Karcheva, *Soobsch. Akad. Nauk. Gruz. SSR* **73**, 97 (1974); *Chem. Abstr.* **80**, 127460.

Thin film galvanic cell Pb/PbF$_2$/CuF$_2$/Cu.

J. H. Kennedy and J. C. Hunter, *J. Electrochem. Soc.* **123**, 10 (1975).

Effect of Sb and various anions on potential of Pb.

E. M. Khairy, *J. Electroanal. Chem.* **12**, 27 (1966).

Hydrogen evolution from alkaline solutions of Pb amalgams.

K. B. Khlystova, *Fiz. Khim.* 85 (1974); *Chem. Abstr.* **83**, 67828.

Open-circuit decays on Pb.

Ya. Kolotyrkin, *Zh. Fiz. Khim.* **31**, 581 (1947).

E.M.F. measurements of Pb–Hg amalgams.

L. F. Kozin, *Tr. Inst. Khim. Nauk. Akad. Nauk Kazak SSR* **9**, 81 (1962); *Lead Abstr.* 1964 2252.

Effect of impurities on the electrochemistry of Pb–Sb–As alloy.

G. V. Krivchenko, *Sb. Rab. Khim. Istochnik Toka Uses Nauchno Issled* **17** (1973); *Chem. Abstr.* **82**, 9313.

Pb/Pb^{++} on Hg electrode in tri-butyl phosphate.

D. Krznaric, *J. Electroanal. Chem.* **44**, 401 (1973).

Determination of p.z.c. of PbO$_2$ electrode.

F. I. Kukoz, *Sov. Electrochem.* **2**, 74 (1966).

Exchange reaction of Pb^{++} with Hg$_2$S monolayer.

H. A. Laitinen, *Anal. Chem.* **42**, 473 (1970).

Anodic deposition, cathodic stripping of PbO$_2$ on SnO$_2$.

H. A. Laitinen, *J. Electrochem. Soc.* **123**, 804 (1976).

Surfactants and Pb polarography.

S. Lal and G. D. Christian, *Monatsh. Chem.* **10**, 1214 (1973).

Pb/Pb^{++} in liquid ammonia.

D. Larkin, *J. Electrochem. Soc.* **119**, 189 (1972).

PbO solid electrolyte voltaic cell.

K. L. Laws, *J. Chem. Phys.* **39**, 1824 (1963).

Cathodic passivation of PbO$_2$ in aqueous media.

V. I. Lazarev, *Izv. Vyssh. Ucheb. Zaved. Khim. Khim. Tekhnol.* **18**, 838 (1975); *Chem. Abstr.* **83**, 123103.

Effect of surface–active materials on passivation of Pb electrodes.

V. F. Lazarev, *Zh. Prikl. Khim.* **38**, 1305 (1965); *Lead Abstr.* 1966 3929.

Hydrogen evolution on Pb in alkali.

T. S. Lee, *J. Electrochem. Soc.* **118**, 1278 (1971).

Potential of zero charge of Pb electrode.

D. I. Leikis, *J. Electroanal. Chem.* **46**, 161 (1973).

Impedance measurements of the Pb/Pb^{++} system.

D. I. Leikis, *Sov. Electrochem.* **11**, 317 (1975).

Reduction of PbO$_2$ in silicofluoric acid, effect of additives.

A. I. Levin, *Tr. Ural Politekhn. Inst.* No. 170,83 (1968); *Chem. Abstr.* **71**, 18145.

Cathodic passivation of PbO$_2$.

A. I. Levin, *Izv Vyssh. Uchebn Zaved.* **18**, 838 (1975); *Chem. Abstr.* **83**, 123103.

Expanders and discharge of Pb microelectrode.

A. Le Mehaute, *J. Appl. Electrochem.* **6**, 543 (1976).

Electrochemical acceleration of creep in Pb single Xtals.
V. I. Likhtman, *Sov. Electrochem.* **5**, 729 (1969).
Electro-oxidation/reduction of PbO suspension.
M. A. Loshkarev, *Ukr. Khim. Zh.* **29**, 287 (1963); *Lead Abstr.* 1964 2260.
Cathodic reduction mechanism of PbO_2.
L. I. Lyamina, *Sov. Electrochem.* **10**, 841 (1974).
Redox reactions on PbO_2 reduced PbO_2 electrodes.
L. I. Lyamina, *Sov. Electrochem.* **11**, 238 (1975).
Effect of temperature on anodic passivation of Pb.
M. Maeda, *Acta Metal.* **6**, 66 (1958).
Behaviour of Pb-ion sensitive electrodes.
V. Majer and K. Stulik, *Anal. Lett.* **6**, 577 (1973).
Electrochemistry of surface oxides.
A. C. Makrides, *J. Electrochem. Soc.* **113**, 1158 (1966).
Electrode behaviour of film membranes based on Pb di-isobutyldithio-phosphate.
E. A. Materova, *Ionnyi Obmem. Ionometria* **1**, 137 (1976); *Chem. Abstr.* **87**, 124561.
Coulometric determination of galenite from electrodeposited PbO_2.
I. F. Mazalov, *Zh. Anal. Khim.* **29**, 1427 (1974).
Electrochemical behaviour of galena in acid media.
I. F. Mazalov, *Khim. Khim. Tekhnol.* **18**, 70L (1975); *Chem. Abstr.* **87**, 55836.
Surface changes on Pb electrodes in electrolysis.
M. Metzler, *Electrochim. Acta* **11**, 111 (1966).
Electrochemistry of PbO_2.
K. Micka, *Chem. Listy* **65**, 449 (1971); *Lead Abstr.* 11-1400.
$Pb-HClO_4$ electrochemistry.
N. K. Mikhailova, *Sb. Rab. Khim. Istokhim. Toka* **9**, 6 (1974); *Chem. Abstr.* **83**, 100738.
Fracture and re-passivation of anodic films during deformation (Pb/H_2SO_4).
N. K. Mikhailova, *Zasch. Metal.* **10**, 57 (1974).
Ionic transport in Pb glasses.
G. C. Milnes, *Phys. Chem. Glasses* **9**, 43 (1968); *Chem. Abstr.* **68**, 107450.
Infra-red study of Pb patination.
T. Monk, *J. Less. Comm. Metals* **9**, 222 (1964); *Lead Abstr.* 17–4915.
Electrocapillary phenomena at Pb electrodes.
I. Morcos, *J. Electrochem. Soc.* **121**, 1417 (1974).
Suspensions of slightly soluble Pb compounds on Hg electrode.
A. S. Musina, *Izv. Akad. Nauk. Kaz. SSR Ser. Khim.* **26**, 60 (1976); *Chem. Abstr.* **85**, 150361; **85**, 150861; **85**, 183956.
Corrosion, cathodic protection, passivation of Pb.
J. van Muylder and M. Pourbaix, CEBELCOR Tech. Rept No. 13, Aug. (1953).
Cathodic corrosion of Pb and protection.
J. van Muylder and M. Pourbaix, CEBELCOR Tech. Rept No. 14, July (1954).
Kinetics of $PbSO_4$ crystallisation/dissolution.
G. H. Nancollas, *Trans. Far. Soc.* **66**, 3103 (1970).

Cathodic reduction of O_2 and relation to Xtal structure on Pb.
V. N. Nikulin, *Zh. Fiz. Khim.* **6**, 1360 (1959).
Effect of O_2 and surfactants on Pb electrorefining with Pt electrode.
F. M. Niyashita, *Nippon Kogyo Kaishi* **2**, 431 (1976); *Chem. Abstr.* **86**, 147634.
Pourbaix diagrams for $PbS–H_2O$.
M. Novak, *Corrosion Sci.* **4**, 159 (1964).
Dissolution of Pb from PbS and graphite anode.
S. Ono, *Denki Kagaku* **33**, 358 (1965); *Lead Abstr.* 1965 3702.
Anodic oxidation of Pb–Sb alloys.
K. Okada, *Yuasa Jiho* **44**, 16 (1977); *Chem. Abstr.* **87**, 138582.
Hydrogen evolution from Pb in buffered media.
S. E. Ostrovskaya, *Sov. Electrochem.* **9**, 1347 (1973).
Potentiostatic study of organics on $Pb/PbSO_2$ system.
N. N. Ozhiganova, *Sb. Rab. Khim. Istochnikam. Toka* **9**, 28 (1974); *Chem. Abstr.* **83**, 100739.
Double layer studies on Pb electrodes.
H. V. Palm and V. E. Past, *Sov. Electrochem.* **1**, 527 (1965).
Potentiodynamic examination of Pb in sulphuric acid.
H. S. Panesar, *7th Int. Power Sources Symp., Brighton* 15 Sept. (1970); *Lead Abstr.* 1971 925.
Adsorption of peptone and electrodeposition of Pb–Sn.
M. Paunovic, *Plating* **58**, 599 (1971); *Lead. Abstr.* 11-1445.
Oxygen evolution on Pb.
D. Pavlov, *Werkst. Korros.* **19**, 671 (1968).
Formation mechanism of Pb(II) compounds by Pb anodisation.
D. Pavlov, *Electrochim. Acta* **13**, 2051 (1968).
Mechanism of passivation processes on Pb.
D. Pavlov, *Electrochim. Acta* **15**, 1483 (1970).
Photoelectrochemical properties of Pb electrode in H_2SO_4.
D. Pavlov, *J. Electrochem. Soc.* **124**, 1522 (1977).
Photochemical effects on anodic Pb oxidation.
D. Pavlov, *Electrochim. Acta* **23**, 845 (1978).
Thermodynamic study of $Pb–PbSO_4$ electrode.
K. S. Pitzer, *J. Phys. Chem.* **80**, 2863 (1976).
Polarisation characteristics of battery plates during discharge.
S. P. Poa, *J. Chin. Int. Chem. Eng.* **6**, 93 (1975); *Lead Abstr.* 17-4885.
Kinetic study of PbO_2 electrode with ring-disk.
J. P. Pohl and H. Rickert, *10th Int. Power Sources Symp. Brighton* (1976).
Electrochemical behaviour of PbO_2 electrode.
J. Pohl and H. Rickert, *in* "Power Sources 5" (D. H. Collins, ed.). Academic Press, London and New York (1975).
Oxygen evolution on Pb in alkali.
S. S. Popova, *Sov. Electrochem.* **4**, 504 (1968).
Comparison of electrochemical corrosion tests with actuality.
M. Pourbaix, *Werkst. Korros.* **15**, 821 (1964); *Lead Abstr.* 1965 3002.

Half-wave potentials of Pb^{++}/Pb in chloride media.
H. P. Raaen, *J. Electroanal. Chem.* **8**, 475 (1964).
Solubility of Pb salts in organic electrolytes.
M. L. B. Rao, *Electrochem. Technol.* **6**, 105 (1968).
Raman study of $Pb-PbCl_2$ corrosion products.
E. S. Reid, *J. Electroanal. Chem.* **80**, 405 (1977).
Electrochemical and structural behaviour of alpha PbO_2.
P. Reinhardt, *Z. Phys. Chem.* **257**, 193 (1976); *Chem. Abstr.* **84**, 142236.
Powder (galena) electrodes.
A. Reyman, *Przemysl Chem.* **41**, 313 (1962); *Lead Abstr.* 1963 1355; also *Przemysl Chem.* **41**, 390 (1962); *Lead Abstr.* 1963 1534.
Co-existence of active and passive zones on the $PbO_2-PbCrO_4$ electrode.
H. Rickert, *Werkst. Korros.* **17**, 376 (1966); *Lead Abstr.* 1966 4247.
Kinetics of PbO_2 electrode—Part I and II.
H. Rickert, *Z. Phys. Chem.* **95**, 47 and 59 (1975).
Photoemission study of double layer on Pb electrode.
Z. A. Rotenburg, *Sov. Electrochem.* **10**, 682 (1974).
Knife-cut (fresh surface) Pb electrode.
P. Ruetschi, *J. Electrochem. Soc.* **104**, 406 (1957).
Double layer studies on Pb electrode.
P. Ruetschi, *J. Electrochem. Soc.* **107**, 325 (1960).
Anodic oxidation of Pb at constant potential.
P. Ruetschi, *J. Electrochem. Soc.* **111**, 1323 (1964).
Ion-selectivity and diffusion potentials in corrosion layers of $PbSO_4$.
P. Ruetschi, *J. Electrochem. Soc.* **120**, 331 (1973).
Double layer structure on Pb electrode.
K. V. Rybalka, *Sov. Electrochem.* **4**, 1360 (1968); **3**, 332, 1013 (1967).
BNF extender and kinetics of $Pb/PbNO_3$ system.
K. V. Rybalka, *Sov. Electrochem.* **13**, 79 (1977).
Half-cell potential of PbS in aqueous solution.
M. Sata, *Electrochim. Acta* **11**, 361 (1966).
Side reactions in anodic oxidation of Pb(II) in anhydrous acetic acid.
M. S. Sataev and N. G. Bakchisteraistyan, *Tr. Mosk. Khim. Tekhnol. Inst.* **71**, 218 (1972); *Chem. Abstr.* **80**, 103152.
PbO_2 as a specific electrode for H_2O_2.
K. G. Schick, *Anal. Chem.* **48**, 2186 (1976).
Pitting and formation of surface films on Pb.
E. L. Schmeling and B. Roeschenbleck, *ZDA/LDA Transl.* 71/287.
Pb^{++} adsorption on Ag by thin-layer method.
E. Schmidt, *J. Electroanal. Chem.* **28**, 349 (1970).
Pb(II) adsorption on Au electrodes.
E. Schmidt, *J. Electroanal. Chem.* **34**, 377 (1972).
Optical/electrochemical studies of Pb underpotential deposition on Au.
E. Schmidt, *J. Electrochem. Soc.* **121**, 1610 (1974).
Potentiodynamic desorption spectrum of Pb monolayers on Au.
J. W. Schultze, *Surface Sci.* **54**, 489 (1976).

E.M.F. of $PbCl_2$ in presence of impurities (activity coefficient measurement).
    K. Schwabe, *Electrochim. Acta* **13**, 1837.(1968).
Cathodic polarisation of PbO_2 in $PbNO_3$.
    T. Sekine, *Denki Kagaku* **41**, 435 (1973); *Chem. Abstr.* **80**, 77492.
Double layer of Pb in presence of MeOH.
    T. Sekine, *Indian J. Chem.* **12**, 1180 (1974); *Lead Abstr.* 15-1213.
Ozone formation on PbO_2.
    D. P. Semchenko, *Izv. Ser Nauk. Nauchn. Tsentra. Vyssh.* **3**, 98 (1975); *Chem. Abstr.* **83**, 105409.
Effect of F^- on oxygen evolution (inhibition).
    D. P. Semchenko, *Sov. Electrochem.* **7**, 932 (1971).
Electrochemistry of $PbCl_2$ in DMSO.
    M. Shaikh, *Electrochim. Acta* **19**, 541, 545 (1974).
Effect of humic acid on double layer and overvoltage of Pb cathodes.
    S. A. Shapiro, *Khim. Tverd. Topl.* **112** (1967); *Chem. Abstr.* **67**, 17153.
Differential capacitance and cyclic voltammetry of Pb in acid.
    T. F. Sharpe, *J. Electrochem. Soc.* **116**, 1639 (1969).
Adsorption of lignosulphonates on Pb.
    T. F. Sharpe, *Electrochim. Acta* **14**, 635 (1969).
Slow cyclic voltammetry on anodised lead.
    T. F. Sharpe, *J. Electrochem. Soc.* **122**, 845 (1975).
Lead alloys as PbO_2 electrodes.
    T. F. Sharpe, *J. Electrochem. Soc.* **124**, 168 (1977).
Electropolishing of Pb.
    B. A. Shenoi and K. S. Indira, *Electroplat. Metal Finish.* **22**, 19 (1969).
Structure of $Pb/PbSO_4$ electrode in reduced state. Effect of additives.
    A. C. Simon and S. M. Caulder, *Electrochim. Acta* **19**, 739 (1974).
Cl_2 evolution on Pt and PbO_2 anodes.
    V. I. Skripchenko, *Trudy Novorherkass. Politekhn.* **322**, 95 (1976); *Chem. Abstr.* **86**, 179351.
Impedance of $Pb^{++}/Pb(Hg)$ electrode.
    J. H. Sluyters, *J. Electroanal. Chem.* **15**, 151 (1967). Also *J. Electroanal. Chem.* **18**, 93 (1968).
Adsorption of Pb(II) at Hg electrode.
    M. Sluyters Rehbach, *J. Electroanal. Chem.* **38**, 17 (1972).
Permeation of electrolytic hydrogen through Pb.
    F. R. Smith, *Can. J. Chem.* **48**, 1789 (1970).
Study of $Pb–PbSO_4$ system by impedance method.
    E. M. Smrotskova, *Sov. Electrochem.* **11**, 1439 (1975).
Technique for electron microscopy of battery electrodes.
    K. M. Solov'eva, *Sb. Rab. Khim. Istochnikam. Toka* (1973) (8), 198; *Chem. Abstr.* **81**, 52183.
Leaching of Pb from platinum black.
    M. Spiro, *J. Electroanal. Chem.* **28**, 151 (1970).
Pb in platinised Pt.
    M. Spiro, *Chem. Rev.* **71**, 177 (1971).

Cyclic voltammetric studies on pure Pb.
    J. G. Sunderland, *J. Electroanal. Chem.* **89**, 343 (1976).
Effect of Pb impurity on h.e.r. at Ga cathodes.
    K. Szabo, *Magyar Kem. Foly* **79**, 373 (1973).
Specular reflectance of Pb(II) on Au electrodes.
    T. Takamura, *J. Electroanal. Chem.* **41**, 31 (1973). 31.
Electro-optical studies of Pb submonolayers on Au by Faradaic adsorption.
    T. Takamura, *Electrochim. Acta* **19**, 933 (1974).
Voltammetry at PbO_2 electrode.
    D. R. Tallant, *J. Electroanal. Chem.* **18**, 413 (1968).
Oxidation of iodides studied on PbO_2 anodes.
    S. P. Tevosov, *Uch. Zap. Azerb. Un-T Ser. Khim.* **4**, 64 (1973); *Chem. Abstr.* **82**, 177194.
Electron transfer reactions at Pb electrodes.
    H. R. Thirsk, *J. Chim. Phys.* **49**, C 131 (1952).
Electrode process on continuously renewed Pb electrode.
    N. D. Tomashov, *Electrochim. Acta* **15**, 501 (1970).
Electrosynthesis of cyanoethyl-Pb compounds.
    A. P. Tomilov, *Zh. Obsch. Khim.* **35**, 391 (1965); *Lead Abstr.* 1965 3252.
Electrode reactions of $PbCl_2$ in non-aqueous media.
    S. H. Toshima, *Denki Kagaku* **42**, 508 (1974); *Chem. Abstr.* **82**, 91679.
Alternate anodic–cathodic charging of Pb in NaOH.
    A. R. Tourky, *J. Chem. (U.A.R.)* **11**, 177 (1968); *Lead Abstr.* 1970 302.
Anodic precipitation of PbO_2 from nitrates.
    I. B. Tsvetkovsky, *Bull. Lening. Univ. (Phys. Chem.)* 110 (1974).
Reduction of PbO and PbO_2.
    H. V. K. Udupa, *Trans. Indian Metal.* **19**, 99 (1966).
Electrochemical preparation of films of Pb and Cd sulphides.
    Ya. A. Ugai, *Sov. Electrochem.* **12**, 835 (1976).
Reduction of alkyl halides at Pb electrodes.
    H. E. Ulery, *J. Electrochem. Soc.* **116**, 1201 (1969).
Double layer of Pb electrode in MeOH.
    Z. N. Ushakova, *Sov. Electrochem.* **12**, 485 (1976).
Double layer at Pb concentrated solutions.
    Z. N. Ushakova and V. F. Ivanov, *Sov. Electrochem.* **9**, 753 (1973).
Film formation on anodically polarised Pb.
    H. Vaidyanathan, *J. Electrochem. Soc.* **121**, 876 (1974).
Kinetics of low temperature $Pb/PbSO_4$ electrode.
    E. M. L. Valeriote, J. S. NTIS AD-A Rept. 011028 NTIS (1975); *Chem. Abstr.* **83**, 185482.
Kinetics of $PbSO_4/PbO_2$ electrode at low temperatures.
    E. M. L. Valeriote, *in* "Power Sources 5" (D. H. Collins, ed.). Academic Press, London and New York (1975).
Kinetics of potentiostatic oxidation of $PbSO_4$.
    E. M. L. Valeriote, *J. Electrochem. Soc.* **124**, 370 and 380 (1977).

New method for electropolishing and etching of Pb.
   M. Verhaege, *Prakt. Metal.* **10**, 220 (1973).
Dissolution of Pb in $PbCl_2$ by chronopotentiometry.
   M. M. Vetyukov, *Sov. Electrochem.* **5**, 362 (1969).
Role of semiconducting films in electropolishing of Pb.
   A. K. Vijh, *Electrochim. Acta* **16**, 1427 (1971).
Cyclic voltammetry of Pb in sulphuric acid.
   W. Visscher, *J. Power Sources* **1**, 257 (1977).
Electrode potentials of Pb amalgam in Pb^{++} solution.
   O. S. Vitorovic, *Serbian Glas. Chem. Drus.* **40**, 397 (1975); *Chem. Abstr.* **85**, 84653.
Cyclic voltammetry for accelerated Pb corrosion test.
   J. L. Weininger, *J. Power Sources* **2**, 241 (1978).
Anodic oxidation of ethyl xanthate on galena electrodes.
   R. Woods, *Aust. J. Chem.* **25**, 2329 (1972).
Zeta potentials of galena–xanthate–oxygen.
   B. Yarar, *Trans. Inst. Min. Metal. Sect. C* **83**, C96 (1974); *Lead Abstr.* 14-3075.
Optical/electrochemical studies of Pb underpotential deposition on Au.
   E. Yeager, AD 766,821 (1973); *Chem. Abstr.* **80**, 55235. Also *J. Electrochem. Soc.* **121**, 474, 1611 (1974).
Effect of metal cations on potential of Pb electrode in H_2SO_4.
   A. I. Zaborovsky and E. A. Kalinovsky, *Ukr. Khim. Zh.* **39**, 28 (1973).
Electrosynthesis of sodium plumbide.
   O. G. Zarubitskii, *Khim. Tekhnol.* (6) 61 (1975); *Chem. Abstr.* **84**, 186587.
Coulometric measurement of Pb thickness.
   Z. Zika, *Koroze Ochrana Mater.* **14**, 29 (1970); *Lead Abstr.* 1970 768.
Reference electrode (Pb amalgam/$PbCl_2$).
   British Patent 894,289 (1960).
Sacrificial anode (Pb) cell for organometallics.
   British Patent 923,807 (1961); *Lead Abstr.* 1963 1762.
Pb–Pt bielectrode.
   German Patent 1,166,586 (1959); *Lead Abstr.* 1964 2665.
Solid electrolyte $K_{0.25} \cdot Pb_{0.75}$, $F_{1.75}$ etc.
   U.S. Patent 3,973,990 (1976).
Catalytic decomposition of hydrogen peroxide by Pb oxides.
   AD138,643, 1 Dec. (1946).

Index